THE ROCKET AND THE REICH

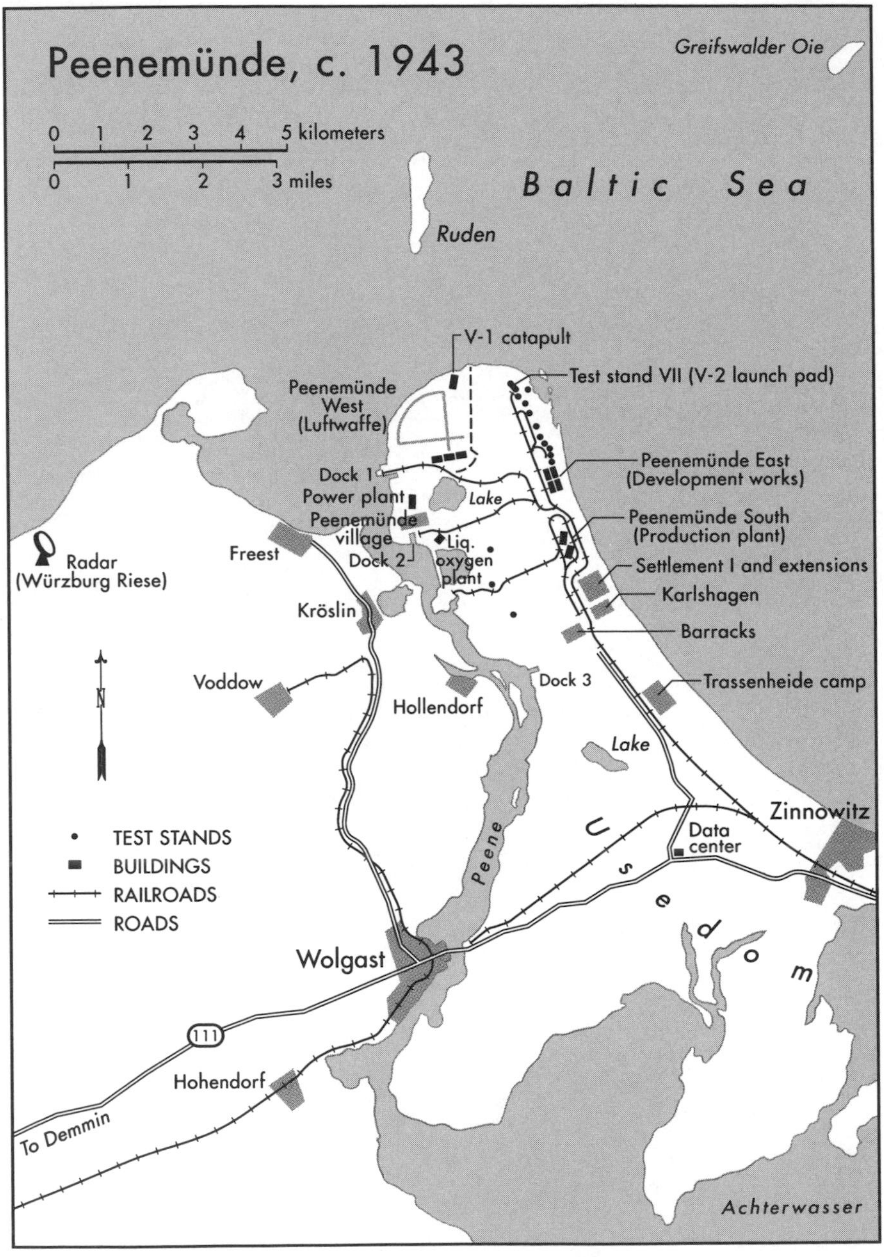

Peenemünde, c. 1943
0 1 2 3 4 5 kilometers
0 1 2 3 miles
Greifswalder Oie
Baltic Sea
Ruden
V-1 catapult
Test stand VII (V-2 launch pad)
Peenemünde West (Luftwaffe)
Dock 1
Power plant
Lake
Peenemünde village
Dock 2
Liq. oxygen plant
Peenemünde East (Development works)
Peenemünde South (Production plant)
Settlement I and extensions
Karlshagen
Barracks
Radar (Würzburg Riese)
Freest
Kröslin
Voddow
Hollendorf
Dock 3
Trassenheide camp
Lake
N
Zinnowitz
Data center
Usedom
Peene
TEST STANDS
BUILDINGS
RAILROADS
ROADS
Wolgast
111
Hohendorf
To Demmin
Achterwasser

Michael J. Neufeld

The ROCKET and the REICH

Peenemünde and the Coming of the Ballistic Missile Era

HARVARD UNIVERSITY PRESS
Cambridge, Massachusetts

Printed in the United States of America

Third printing, 1999

Published by arrangement with The Free Press, a division of Simon & Schuster, Inc.

First Harvard University Press paperback edition, 1996

Library of Congress Cataloging-in-Publication Data

Neufeld, Michael J.
The rocket and the reich : Peenemünde and the coming of the ballistic missile era. / Michael J. Neufeld
p. cm.
Includes bibliographical references and index.
ISBN 0-674-77650-X
1. Rockets (Ordnance)—Research—History—20th century. 2. Liquid propellant rockets—Research—History. 3. Ordnance—Manufacture—History—20th century. 4. Germany. Heer. Heeresversuchstelle Peenemünde—History. 5. Peenemünde (Germany)—Manufactures—History. I. Title.
UF535.G3N48 1995
355.8'25195'094309044—dc20 94-30088
CIP

To the tens of thousands of prisoners

who suffered, died—or survived—

Dora and the other concentration camps of the V2 program

May their sacrifice never be forgotten

Contents

Preface

I first became interested in the German rocket engineers as a small boy in Canada when my father discussed the space race. Did the Russians get the better Germans, or did the Americans fail to exploit their advantage when they got Wernher von Braun's group after the war? That was one of the questions bandied about in the media and around the dinner table at the time. In the aftermath of those early discussions—and televised launches—I lived and breathed spaceflight as a teenager in the 1960s. My ambition to become an aerospace engineer or an astronomer was not realized, however, so I followed other interests and became a historian specializing in modern Germany. Only after revising my dissertation in labor history did I turn back to my old avocation in search of a field that would sustain my enthusiasm. The history of science and technology in general, and the story of German rocketry in particular, seemed to offer an ideal combination of my new and old interests.

In mid-1988 I came to the National Air and Space Museum as a visiting fellow. I soon concluded that a new book on Peenemünde and the V-2 would be far from redundant, that indeed it was a necessity. Although the development of the first large ballistic missile was one of foundations of the nuclear arms and space races, the topic's history had not been well researched. The books on this subject, especially those in English, tended to cite the memoirs of participants uncritically, while giving a less-than-frank treatment of the Nazi records of prominent leaders like von Braun. With two exceptions, Heinz Dieter Hölsken's *Die V-Waffen (The V-Weapons)* and David Irving's *The Mare's*

Nest, serious archival research was entirely neglected and even Irving's research was incomplete and unfootnoted.

In the mid-1980s a new genre began to appear: books by investigative journalists on the scandals of Project Paperclip, which brought German scientists and engineers to the United States after the war. Those works were largely engendered by the Rudolph affair. In 1984 the Justice Department forced Arthur Rudolph, the former project manager of NASA's Saturn V moon rocket, to leave the United States because of his involvement with concentration camp labor during V-2 production. While the resulting exposés uncovered much new information about the Nazi records of the German rocketeers, all too often they combined sensationalism with a simplistic view of life in the Third Reich.

The present work thus aims to provide a balanced and readable history of the German Army liquid-fuel rocket program based on archival research. The symbolic center of the book is the rise and fall of the Army rocket facility at Peenemünde as a major research and development institution. Because the stunning technological revolution effected at Peenemünde is so central to understanding the shape of the institution and the character of the program, I do not shy away from discussions of the technology. But, for the sake of readability, I eschew jargon or a more theoretical examination of the place of Peenemünde in the history of science, technology, and the military in the twentieth century. The epilogue will draw attention to some interesting aspects of Peenemünde's legacy which emerge from the rocket program's relationship with Nazism and its transfer to the United States and elsewhere. Readers looking for a more theoretical or historiographic examination of those topics are invited to consult my articles listed in the bibliography. Those who want more detail about the organizational history of the rocket program should turn to Appendix 2.

This book would not have been finished without the National Air and Space Museum (NASM) and the Smithsonian Institution. The project was supported in its first two years by a NASM Verville Fellowship and by a Smithsonian Institution Postdoctoral Fellowship. The National Science Foundation also made a large contribution through a Scholar's

Award (DIR-8911103) from the Science and Technology Studies Program. During my fellowship period in the NASM Department of Space History, David DeVorkin, Allan Needell, Robert Smith, Martin Collins, John Mauer, Frank Winter, Ron Doel, Michael Dennis, Paul Ceruzzi, Cathy Lewis, Gregg Herken, Joe Tatarewicz, Mandy Young, Joanne Bailey, and Pat Fredericks provided me with a fruitful intellectual climate and helpful administrative support. Frank Winter in particular gave me the benefit of his extensive knowledge of the early rocket societies.

Since my move to the Aeronautics Department as a curator in 1990, its chairman, Tom Crouch, has not only given me much time to finish the book but also provided me with the fruits of his own deep thinking about Peenemünde and Wernher von Braun. I have learned much about the Luftwaffe from Tom Dietz, while Anita Mason and Collette Williams have given cheerful and helpful administrative assistance. I have also benefited from discussions with Von Hardesty, Dom Pisano, Peter Jakab, Jacob Vander Meulen, and Claudio Segrè. I would further like to thank the NASM branch of SI Libraries, especially its current and former branch chiefs, Martin Smith and Dave Spencer, for allowing me to borrow a microfilm machine, plus Mary Pavlovich for her help with interlibrary loans. I am also grateful to the Archives Division, above all to Paul Silberman for turning up the long-forgotten and mislabeled "FE" microfilm of Peenemünde documents.

Every chapter was carefully read by two close friends and colleagues from my days in Space History: David DeVorkin, curator of the history of astronomy, and Ron Doel, twice a visiting fellow in the Department. Their good advice has been incorporated everywhere in the book. Robert Smith also gave me valuable comments on early chapters. They can take no responsibility for my errors and misjudgments, however, nor do the opinions expressed in this book necessarily represent the views of the National Air and Space Museum or the Smithsonian Institution.

From outside the museum, I would especially like to thank Michael Hubenstorf of the Free University of Berlin, who provided me with comfortable accommodations, warm hospitality, and helpful research hints during my stays in that city. Richard Muller of the Air Command and Staff College, a former NASM fellow, sent me material about the Luftwaffe and gave me the benefit of his encyclopedic knowledge of

German military history. Guillaume de Syon of Boston University, Richard Breitman of American University, Donald MacKenzie of the University of Edinburgh, and Wolfgang Rüdig of the University of Strathclyde have kindly supplied me with archival and secondary information relevant to the book, while Yves Béon and Roland Hautefeuille, both of Paris, generously gave me copies of their books. Joyce Seltzer of Harvard University Press, my former editor at The Free Press, made many helpful suggestions for improving the writing of this study. Sheila Weiss of Clarkson University encouraged me to take the plunge into the new field in the first place.

My debts to archives and libraries are considerable. I would like to thank Robert Wolfe, Harry Riley, Ed Reese, and other members of the National Archives staff; David Marwell and Frau Wolf of the Berlin Document Center; Philip Reed of the Imperial War Museum; James Hagler of the Space and Rocket Center; Rudolf Heinrich, former archivist of the Deutsches Museum; and Helmut Trischler, its current research director. I would also like to express my gratitude to the staffs of the Bundesarchiv Koblenz, the Bundesarchiv/Militärarchiv Freiburg, the Humboldt University Archive, the Redstone Scientific Information Center, the Kz-Gedenkstätte Mittelbau-Dora and the Peenemünder Informationszentrum. The *Journal of Military History* kindly allowed me to reprint material that first appeared in a different form in that publication.

For the kind granting of their time and permission to conduct oral history interviews, I would like to thank the former Peenemünders whom I visited in Huntsville and elsewhere, especially Gerhard Reisig, who was very helpful at an early stage of this project. I know that some of the things I have said in this book will anger them, but I have always tried to be faithful to the truth as I saw it. I do not think that there is such a thing as collective guilt; each case must be evaluated individually. The German rocket engineers were neither uniformly innocent nor uniformly guilty of Nazi enthusiasm or the abuse of slave labor. It is time that they stop being lumped together. I understand that the majority were merely doing their job in wartime in a totalitarian society over which they had no control, but neither is that a reason to exempt every last one from responsibility. On those questions, I am grateful to Eli Rosenbaum of the Office of Special Investigations, U.S. Justice Department, and to a freelance journalist, Linda Hunt, for supplying me with valuable information. Other interviews were kindly granted by Dr. Hermann Kurzweg, Rolf Engel, Dr. and Mrs. Hans Geipel, Mr. and Mrs. Manfred Schubert, and Mr. and Mrs. Gerhard Zanssen.

Finally, Karen Levenback gave me much love and support during the writing of this book, as well as the benefit of her professional expertise in editing and writing. Nearly every chapter is better written because of her insightful advice. Her encouragement during the long hours at the computer have been invaluable as well. I would also like to thank our cats, Newt, Birch, and Kepler, for the suggestions they made by jumping on my keyboard. It would not have been the same without them.

Prologue

Summer 1943

In early August 1943 Dieter Huzel, a newcomer to the secret German Army rocket center at Peenemünde, at last got his chance to see the A-4 missile. As a friend showed him into the launch checkout building for the vehicle that would later become famous as "Vengeance Weapon 2" or V-2, his eyes slowly adjusted from the glare of the summer sun. Finally,

> I saw them—four, fantastic shapes but a few feet away, strange and towering above us in the subdued light. I could only think that they must be out of some science fiction film—*Frau im Mond [The Woman in the Moon]* brought to earth.
>
> I just stood and stared, my mouth hanging open for an exclamation that never emerged. Then, slowly, I walked around them. They fitted the classic concept of the space ship—smooth, torpedo-shaped—giving no hint of the mechanisms within, and resting tip-toe on the points of four swept cruciform fins. By today's standards the A-4 was a small missile, but these were 46 feet tall and by all odds bigger than anything I had ever dreamed of. They were painted a dull olive green, and this, said Hartmut, as well as their shape, had won them the nickname of cucumber. I laughed, and the spell was broken.[1]

It was indeed a remarkable sight. Despite the A-4's utopian origins in the spaceflight movement of the 1920s, the German Army had committed massive resources to build this exotic weapon, and had succeeded. But the missile was only the most spectacular product of this investment. What the money had purchased, first and foremost, was the Peenemünde Army Center, which had created this astonishing technological achievement.

Huzel had arrived just as the institution was reaching the apogee of its trajectory. Up to 12,000 people worked—or were forced to work—for the Army on the Baltic island of Usedom, where Peenemünde was located. Since his arrival, Huzel had passed through the new "works train" station, where the hordes of commuters reminded his friend "of the Berlin U and S stations at rush hour." From the the modern electric train that coursed through the center, Huzel had viewed massive facilities, including a camp of mostly Russian and Polish forced laborers and the F-1 assembly building at the Pilot Production Plant (Peenemünde-South). That building, more than 250 meters (800 feet) long, held an assembly line that would produce three hundred A-4s a month in the fall. Unbeknownst to him, F-1 also contained a small concentration camp, where six hundred mostly Russian and French-speaking prisoners were held under SS guard as the first installment of the labor force.[2]

The newcomer's journeys also took him to the older Development Works at Peenemünde-North (or -East, the original name used to distinguish it from the much smaller Luftwaffe facility at Peenemünde-West). At the Development Works Huzel found, scattered among the trees, a large campus of administrative buildings, laboratories, test stands, and workshops. About six thousand engineers and skilled hands worked in this facility alone as a result of concentrating "everything under one roof" instead of contracting out most work to companies. Here the first A-4s had been designed and built since the center opened in 1937.[3]

How had this remarkable institution come about? Why had the Third Reich invested hundreds of millions of marks in guided missile research and development? And why had the leaders of the Army rocket program incorporated into their project, willingly or unwillingly, one of the worst features of the Nazi regime for which they worked—concentration camp labor? To answer those questions, we must first turn to the late 1920s and early 1930s. During that period key officers in artillery development came to believe that liquid-fuel rocketry could be the basis for a potentially war-winning secret weapon: the ballistic missile. Supersonic projectiles like the A-4 would rain down on potential enemies, causing physical destruction and psychological shock. If the surprise was great enough, opposing leaders might well concede victory to the technologically superior German forces. But the Army's interest in the technology had been awakened by quite another scenario, one popularized by a small but determined band of spaceflight enthusiasts. To those true believers, the liquid-fuel rocket would be the means for liberating the human race from the bonds of the earth.

Chapter 1
The Birth of the Missile

Berlin, October 1929. In the fashionable west end, near the Kurfürstendamm, the entire façade of the UFA Palace movie theater was "covered with a gigantic animated panel, showing the earth and the moon against a starry sky and a projectile-like moon rocket making round trips between them."[1] Playing in the theater was the new science fiction movie *Frau in Mond (The Woman in the Moon),* by Fritz Lang, the renowned director of *Metropolis.* The newspapers carried notices of the impending launch of a stratospheric rocket by Hermann Oberth, the film's scientific adviser and the father of the Weimar spaceflight movement. That movement had arisen in response to Oberth's writings about the feasibility of space travel, but it was also an expression of the Weimar Republic's forward-looking and innovative culture. The upheavals of 1918—the loss of World War I, the abdication of the Kaiser, and the founding of a democratic republic—had created unprecedented freedom but had also exacerbated deep social and political tensions in German society. Those conditions fostered original art and original thinking.

Not far from the theater, in the quarters of the Army Ordnance Office, Lieutenant Colonel Karl Emil Becker (1879–1940) had begun to investigate the revival of the rocket as a weapon. Since the mid-nineteenth century, when rifled, breech-loading artillery guns had greatly improved accuracy and range, the black-powder rocket had fallen out of favor as a bombardment weapon. Military experts and lay people alike came to regard this traditional form of the rocket, which burned a gunpowder-like fuel in a metal or paper casing, as little better than a toy. In World War I rockets had been used only for signal or illuminating flares and other minor applications. In the interwar period, however, improved possibilities for the safe manufacture and storage of solid propellants, including new smokeless powders, made rocketry again more interesting to the military. Becker, who had a doctorate in engineering and headed Section 1 (ballistics and munitions) of Army Ordnance's Testing Division, was especially interested in solid-fuel rockets as a means of launching poison gas against enemy troops on the battlefield.[2]

The fact that the Allied-imposed Versailles Treaty of 1919 omitted any mention of rocket development reinforced Becker's interest in the technology. Like most of his fellow officers, he was an ultranationalist who yearned for the day when a new right-wing authoritarian regime could overthrow the treaty's onerous limits on German military power. Until that day, however, Becker and his compatriots would use all available means to circumvent the treaty, which restricted the Reich to an Army of 100,000 lightly armed men, a tiny Navy, and no air force at all. Not only had the Army maintained hidden units to violate its size limit, it had conducted covert research into poison gas, aircraft, tanks, and other banned weapons at home and abroad, most notably in the Soviet Union. The investigation of legal technologies like the rocket was yet another way to prepare for rearmament, an increasing concern of the Weimar military in the relatively stable years of the late 1920s. But Becker's interest in rocketry as a means of illegal chemical warfare shows that finding a loophole in the treaty was not central to his decision to look into the technology. Of more significance was the Versailles ban on heavy artillery, an important class of weapons that was Becker's specialty. Provided that rockets could be made sufficiently powerful, they could replace not only short-range battlefield weapons but also long-range heavy guns.[3]

If Becker had military reasons for taking up rocketry, the Weimar spaceflight fad also undoubtedly had a crucial impact. The fad had begun with Hermann Oberth's seminal 1923 book, *Die Rakete zu den Planetenräumen (The Rocket into Interplanetary Space).* Oberth (1894–1989), a member of the German minority of Transylvania, was an unwilling Rumanian citizen after the Hapsburg Empire's collapse in 1918. His slim volume had defended the radical concept of manned spaceflight and had made concrete suggestions for overcoming the technical difficulties involved. Most notably, Oberth's book showed that by mixing and burning a liquid fuel like alcohol with an oxidizer like liquid oxygen, one could dramatically improve performance over the traditional black powder rocket. At first the book attracted little notice, but in 1924 Oberth's cause was taken up by the irrepressible Max Valier, an Austrian writer and self-proclaimed astronomer residing in Munich.[4]

Valier's articles, books, and speeches did much to popularize the idea of spaceflight with the Weimar public. Although most of Oberth's ideas had been anticipated by, among others, Konstantin Tsiolkovsky in Russia and Robert Goddard in the United States, their insights were inaccessible to lay and specialist readers alike. Tsiolkovsky's publications went back as far as 1903 but were buried in obscure Russian periodicals. Goddard's "A Method of Reaching Extreme Altitudes" (1919–20) had avoided explicit references to liquid-fuel rocketry and manned spaceflight. Even so, his discussion of a staged powder rocket to hit the moon had unleashed a wave of sensationalism and ridicule in the newspapers that made the shy physicist even more secretive than before. His impact in Europe was largely confined to wild rumors in the popular press about his activities. The fact that he had launched the world's first liquid-fuel rocket in 1926 remained virtually unknown for a decade afterward.[5]

Oberth's intellectual boldness and Valier's knack for publicity, in contrast, made the spaceflight idea more visible and respectable in Germany than almost anywhere else. In 1927 Valier participated in the formation of the Society for Space Travel, often known by its German abbreviation, VfR. Until 1930 the VfR was headquartered in Breslau (now Wroclaw, Poland), because its first president was Johannes Winkler, a church administrator and frustrated engineer there. Winkler's new journal, *Die Rakete* (The Rocket), became the organ of the society. But it was Valier's alliance with Fritz von Opel, heir to the car manufacturing fortune, that finally put rocketry on the front pages. In order to generate publicity, Opel and Valier used commercial black-powder rockets to power spectacular race car demonstrations in April and May 1928. Those experiments unleashed a wave of publicity in the media, and other stunts followed with rail cars, gliders, bicycles, and even a

Valier rocket ice sled. Their visibility also strengthened Fritz Lang's resolve to make the moon flight movie he had been thinking about since *Metropolis*.[6]

Although some skepticism and ridicule had accompanied all that activity, especially speculations on the subject of spaceflight, the 1928–29 popular fad showed that, with the possible exception of Soviet Russia, Germany responded more enthusiastically to the potential of the rocket than any other country. Nationalism no doubt played a key part here. Germans tended to seize on almost any sign of their technological superiority or their rapid recovery from the humiliations of the war and Versailles. Despite bitter political and ideological divisions in the country, technological progress was desired by almost everyone, and the rocket fad provided escapist entertainment for the new mass culture of the 1920s.[7]

Thus, when Becker began to investigate the rocket in 1929, he did so against a background of media publicity and highly visible demonstrations of powder rockets in action. His curiosity may also have been piqued by discussions, mostly on the margins of the spaceflight movement, of the possibility of a large ballistic missile based on liquid fuels. Oberth, for one, had discussed the possibility of launching poison-gas attacks on enemy cities with intercontinental rockets in the enlarged 1929 version of his book, *Wege zur Raumschiffahrt* (Ways to Spaceflight), apparently because he had received so many queries from the public about the idea. He considered it impractical for the next decade or two, however, because of the difficulty of accurately guiding the missile to its target.[8]

At the end of 1929 Becker asked for and received permission from the Reich Defense Minister for a small solid-fuel rocket program. Testing of commercial black-powder units began shortly thereafter. Assisting the fifty-year-old Becker were a small number of junior officers with engineering training. His second in command in the ballistics and munitions section bore the impressive aristocratic moniker d'Aubigny von Engelbrunner Ritter [Knight] von Horstig. Captain von Horstig (b. 1893) also had an engineering doctorate and shared his superior's World War I experience in the artillery. Three slightly younger veterans would soon emerge as the central figures in the administration of the early program: Erich Schneider, Walter Dornberger, and Leo Zanssen.[9]

All three were products of a "study officer" program that Becker had been instrumental in founding. Appalled by the antitechnological attitudes of the old Imperial officer corps and dismayed over personal experiences with poorly organized procurement in wartime heavy artillery development, Becker successfully pushed engineering training for selected individuals in the Army. He was aided in that endeavor by the new Army leadership and by his mentor, Professor Carl Cranz, author of a famous ballistics textbook, which Becker helped to revise in the 1920s. Cranz's Prussian Army artillery laboratory had been converted into an institute of applied physics at the Technical University of Berlin after the war, in order to prevent its dissolution under the Versailles Treaty. Not only did Becker receive his doctorate from Cranz's institute, but it also became the center of a regular "diploma engineer" program (equivalent to a master's degree) for study officers. Schneider graduated from the University in 1928, Dornberger in 1931, and Zanssen in 1933, all as mechanical engineers with special expertise in artillery ballistics.[10]

Of the three, Dornberger (1895–1980) would become the most important. The son of a pharmacist from the southwest German city of Giessen and a veteran of heavy artillery units on the Western Front, Dornberger would become a masterful salesman, administrator, and political infighter for the rocket program. A spaceflight enthusiast, he read Oberth's *Wege* around the time of its appearance in 1929. He began work in Becker's section in 1930, purportedly with the assignment of looking into liquid-fuel rocketry, but until 1936 his main area of concentration was small battlefield solid-fuel rockets. Zanssen, another middle-class officer from western Germany and a close friend from the University, was Dornberger's alter ego and served under him through much of the history of the program.[11]

THE RISE OF AMATEUR ROCKETRY

At the same time as Army Ordnance began its small-scale investigations in 1929–30, liquid-fuel rocket development began in earnest among the spaceflight fanatics in the VfR and outside of it. It had been apparent for some time that a move from theory to practice was necessary. As early as 1924 Oberth and Valier had been looking for a funding source, such as a millionaire or a corporation, to make that possible. Valier's search eventually led him to his short-lived alliance with Opel and to the idea of using commercially available black-powder rockets to put on a series of stunts with cars and other vehicles. That publicity-seeking approach, which did nothing to advance rocket engine development in the short run, proved to be the last straw for the already strained relationship between the querulous and suspicious Oberth and the technically untutored Valier. Like almost everyone else in the spaceflight movement, however, the two looked to the same models: the heroic independent inventors of the late nineteenth and early twentieth centuries, like Edison, Diesel, and Ford. They expected some far-sighted, wealthy investor to finance their rocket development and did not foresee that such an enormously expensive technology could only be created by a government-financed military-industrial complex. Motivated by a burning vision of travel to the moon and the planets, the spaceflight pioneers also grossly underestimated the complexity and difficulty of the technology.[12]

Some of the early pioneers did receive limited corporate support. Johannes Winkler was the first to begin more serious work, with preliminary experiments in Breslau in 1928–29 and further work at the Junkers Aircraft Company in Dessau in 1929–31. The head of the company, Hugo Junkers, a well-known airplane designer, hoped that rocket engines could be used to assist the takeoff of heavy airplanes and could serve as a propulsion system for high-speed aircraft. Winkler made preliminary experiments using various propellants, such as ethane and nitrogen monoxide, but settled on methane and liquid oxygen as his main fuels. Liquid oxygen was the ideal oxidizer, but that entailed all the difficulties of handling a fluid that boils at a temperature of –183°C (–297°F), and one which had a distressing tendency to set off explosions if it came into contact with grease and organic materials. Winkler nonetheless succeeded in making the first verifiable launch in Europe of a liquid-propellant rocket in March 1931, immediately after quitting Junkers and obtaining private money. Winkler's rocket engine generated only 7 kg (14 lb) of thrust. (Thrust is the force on the rocket created by the gases exiting the engine nozzle, as per Newton's third law of motion: Every action produces an equal and opposite reaction.)[13]

Valier, meanwhile, had secured the support of a manufacturer of liquid-oxygen equipment, Paul Heylandt. One of Heylandt's firms, the Industrial Gas Utilization Company in south Berlin, became the site of Valier's attempts to develop a rocket car using liquid fuels beginning in late 1929. Assisted by one of Heylandt's engineers, Walter Riedel (1902–68), later chief of the design bureau at Peenemünde, Valier designed an engine using kerosene and liquid oxygen. Its performance was unstable because of problems with the injection and atomization of the fuel, one of the most critical difficulties experienced by all the early experimenters. Arthur Rudolph (b.1906), another young engineer at Heylandt and a future branch chief at Peenemünde, witnessed the accident that killed Valier on Saturday evening, May 17, 1930. As the three were making engine runs on a primitive test stand, the motor suddenly exploded in a hail of metal. Riedel caught the staggering Valier and then ran for help. Rudolph, who was knocked flat by the explosion, finally reached Valier, but a piece of shrapnel had punctured the Austrian's aorta. Within a minute Valier was dead, the first victim of a dangerous trade. There was a minor public uproar, and a bill was introduced in the Reichstag to ban rocket experiments, but it did not pass. Heylandt decided to discontinue his involvement, but Rudolph would not quit so easily.[14]

The most important rocket group of the early 1930s—*Raketenflugplatz* (Rocketport) *Berlin*—arose, however, as a byproduct of Hermann Oberth's involvement with the film *Frau im Mond.* Oberth came to Berlin from Rumania in late 1928 to work as scientific adviser for the movie, which promised to be a historic breakthrough for the spaceflight movement. Fritz Lang was the most famous and powerful German film director of his era. Once in Berlin, Oberth asked Lang to help him obtain money for rocket development. Lang persuaded the UFA film conglomerate to bankroll the launching of a stratospheric sounding rocket during the film's premiere. But Oberth, an impractical physics teacher from a small town in Transylvania, had no engineering experience. He advertised for assistants, and a World War I fighter pilot with dubious engineering credentials showed up. Rudolf Nebel was more of a salesman and a con artist than an engineer. It is appropriate that his last name can be translated as "fog." Oberth found a second assistant in a freelance aviation and space writer, Alexander Sherchevsky, "a Russian emigrant . . . who lived," Oberth wrote a few years later, "completely in filth. And fairly literally at that. I had the impression that, if one threw him against the wall, he would stick there." On another occasion Oberth described him as "the second laziest man I ever met."[15]

The three set out to build the rocket, but the project turned into a fiasco. Scherchevsky was useless and had to be let go, Oberth was injured in an explosion, and the film company issued exaggerated and misleading press releases about the rocket's performance. After suffering a nervous breakdown, Oberth left for Rumania even before the movie's star-studded premiere on October 15. He had lost most of his money in the venture, because the film company refused to reimburse him. The ill-fated project was left in the hands of Rudolf Nebel, who delayed the announced launch until November and then canceled it altogether. About the same time—the end of 1929—Winkler had to stop publication of *The Rocket* because the journal's finances had been poorly managed. That cut off much of the membership of the VfR from contact with the society, with the result that the number of members dropped significantly. The Berlin leadership regrouped and decided to form a liquid-fuel rocket group, starting with the leftover materials from the Oberth rocket. Because Nebel had been empowered as Oberth's representative and was energetic and unscrupulous, he came to dominate this effort.[16]

Nebel began making the rounds of government ministries, scientific institutions, and corporations to look for funds to continue Oberth's work. In the first months of 1930 he may even have talked himself into the offices of Albert Einstein and Reich Interior Minister Carl Severing. Severing allegedly promised support, but soon thereafter the coalition cabinet he was in collapsed under the economic strain of the Great Depression, ushering in an era of weak right-wing governments dependent on the decree powers of Reich President Paul von Hindenburg, the retired Field Marshal. Nebel's campaign nevertheless produced one success. After he met Becker, Army Ordnance gave him a fairly large sum of money; Nebel's memoirs say 5,000 marks (about $1,200). The money was supposed to be at the disposal of Oberth for the launch of his rocket on the Baltic coast. Through Severing or Becker, Nebel was referred to the Reich Institution of Chemical Technology, which performed some of the same functions as the U.S. Bureau of Standards. Its director agreed to provide workshop space and a certification of Oberth's rocket engine, which would be useful in further fundraising.[17]

Nebel wired Oberth to join him in Berlin, which Oberth eventually did. After some work they finally succeeded in constructing a small 7-kg-thrust gasoline–liquid-oxygen engine. A famous picture taken on the day of the official test, July 23, 1930, shows Oberth and Nebel with the institute director and a number of helpers. Among them were two whose future roles at Peenemünde would be crucial: Klaus Riedel (1903–44) and Wernher von Braun (1912–77). Riedel (no relation to his namesake at Heylandt), a heavy-set young engineer and spaceflight enthusiast, would be Nebel's primary designer in the rocket group. Von Braun, in a suit with knee breeches, looks both his age and his status: eighteen years old and wealthy. He came from venerable Prussian Junker stock and possessed the title of *Freiherr* (baron), although he did not consistently use it. His father had been a high-ranking civil servant in Imperial Germany, but was forced out in 1920 for not distancing himself sufficiently from a far-right coup attempt against the new Weimar Republic. The elder von Braun became a banker with close connections to President Hindenburg and the old reactionary elites. His son's enthusiasm for engineering and advanced technology was mysterious to him. Wernher had become a teenage spaceflight fanatic after encountering Oberth's works in 1926. By 1930 he was preparing to go to engineering school at the Technical University of Berlin, an unusual career choice for an aristocrat.[18]

After the test Oberth once again returned to Rumania, but Nebel founded the famous Raketenflugplatz Berlin. Nebel and Riedel had already begun working on what they called the *Mirak,* for "minimum rocket," a small rocket based on a modified Oberth engine. In his search for an appropriate testing ground, Nebel found an abandoned ammunition dump in Reinickendorf, a nondescript working-class district near the northern edge of Berlin. A number of massive concrete storage bunkers surrounded by earthen blast walls were situated in the midst of a hilly and wooded area. The only access road was poor, and the lowlands were swampy. Nebel was able to obtain permission from the city and the Reich Defense Ministry to use the land, and he, Riedel, and others opened the Raketenflugplatz on September 27, 1930.[19]

Unbeknownst to anyone in the group except Nebel, the ballistics and munitions section had played a key role in securing him a lease for three years on the facility. Becker's section may even have suggested the location. In short order, however, Becker became disgusted with Nebel. In a May 1931 memorandum to Ordnance's aviation section, he denounced Nebel's dishonesty; his lack of the "necessary practicality, quietude, and secrecy"; and his tendency to write "sensationalistic articles in newspapers and magazines" merely for the purpose of raising money. The issue of secrecy was doubtlessly crucial, but there was also a culture clash between Nebel's blatant self-promotion and the mentality of the officer corps. Ordnance had thus cut off all contact with him before the spring of 1931.[20]

Nebel's *modus operandi* was later described by von Braun:

> One day, Nebel took me out on one of his "acquisition trips." We visited a director of the large Siemens corporation. Nebel told him eloquently about his plans—the liquid[-fuel] rocket motor, the stratosphere, lightning voyages across the ocean, the moon. The man was half amused, half impressed. The result was a trunk full of welding wires. With the wires we proceeded to a welding shop in town. Nebel told the shop superintendant that we had plenty of aluminum welding to do, but suffered from lack of a skilled man. Soon a deal was worked out, according to which Nebel would supply the shop with welding wires, while they would weld our tanks and rocket motors—all on a cash-free basis.[21]

Nebel made it a point never to pay for anything. Shell Oil provided free gasoline, Siemens free meals. He acquired skilled labor by giving unemployed craftsmen free housing in the bunkers in return for work on the rockets; many of them became true believers. Dimitri Marianoff, Einstein's "stepson-in-law" and a visitor at Raketenflugplatz, said: "The impression you took away with you was the frenzied devotion of Nebel's men to their work. . . . Not one of these men was married, none of them smoked or drank. They belonged exclusively to a world dominated by one single wholehearted idea." But they were not so single-minded that they could not celebrate. Von Braun reports that successes were often followed by drinking parties at a "downtown nightclub." If so, he must have been footing much of the bill, since most of the others were receiving only miserly sums from unemployment insurance.[22]

With Klaus Riedel primarily responsible for design and Rudolph Nebel concentrating on raising funds and materials, the Raketenflugplatz worked toward a flying version of the Mirak. After its launching in May 1931 it was rechristened the *Repulsor,* for the space vehicles in a popular German science-fiction novel. Throughout the rest of 1931 and into 1932, Nebel's group launched various versions dozens of times, including many demonstrations for which spectators were charged admission. The Raketenflugplatz also publicly demonstrated the burning of larger rocket motors with thrusts of up to about 50 kg (110 lb).[23]

The Raketenflugplatz's many different engine and vehicle configurations embodied certain common principles. Liquid oxygen was the oxidizer, and the fuel was easily obtainable gasoline. In line with Oberth's original suggestions, however, alcohol was later substituted, because that made it possible to add water, lowering the combustion temperature. Cooling the engine was a difficult problem; burnthroughs of nozzle walls, leading to explosions or erratic performance were common. The group first tried surrounding the engine with the liquid oxygen tank for cooling, then putting a jacket of water around the combustion chamber, and finally circulating watered alcohol through the cooling jacket before injection. The technique of using fuel circulation through the engine and nozzle walls, foreseen by Oberth and other pioneers, is called "regenerative cooling" and is a central feature of almost all large rocket engines.

All the early engines fed the fuels into the combustion chamber under pressure. The Raketenflugplatz at first used the liquid oxygen's own evaporation to build pressure in that tank and employed a carbon dioxide cartridge in the gasoline tank, but in the end it adopted a better solution: Compressed nitrogen expelled the propellants in both tanks.

The most distinctive feature of the vehicles produced by the Berlin group, and by most other groups at the time, was the "nose drive" configuration. In contrast to the stereotypical image of the rocket–with engine and tail fins at the rear—these rockets had the engine at the top and the tanks trailing behind. According to Willy Ley, a VfR member and freelance science writer who came to the United States in 1935 to escape Hitler's regime, the Raketenflugplatz had begun by consciously imitating the classic black-powder rocket. The first "One-Stick Repulsor" had its gasoline tank in a long tube attached to the side of the engine head. For centuries a stick had been used to give powder rockets a crude stability, but the aerodynamic principles had not been understood at the time, and even if the amateur experimenters did understand them, they had lacked the resources and systematic approach to exploit that knowledge. The bizarre-looking vehicles that resulted from the "nose drive" showed that the Raketenflugplatz never mastered stability and control in flight. But that was less important than getting a vehicle off the ground without endangering the onlookers too much. The endless problems with propulsion—burnthroughs, leaks, explosions, and valves and lines frozen by liquid oxygen—were much more pressing.[24]

ARMY ORDNANCE AND THE BALLISTIC MISSILE

As the enthusiastic amateur groups stumbled forward into the new territory of liquid-fuel rocketry without a map, but with the goal of spaceflight on the distant horizon, Becker and his subordinates began to chart a path toward their own objective: the ballistic missile. Long-range artillery had approached its limits with the Paris Gun, a special 21-centimeter (8.25-inch) howitzer used by the Germans to shell the French capital from 130 kilometers (80 miles) away in the spring of 1918. Becker had worked as an assistant on that spectacular project. After lobbing only 320 shells, however, each with 10 kilograms (22 pounds) of high explosives, the gun wore out its main and reserve barrels and had made little impact on French morale. By replacing conventional gunnery with liquid-fuel rocket engines, one could eliminate not only barrels and their massive supporting equipment but also all limits on range and payload. Moreover, Becker believed, the surprise deployment of stunning new weapons could have a dramatic effect on the enemy's psychology. A rain of fairly accurate long-distance projectiles might even cause the collapse of enemy morale. That idea had failed with the Paris Gun, but the sudden deployment of a much larger projectile based on a revolutionary technology could be effective. To produce the necessary shock and surprise, it would be imperative to develop the ballistic missile in absolute secrecy, even though it was not outlawed by Versailles. Secrecy would have the added benefit of concealing the missile's potential from the other powers.[25]

Such were the concepts that stood behind Army Ordnance's growing commitment to liquid-fuel rocketry from 1930 to 1932. Becker's group nevertheless moved only haltingly toward an investment in the infant and unproven technology. After the abortive attempt to support Oberth and Nebel in 1930, the ballistics and munitions section did little with liquid fuels for nearly a year. The cash-strapped Ordnance rocket project, itself only a small part of artillery development under Becker, focused instead on the feasibility of unguided, solid-fuel battlefield weapons with poison gas or high-explosive warheads. The first sign of renewed interest in the more advanced technology came on October 16, 1931, when Becker wrote to the Heylandt company requesting a confidential meeting between Captain von Horstig and Paul Heylandt. Becker expressed interest in the company's "liquid-fuel blow-pipe." His awkward use of the term "blow-pipe," instead of "rocket" or "motor," shows that he was unfamiliar with the technology.[26]

Becker's inquiry was sparked by a new rocket car that the Heylandt company had finished and tested in April–May 1931. After the death of Valier, Arthur Rudolph had continued to experiment with the engine to determine why its combustion was so unstable. He did so against the express orders of Heylandt and almost lost his job as a result. Sometime in late 1930 or early 1931, Paul Heylandt started a new rocket-car project. He was still fascinated with advanced technology and was interested in recouping his investment of more than twenty thousand marks in Valier's experiments. Most of the work on the car was done by Riedel and his superior, Alfons Pietsch, and Rudolph helped produce a much improved, regeneratively cooled engine with about 160 kg of thrust. Heylandt returned from a trip to the United States in time for the public trials, but the impact of the Depression plus the fading of enthusiasm for rocket stunts made the new car a financial and public relations flop. Pietsch lost his job shortly thereafter in one of the company's many layoffs. The car was shown again during the summer months, but with no better result—at least until Becker inquired about its engine.[27]

A number of things may have sparked Ordnance's renewed attention to liquid-fuel rocketry, beyond a general interest in the futuristic ballistic missile concept, which most military officers of that era would have regarded as utopian or impossible. During the spring and summer of 1931, the activities of the Raketenflugplatz, the Heylandt group, and Johannes Winkler had made the technology more visible and viable. Army Ordnance may also have been influenced by the political climate. During the proceeding year the country had begun to slide into chaos. The mass unemployment of the Great Depression, combined with the Weimar Republic's already weak popular support, had led to political polarization and street fighting. On the far left, the Communists gained much ground, but their gains were eclipsed by the extreme right-wing National Socialists (Nazis), who leaped from marginality to major party status in the national elections of 1930. The weak Weimar cabinet of Chancellor Heinrich Brüning also became increasingly conservative and authoritarian. In this poisonous environment, nationalist interest in new weapons technologies and rearmament grew. The military possibilities of rocketry were mentioned more often in the press, among others by Nebel, who tried jingoistic appeals for funds. Nebel was a supporter of the ultraconservative German National People's Party and a member of its massive veterans' organization, the Stahlhelm (Steel Helmet), but he was not opposed to the Nazis. He had written to Adolf Hitler to ask for support for rocketry as early as January 1930, and to Hermann Göring and Josef Goebbels thereafter—to no avail.[28]

Becker and his assistants—von Horstig, Schneider, and Dornberger—shared Nebel's extreme right-wing politics even as they despised his slippery, self-promotional character. But the budget and the ambitions of the ballistics and munitions section were still modest. When Becker approached Heylandt in October 1931, his first interest was fundamental research into the technology. At the outset of the negotiations, Becker wanted to purchase the rocket-car engine for testing at Ordnance facilities, but in November he decided only to issue Heylandt a study contract on the correct form for the nozzle and combustion chamber. Using compressed air would eliminate the dangers and difficulties of measuring a burning engine, and the results would be equally valid for solid or liquid fuels. The company had already begun those experiments when it received the contract in December. The report, which appeared at the end of April 1932, showed that the commonsense assumption that a long and narrow nozzle was superior was not borne out. The best results were obtained with the largest angle of opening between the sides of the nozzles tested: fifteen degrees. Later experience would show that such an angle was still too small by a factor of two, but the study was the beginning of the Army's much more scientific—and secret—approach to liquid-fuel rocketry.[29]

After another long and unexplained delay, in October–November 1932 Becker's Section 1 negotiated a new contract with the Heylandt company for a small 20-kg-thrust liquid-oxygen/alcohol engine. It would be based on experiments done by the company at its own expense in the preceding months. Meanwhile, Ordnance had not finished dealing with the rocket groups and inventors. Becker and von Horstig had been distracted over the winter of 1931–32 by the claims of Wilhelm Belz of Cologne, who was said to have launched a liquid-fuel rocket to several hundred meters in altitude and six kilometers in range. Reflecting the increasing influence of the far right, Belz's claims had been strongly supported by a heavy artillery veterans' organization in Munich, one of whose members was in contact with the Nazi leader Rudolf Hess. It soon turned out, however, that Belz was a fraud. He had apparently used a Sander black-powder rocket to achieve an ascent much more modest than that claimed.[30]

The time wasted on Belz notwithstanding, Ordnance had made an important step toward building up its own liquid-fuel rocket program in its work with Heylandt. An even more important step came from renewed contacts with the Raketenflugplatz. Although earlier he had declared Nebel beyond the pale, Becker reversed himself in early 1932. He, von Horstig, and Dornberger visited the Raketenflugplatz more than once dressed in civilian clothing to look less conspicuous. After negotiations, Becker officially wrote to Nebel on April 23, 1932, inviting him to make a secret demonstration launch at the Kummersdorf weapons range, 40 km southwest of Berlin. The rocket was to eject its parachute and a red flare at the peak of its trajectory. The terms were very strict: If they were fulfilled Nebel would be paid 1,367 marks for expenses; if not, he would receive nothing.[31]

The demonstration was held in the early morning on June 22. In order to maintain secrecy, Nebel was to appear with his car at 4 A.M. outside the Kummersdorf range. A long aluminum launch rack with the rocket inside was mounted on the open-topped car. Traveling with Nebel were Klaus Riedel and Wernher von Braun. After a long roundabout trip on poor roads that may have damaged the fragile rocket, the group and and its Army hosts arrived at the launch site, which was, von Braun recalled, "covered with photo-theodolites, ballistic cameras, and all sorts of equipment we then never knew existed." Among the Army participants were Becker, von Horstig, Schneider, Dornberger, and Dr. Erich Schumann, a physicist who directed a small research branch of Section 1 and held a professorship at the University of Berlin. Schumann was close to the Nazi Party and would become a key administrator in the science policy of the Third Reich.[32]

The unlikely looking vehicle that Nebel and his assistants launched was 4 meters (13 feet) long, and the main body was only 6 centimeters (2.4 inches) in diameter. It weighed about 12 kilograms when fueled, had an engine with a water-cooling jacket in the nose, and the parachute and flare in a tail compartment with ineffective little fins. Around 6:30 A.M. the rocket was ignited and rose rather too slowly from its rack, swung lightly back and forth and then turned over into an almost horizontal trajectory. It reached no more than 600 meters in height after piercing the low cloud deck and crashed 1,300 meters (less than a mile) away without ever opening its parachute.[33]

Ordnance's observers made known their displeasure on the spot. The conclusion to Captain Schneider's launch report expresses their renewed distaste for Nebel:

> Once again it is apparent that Nebel works unreliably and that his assertions must be treated with the greatest skepticism, since in his meeting with our office he described the promised maximum altitude, 8 km, as no problem, yet at the launch he no longer would speak of this figure. Even the altitude that he guaranteed there, 3.5 km, was not reached in the actual test. For Testing Division, the conclusion must be reached that closer cooperation with Nebel is out of the question, even

> though he was able to produce liquid-fuel rocket with an engine that worked well for a duration of many seconds, because he makes assertions against his better judgment.[34]

In short, the Army thought he was a liar and refused to pay.

In the aftermath, Nebel repeatedly visited the offices of Army Ordnance to argue over the outcome, but it was a waste of time. Eventually the twenty-year-old von Braun visited Ordnance and found Becker, in contrast to Nebel's description, warm, knowledgeable, and scientific. The two established an immediate personal connection. Becker outlined Ordnance's objections to the Raketenflugplatz's unsystematic approach: "What we need first is accurate measurements and data. . . . How do you measure your propellant consumption, your combustion pressure, your thrust?" Becker also criticized the publicity-seeking approach of the group, which deeply offended his desire for secrecy. He offered the rocketeers a chance to work for him, but only "behind the fence of an Army post."[35]

The failed demonstration was a crucial turning point; from then on, Ordnance concentrated on building up its own in-house liquid-fuel rocket program. A drawing dated June 24, only two days after the launch, shows a proposed liquid-fuel test stand to be built at the Kummersdorf weapons range. The three-sided reinforced concrete structure was to be more than 6 meters wide and 7 meters long. Becker forwarded the proposal to Schumann and others on June 25. Although a marginal notation says that it was rejected, it was built by November. Thus, when Becker saw von Braun in July, he may well have been considering recruiting Raketenflugplatz people for Kummersdorf, although it is highly unlikely that he ever wanted Nebel.[36]

The young engineering student took the offer back to his companions, and lengthy debates followed. Nebel heatedly rejected Army bureaucracy and red tape—he was far too much of a loose cannon to stand for that. Nor did he ever comprehend the need for a complex military-industrial organization to force the development of rocket technology. In his 1972 memoirs he states, laughably, that if the Army had given him the money he could have built the V-2 by 1939. Klaus Riedel was also skeptical of the military; he wanted to found a rocketry and spaceflight corporation. That romantic idea was rooted in the movement's traditions, springing from powerful images of inventors and invention in popular culture. Von Braun, on the other hand, was more practical. He foresaw the need for military funding to master the daunting engineering task of building a complicated liquid-fueled, gyroscopically guided missile.[37]

One issue not discussed was the morality of working for the Army. As a rabid nationalist, Nebel obviously could not object on moral grounds, and although von Braun appears to have been apolitical and interested in little but space travel, he had been brought up in a very conservative family. At the beginning of June 1932 his father had been appointed Minister of Agriculture in the new reactionary cabinet of Chancellor Franz von Papen. The elder von Braun was one of the barons in the "Cabinet of Barons"—a government close to the Army and the old elites but lacking almost all popular support. With such a background, his son was reflexively nationalistic but not automatically sympathetic to the "vulgar" Nazis. The Reichswehr thus presented no political problem for the younger von Braun, and war in any case seemed very far away in the politically chaotic summer of 1932. The debates at the Raketenflugplatz were solely about how to exploit the Army's offer. In the unpublished version of von Braun's memoir article, he states:

> There has been a lot of talk that the Raketenflugplatz finally "sold out to the Nazis." In 1932, however, when the die was cast, the Nazis were not yet in power, and to all of us Hitler was just another mountebank on the political stage. Our feelings toward the Army resembled those of the early aviation pioneers, who, in most countries, tried to milk the military purse for their own ends and who felt little moral scruples as to the possible future use of their brainchild. The issue in these discussions was merely how the golden cow could be milked most successfully.[38]

It is a depressingly frank statement of an attitude common among inventors, engineers, and scientists in the modern era.

The upshot of the debates was that only von Braun would go over to the Army immediately. For the officers in Ordnance, his class background and parentage counterbalanced his youthfulness, but it was his intellectual ability that really won them over. Dornberger was "struck . . . by the energy and shrewdness with which this tall, fair, young student with the broad massive chin went to work, and by his astonishing theoretical knowledge." After completing only the first half of his mechanical engineering program at the Technical University of Berlin, von Braun was made a doctoral candidate in applied physics under Schumann's supervision at the University of Berlin. At the same time—on or about December 1, 1932—he began work at Kummersdorf, with liquid-fuel rocketry as his dissertation topic.[39]

He was not yet a regular civil servant; his position fitted into a pattern already established by Becker and Schumann. As a subdivision of Section 1, Schumann's "Center for Army Physics and Army Chemistry" worked on chemical weapons and other secret advanced research. With monies already limited in the 1920s and with further stringency coming from the desperate budget situation of the Depression years, the research was largely done by graduate students. As a doctoral candidate, von Braun did not receive a direct salary from the Army. Instead he was provided with a monthly stipend of 300 marks under a contract to continue "experimental series B (research on the liquid-fuel rocket)." Whatever his official status, when von Braun began to work at Kummersdorf, Ordnance's own liquid-fuel rocket program can fairly be said to have begun. Less than five years later he would be technical director of hundreds of people at Peenemünde.[40]

THE SUPPRESSION OF THE ROCKET GROUPS

Only two months after von Braun began work at Kummersdorf, Hitler came to power. On January 30, 1933, the leader of the National Socialist German Workers' Party was appointed Chancellor in a coalition cabinet dominated by members of the old elites—Prussian landowners, Army officers, bankers, and representatives of heavy industry. Von Braun's father was out of a job with the organization of the new government, although he would have been willing, by his own account, to serve in a Hitler cabinet if asked. It was not that he was enthusiastic for the Nazis—he was not—but he shared the catastrophic illusion of his colleagues that they had no choice but to try to use the Nazis' mass base to install a right-wing authoritarian regime. Within months, Hitler's minions ruthlessly eliminated other parties and considerably reduced the power of the old elites in the Nazi system. But the Army still retained some autonomy from political interference, and the coalition or "polycratic" (multiple power center) character of the National Socialist regime continued. Although the Third Reich successfully projected to the world the image of a monolithic totalitarian state, it was closer to a collection of warring bureaucratic empires. The resulting political battles would play a crucial role in the history of the rocket program and Peenemünde.[41]

The consolidation of a fascist government committed to the rearmament of Germany and to the elimination of internal dissent presented Army Ordnance with an opportunity to suppress the amateur groups. Even before 1933 Becker and his associates had attempted to keep rocketry secret in order to preserve the element of surprise against foreign powers. The Weimar constitution made it impossible, however, to place any controls over the amateur groups or even to punish Nebel for letting slip his contacts with the Defense Ministry. Of course it is also true that, until mid-1932, the officers in Ordnance hoped that liquid-fuel rocket development would make progress under the aegis of the groups or industrial firms, since they had little money for anything except solid-fuel rockets. But after the establishment of an in-house program and the Nazi seizure of power, they moved quickly to eliminate public discussion and experimentation.[42]

The early phases of Ordnance's campaign are shrouded in obscurity. The first victim may have been Rolf Engel, a rocket enthusiast the same age as von Braun. Engel had lived at the Raketenflugplatz and in 1932 had been Johannes Winkler's chief assistant in a project to build a larger rocket. Toward the end of that year, with the rocket a dismal failure and the money exhausted, Engel organized a government-financed relief project in Dessau for unemployed engineers, many of them from Winkler's former employer, Junkers Aircraft. After the Nazi seizure of power, the new rocket group even received offices in the famous Bauhaus school of architecture and design, whose occupants had fled the country. But Engel's project came to a sudden end on April 4, 1933, when the political police arrested him and a colleague. They were charged with "negligent high treason" for corresponding with prominent space pioneers in other countries. Before the charges were dropped, Engel spent six weeks in prison in the difficult conditions created by the mass arrests of the Nazi takeover. He contracted a case of jaundice and was ill for some time afterward.[43]

According to Engel, Becker and von Horstig had instigated the arrest and had wanted to do likewise against Rudolf Nebel and against Reinhard Tiling, a solid-fuel rocket experimenter on the North Sea coast. But Tiling had friends in the Navy, and Nebel had a high-level political connection in the person of Franz Seldte, leader of the Stahlhelm veterans organization, and Labor Minister in the Hitler coalition cabinet. (Winkler was protected because he had returned to Junkers in 1933, where he worked in secret.) It is certainly true that Nebel was supported by Seldte, but no documents have survived to verify Engel's claim that Army Ordnance ruthlessly tried to suppress all the amateur rocket groups in the spring of 1933, as opposed to a year later. Almost all of Engel's assertions are based on statements allegedly made to him by Nazi leaders in the mid-1930s and by Dornberger in the mid-1950s. Still, his story has an inherent plausibility, especially regarding his own arrest. The secret police came to confiscate all the Dessau group's technical materials after he was in jail, even though those documents had nothing to do with the nominal reason for his arrest. Ordnance must have wanted to put his group out of action.[44]

It is also possible that Becker would have wanted to suppress Nebel's work at the Raketenflugplatz that spring. Until late 1932 Nebel had found it difficult to raise money, but he generated a new wave of publicity in June 1933 with his latest and most bizarre project, the "Magdeburg Pilot Rocket." In August 1932 Franz Mengering, an engineer from the north German city of Magdeburg, had showed up at Raketenflugplatz espousing a crackpot theory (dreamed up by someone else) that the apparent form of the universe was an illusion and the surface of the earth was on the inside of a sphere! By developing a large rocket one could prove this thesis. Typically, Nebel did not send him packing, even though he, Riedel, and von Braun all emphatically rejected the theory. Instead, Nebel saw this idea as a new opportunity for raising money. With Mengering, he succeeded in borrowing 35,000 marks from city officials and local businesses for the launch of the first manned rocket during the Pentecost holidays in 1933. In a crazy stunt, a volunteer was to ascend in a large nose-drive rocket with a 750-kg-thrust engine and then jump out with a parachute. Nebel probably knew from the outset that an engine that large could never be built on time. In any case, the Raketenflugplatz had to settle for a 200-kg-thrust engine, which even so was the most powerful the group ever made.[45]

With that engine, Nebel and his associates attempted a number of times in June to launch a subscale unmanned version at Magdeburg. The result was a series of embarrassing failures, ending with a poor launch that smashed the rocket, but the group received some favorable newspaper and newsreel coverage anyway, which must have galled Ordnance. Afterward the remaining enthusiasts at the Raketenflugplatz gathered up the engine and pieces and reconfigured them into a "four-stick Repulsor," which was launched a few times over the summer of 1933 at lakes around Berlin. The last launch ever made by the group was on September 19.[46]

Meanwhile, Nebel had unleashed another round of his endless appeals for funds. In letters to the adjutant of the Reich Air Minister, Hermann Göring, Nebel argued the military potential of the rocket and played up all of his attempts to contact Nazi leaders since 1930. He clearly hoped to get around the hostility of the Army by going to the new Air Ministry, which served as a cover organization for the creation of an air force banned under the Versailles Treaty. Nebel's maneuver did not work, because the letters were routed to Army Ordnance, which did everything in its power to prevent him from receiving any government support.[47]

The game shortly became even more serious. Nebel wrote to England mentioning something about his previous contacts with the Defense Ministry. When Schneider was alerted to this in mid-October, he called the Gestapo, which replied that Nebel had already been ordered into its office and warned never to speak or write about those contacts again. He must already have been under mail surveillance. The incident probably caused a Gestapo raid on the Raketenflugplatz witnessed by Willy Ley. The Gestapo had also contacted the Air Ministry press spokesman about Nebel. Schneider phoned the ministry and told the spokesman that it "would be ideal if these things were not written about in the press at all," but at the very least all discussion of military applications and new technical advances had to be suppressed. It is the first recorded mention of Ordnance's desire to take rocketry into total secrecy.[48]

At the same time Nebel was in further trouble with his own colleagues and with the state. In late September Ley and retired Major Hans-Wolf von Dickhuth-Harrach, the president of the VfR since 1930, denounced Nebel to the state prosecutor for fraud and expelled him from the society. The VfR and the Raketenflugplatz had existed in an uncomfortable symbiosis; even though Nebel had been the VfR's Secretary, the society had formally kept its distance from some of his dubious projects, such as the Magdeburg rocket. The prosecutor found no legal grounds to charge him, which Ley attributed to Nebel's Nazi connections, but the report shows that he had stayed just inside the law or that the bookkeeping was too ambiguous to allow any conclusions. It did not hurt that Klaus Riedel continued to defend Nebel's actions. This nasty conflict basically reflected the collapse of the VfR and the Raketenflugplatz due to monetary problems and Nebel's personality.[49]

As a countermove, Nebel attempted to register the Raketenflugplatz as a society in its own right. But the group had numerous other problems as well. Its three-year lease on the land expired in July 1933, and access to the old ammunition dump became more difficult. Herbert Raabe, a VfR member and occasional visitor to the Raketenflugplatz, remembers being turned away by a soldier guarding the site when he came to visit in the late summer or early fall. After the lease was up, Army administrators also presented Nebel with a water bill for 497 marks that had accumulated, so it was later claimed, because of dripping taps in the buildings. Ordnance refused to take responsibility for the bill and intervened in December to deny Nebel's petition for the recognition of the Raketenflugplatz as a society.[50]

Meanwhile, Rolf Engel had returned to Berlin and had begun an effort to coordinate the remaining amateur rocket societies in the hope of salvaging something. By Engel's account, Nebel agreed to cooperate with him, even though Engel had quit the Raketenflugplatz two years earlier because he felt that Nebel had embezzeled its funds. They approached Wernher von Braun, then saw Karl Becker, who had been promoted to Brigadier General and appointed chief of the Ordnance Testing Division early in 1933. The meeting turned into a shouting match, Engel recalls, after Becker refused to offer them anything but secret work under Ordnance's control. Klaus Riedel had already contacted von Braun a number of times to arrange a rapprochement between Ordnance and Nebel. Von Braun told the Gestapo in July 1934, during an interrogation about his contacts with Nebel, that he had refused to talk to Riedel on the phone. Instead he had met his old friend about five times and warned him that "if Nebel continued his campaign against Army Ordnance, serious consequences could follow"—that is, arrest.[51]

In early 1934 the VfR folded. Its remaining members were taken into an obscure "Registered Society for Progressive Transportation Technology," which carried a few spaceflight articles in its journal from 1934 to 1937. Later in the 1930s another spaceflight society was founded, and it too published a journal, but the discussion was sustained only among a small band of enthusiasts. From the standpoint of the public, rocketry disappeared in 1934 because of the imposition of censorship. Even before the formal press controls were in place, the Army had arranged for the suppression of publications about the topic. Schneider stated in a letter drafted at the end of 1933: "Testing Division was recently forced to intervene a number of times against undesirable propaganda and press statements by Nebel." After a delay that annoyed Ordnance, Josef Goebbels's Propaganda Ministry finally issued a decree on April 6, 1934, banning all discussions of rocketry that mentioned either military uses or technical details.[52]

Rudolf Nebel was a survivor of truly amazing proficiency; all the harassments and problems failed to stop him. He set about working his connections with Labor Minister and Stahlhelm leader Seldte. In mid-1933 the Stahlhelm had been "coordinated" as a Nazi veteran's organization under the supervision of the SA, or Brownshirts, the Nazi paramilitary wing that had provided the thugs for street battles and the seizure of power. Although there was tension between the leaders of the SA and the Stahlhelm as a result of this enforced amalgamation, Nebel had a chance to use his connections to exploit the growing hostility between the SA and the Army. Under the leadership of Ernst Röhm, the SA was laying claim to being the mass army that would supplant the old military. It also made rumblings about the need for a "second revolution" because Hitler had made too many compromises with the capitalists and was not in favor of the immediate plundering of the Jews.

A full picture of Nebel's contacts with the SA will never be known; he conveniently omitted them altogether from his memoirs. Seldte may or may not have put him in touch with Röhm, but the Nazi Gauleiter (regional party boss) for Hamburg did arrange a meeting between the Nebel and the SA leadership through the intercession of an admirer who accepted Nebel's self-description as a poor inventor abused by the Army. Röhm was not present, but Nebel's cause was taken up by Obergruppenführer (Lieutenant General) von Krausser, who promised to talk to Röhm. According to Rolf Engel, who joined the SA in October 1933 and later became a Nazi student leader and SS officer, he also met von Krausser and, on another occasion, Röhm himself.[53]

In the meantime, Nebel had received an innocent inquiry from a scientific institute in Warsaw about the possibility of building a stratospheric rocket. He immediately wrote to Hitler, Goebbels, the Foreign Minister, and other authorities in an attempt to exploit hatred of Poland to gain support for his activities. After a call from the Reich Chancellery on February 23, 1934, Schneider wrote a memo indicating that he had stopped this Nebel initiative. The memo also reveals that the SA leadership had intervened with the Army on behalf of Nebel not long before.[54]

On March 10 Schneider drafted a letter to Röhm explaining the Army's reasons for rejecting Nebel. The document, which was routed through the Army High Command and the Defense Ministry, had a pleading tone that reflected the tension between the Army and the Nazi movement. It noted that on September 21, 1933, Hitler, Göring, and Interior Minister Wilhelm Frick had seen a demonstration of rocketry during their tour of Kummersdorf and that on February 8, 1934, the same tour had been given to Deputy Führer Hess and "a number of higher SA leaders." The Army wished to overcome the "mistrust that is obviously present" by giving an SA representative full insight into the much more systematic work going on at Kummersdorf under military auspices. Under separate cover an explanation of the Army's dealings with Nebel was to be sent as well.[55]

That expedient seems to have worked temporarily, but Nebel did not give up. In May Seldte wrote to the Reich Post Ministry asking that it support Nebel for the civilian purpose of developing mail rockets. The ballistics and munitions section was able to frustrate that initiative as well. But it had so far been unable to move the Gestapo to arrest Nebel, a matter about which von Horstig—Becker's successor as section head—had inquired in March. Nebel probably avoided arrest because of his connection with Seldte.[56]

At the beginning of June an opportunity finally presented itself. Nebel had arranged for the printing of a booklet entitled "Rocket Torpedo" in which he discussed the possibilities of rocketry for anti-aircraft defense, ballistic missiles, and gas attacks. Schneider and von Horstig immediately requested that Nebel be arrested for violating secrecy. Nothing happened right away. Nebel sent his brochure to the SA leadership, and on June 21 it intervened again on his behalf. Exactly one week later a representative of the Stahlhelm leadership called Ordnance and mysteriously requested an immediate confidential meeting regarding Nebel. Schneider and von Horstig were able to convince him that the Army's position was justified.[57]

Nebel's timing could not have been worse. The tension between the Army and the SA had reached a crisis point, and the generals were pressing Hitler for action. Hermann Göring and the chief of the still small SS, Heinrich Himmler, exploited the situation for their own ends by feeding Hitler false rumors of a planned coup by the SA. Beginning on June 30, 1934, the "Night of the Long Knives," SS execution squads shot much of the SA leadership in Munich and elsewhere, including Röhm and von Krausser. Army units provided logistic and backup support. In Berlin, Nebel was arrested and imprisoned at SS/Gestapo headquarters downtown. By his own account he was saved from a potentially worse fate when he was recognized by a police official who had been a frequent visitor at Raketenflugplatz. It was typical of Nebel that in short order he got better treatment and then was let out of jail. He returned to the old rocket site to find his materials and car confiscated on orders from Army Ordnance. Under the terms of the lease, he was told to vacate once and for all. Presumably he had not been evicted earlier because of his connection with Seldte.[58]

Nebel, astoundingly, did not give up. Already in the fall of 1934 he tried to make some kind of arrangement with the large engineering firm Rheinmetall-Borsig. He would come back to haunt the rocket program a few more times, and the people in Ordnance would just as determinedly frustrate him at every turn. He was not alone in trying. Rolf Engel, another bitter opponent of the Army, led a student group that built a test stand for amateur experiments at Siemensstadt in northern Berlin around the end of 1934. But that group disappeared within a few months.[59]

Since the SA purge had effectively given the Army a rocket monopoly, Ordnance now had the power to eliminate even minor irritants. When a young spaceflight enthusiast named Werner Brügel wanted to give a radio talk on rockets for stratospheric exploration, Section 1 moved to stop all radio discussions of the topic and gave the naïve Brügel a tongue-lashing in their offices. Reflecting the prevailing anti-Semitism, Schneider's record of the August 1934 meeting states that Brügel had "an unpleasantly Jewish way of speaking." The Gestapo showed up at Brügel's residence in Frankfurt shortly thereafter, arrested him temporarily, and confiscated his material.[60]

A different side of the Army's drive for a rocket monopoly was experienced by Arthur Rudolph, who had worked on the Valier–Heylandt engines in 1930–31. He had lost his job in mid-1932 because of the Depression, and then met his old boss, Alfons Pietsch, in the unemployment office. They wanted to start in rocketry again, so they tried going to the local SA leader in Berlin for help. Rudolph had joined the National Socialist party and the SA Reserve in mid-1931, supposedly for anti-Communist reasons, and would become the longest-serving Nazi of all the prominent engineers at Peenemünde. The SA turned Rudolph and Pietsch down in 1932, but in May 1933 they obtained a small contract from Army Ordnance to work in secret on their engine. Pietsch disappeared with much of the money, and Rudolph had to finish the job himself. When he showed up to demonstrate the engine at Kummersdorf in August 1934, Dornberger told him: "You either work for us, or you don't work at all." As part of the conditions for his hiring, Rudolph had to leave the SA, but not the Party.[61]

At least one other obscure group in Hannover survived the Army's campaign, only to be eliminated in 1936, and a minor plague of rocket "inventors" appeared between 1934 and 1937 to waste the time of the officers and engineers in Ordnance. Most turned out to be fraudulent or incompetent. One of the more credible was Hermann Oberth, who resurfaced in the summer of 1934 with a missile proposal sent in from Rumania. Because of his foreign citizenship and difficult personality, he was excluded from any participation in the program until 1941. The suppression of the rocket groups and the exclusion of unwanted personalities both contributed to Ordnance's single-minded aim in 1933–34: to use the mechanisms of the Army and the Nazi police state to concentrate development in its own team at Kummersdorf and to eliminate all possible threats to secrecy. But while all this activity was going on, under von Braun's leadership important strides were being made toward the establishment of liquid-fuel rocketry as a viable technology.[62]

FROM A-1 TO A-2

When von Braun began working on his dissertation at Kummersdorf in late 1932, even the modest program of mid-1934 would have seemed luxurious. The resources he had to work with were minimal:

> One-half of a concrete pit with a sliding roof was at my disposal, the other half being occupied with powder rockets. Also, one mechanic was assigned to me. I was instructed to give my work orders to an artillery workshop, which turned out to be loaded to capacity with other tasks, mostly of a higher priority than mine. The mechanics as to how my purchase orders were processed through the cumbersome administrative machinery remained for a long time an opaque mystery to me. It was a tough start.[63]

The "one mechanic" was an old hand from Raketenflugplatz, the skilled metalworker Heinrich Grünow. He was a help, but von Braun lacked the extensive engineering knowledge that might have made the practical job of constructing rocket engines easier.

While von Braun was grappling with those problems at Kummersdorf, the parallel Heylandt program continued. The clear intent of the autumn 1932 contract for a 20-kg-thrust engine had been to produce a laboratory instrument, since its thrust was only one-eighth of that of the 1931 Heylandt rocket-car. The small engine's weight in comparison to its thrust was such that the ballistics and munitions section had to fend off a serious challenge from an unnamed leader of Ordnance. He declared liquid-fuel rocket technology worthless, because obviously this engine could never lift itself off the ground![64]

Although it was clear that engines of higher performance characteristics could be built by Heylandt, Ordnance did not energetically pursue that option, presumably because of the Kummersdorf work. Von Horstig inquired about a 200-kg-thrust engine in December 1932, but six months later Ordnance submitted an order only for a 60-kg engine that Heylandt had proposed. The primary motivation appears to have been to prevent the layoff of Heylandt's rocket group. That engine was successfully tested on company grounds in September 1933. When the company offered in November to build an engine with up to 400 kg of thrust, however, Section 1 turned it down, and told Heylandt to expect no further contracts.[65]

Satisfied that the technology developed at Kummersdorf was superior, Ordnance further consolidated liquid-fuel rocket development in January 1934 by hiring Walter Riedel, the key engineer in Heylandt's group. Riedel, ten years older than the precocious doctoral student, provided the practical design experience von Braun lacked, plus the experience of having worked on rocket engines ever since Valier's original liquid-fuel experiments. It is indicative of how young the rocket group would be that he became known as "Papa" Riedel. As chief of the design office at Peenemünde when it opened in 1937, he was all of thirty-five years old.[66]

Notwithstanding the Heylandt work and the small spinoff contract to Pietsch and Rudolph, the main line of development had always been von Braun's. Starting from Raketenflugplatz designs, he built the alcohol/liquid-oxygen "1W" series (W for water-cooled), with a thrust of about 130 kg. Dornberger's memoirs give a picturesque description of the explosion that supposedly destroyed much of the test stand at the first test on December 21, 1932. That recollection is inconsistent with von Braun's own memory of the first test being a success in January 1933. In any case, explosions, leaks, and burnthroughs did follow. The redesign process was tedious and largely empirical, involving endless variants. Eventually von Braun was able to go to regenerative cooling with the "1B" series (B for *Brennstoff* or fuel) and then the "2B" series with 300 kg of thrust in the autumn of 1933 or thereabouts.[67]

Von Braun's program in 1933 had three main objectives. The first was development of engines based on aluminum alloys. Raketenflugplatz had begun using aluminum for the obvious purpose of saving weight and thus increasing the performance of launched vehicles. In the spring of 1933, troubled by the number of engine failures, von Braun went searching for expertise. "Solidly in Nebel's footsteps, I grabbed the telephone directory and got in touch with welding experts, instrumentation firms, valve factories, and pyrotechnical laboratories." He had learned something from the entrepreneurial methods of his former mentor.[68]

Soon engine parts manufacturing was farmed out to various firms, and von Braun and his superiors made contact in April 1933 with a firm that specialized in aluminum anodizing (*Eloxieren:* surface hardening through the electrolytic formation of an oxidization layer). This proved a crucial breakthrough in increasing the durability of engines. The firm had been working with Nebel, but Ordnance insisted that it cut off all contact with him. In turn that firm led von Braun and his superiors to a small manufacturer who would be the primary contractor for engine and alcohol-tank construction for three years: Zarges, in the southwest German city of Stuttgart. At first it would be a mutually agreeable relationship, but eventually the distance, secrecy considerations, and a desire for greater control over quality would result in a decision to manufacture in-house at Peenemünde.[69]

Von Braun's second objective was the fully automatic operation of ignition and tank pressurization. Proper ignition was a serious problem; if too much fuel or oxidizer reached the engine first and ignition was delayed, an explosion usually resulted. By the end of 1933 the problem was reasonably in hand, but it was never completely solved. Many experiments were also conducted to solve the old problem of how to pressurize the tanks. The weight-saving method of increasing liquid oxygen evaporation with small burning cartridges was tried, but putting gaseous oxygen in the fuel tank led to explosions. It became necessary to use compressed nitrogen or evaporated liquid nitrogen in the alcohol tank, although that meant a separate tank and system. Both forms of nitrogen were tried, but all the problems of tank pressurization remained. As the fuel drained from the tank, the gas would expand and the pressure would drop, resulting in a drop in the rate and pressure of propellant delivery to the engine over time. That meant a slow drop in thrust. Since the pressure of the burning gases in the rocket engine's combustion chamber was about ten atmospheres in the engines of that time, it was necessary to force the propellants into the chamber with a pressure of a few atmospheres higher. That meant the fuel and oxidizer tanks had to withstand at least fifteen atmospheres of pressure (in practice even more), which made them heavy. As rockets got larger, the structural weight problem was magnified exponentially. The limits on tank pressure also limited combustion chamber pressure, which limited performance, because higher-pressure engines are more efficient. It was already clear that complicated turbopumps would have to be developed for larger missiles to get around those problems, a solution already discussed in the works of Oberth and the other pioneers.[70]

Von Braun's third objective was the design and construction of the rocket itself. By June 1933 the drawings were in hand for the first vehicle, the Aggregat-1 ("Aggregate" or "Assembly"), better known as the A-1. It was based on the 300-kg-thrust engine, and its unique feature was its method of stabilization, which derived directly from its origins in an artillery establishment. A liquid-fuel rocket cannot be spun on its axis like an artillery shell or a solid rocket because of the disturbing forces on the propellants in the tanks and lines. As a crude interim solution, Dornberger proposed that only part of the vehicle be spun. Thus the nose of the A-1 was a large gyroscope that stabilized it by brute force. (A gyroscope's axis, like that of a top, will tend to remain fixed in space. If perturbed by an external force it will move or "precess" at a right angle to the force exerted. A gyroscope's resistance to precession is directly dependent on its angular momentum, a product of its mass and rate of rotation.) Before launch, the gyroscope would be spun up to 9,000 rpm by an electric motor on the ground, then left to run solely on its momentum during the rocket's brief flight.[71]

But the A-1 was never to fly. "It took us exactly one half year to build . . .—and exactly one-half second to blow it up," says von Braun, a bit hyperbolically. The late 1933 or early 1934 explosion was due to persistent difficulties with the fuel and oxygen valves, leading to delayed or hard ignition. Eventually the third A-1 was successfully started on the ground, but it was destroyed by the mechanical failure of the liquid-oxygen tank. Ordnance decided on a major redesign, entitled the

A-2 (see Figure 1.1). Von Braun's group separated the tanks and placed the gyro rotor between them. Moving the gyro to the middle had the advantage of bringing the center of gravity backward, thus shortening the moment arm of any deviations of thrust away from the rocket's axis. That increased the stability of the rocket in the early part of the flight, when aerodynamic forces were weakest because of the rocket's low velocity, although stability was actually decreased in the later part of the flight because the rocket's center of gravity was closer to its center of aerodynamic pressure. Separating the tanks also stopped the problem of leakage into the fuel tank caused by vibration-induced cracking of the oxygen tank.[72]

Figure 1.1 also shows a characteristic feature of the early German Army rockets. The engine was actually immersed in the alcohol tank, because it shortened the rocket and helped to cool the engine when the combustion chamber was so long. Difficulties in getting proper atomization and evaporation of the propellant droplets had driven von Braun and his co-workers toward longer and longer combustion chambers to give the propellant mixture more time to burn completely.[73] Incomplete burning was one of the main causes of suboptimum engine performance. The 300-kg-thrust engine had an exhaust velocity of about 1,500 meters per second, whereas the theoretical maximum for a 75-percent-alcohol/liquid-oxygen rocket is a little over 2,000 m/sec at the combustion chamber pressures then feasible—10 to 13 atmospheres. In the equation of the rocket, exhaust velocity is one of the absolutely critical values determining performance; the higher it is, the more efficient the engine. For comparison purposes, the most efficient rocket engine in use today, the Space Shuttle Main Engine, has an exhaust velocity of around 4,500 m/sec using liquid hydrogen and liquid oxygen at a combustion chamber pressure of about 200 atmospheres.

All the problems with the A-1 and the redesign they necessitated meant many delays to the schedule of the program. At the time that von Braun completed his dissertation in April 1934, the A-2 was still months away from being finished. There was nothing unusual about such technical setbacks. In the course of a year and a half, the young physics doctoral student and his few assistants had significantly outstripped the existing amateur rocket technology. The systematic ap-

FIGURE 1.1
The A-2 Rocket

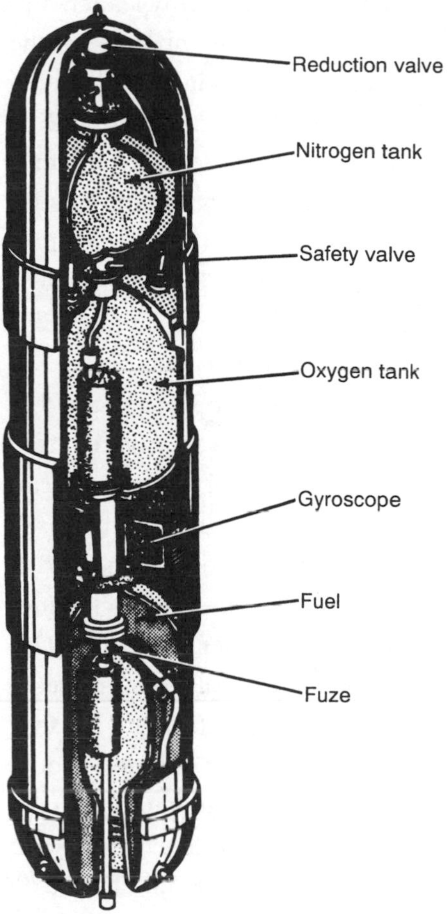

proach imposed by Ordnance had much to do with their success, but von Braun's brilliance was no doubt a factor as well. For his efforts, he received high honors from his dissertation committee, headed by Erich Schumann, when he defended it at the beginning of June 1934. The subject was so secret that even the title was classified. Von Braun's diploma carried a phony title instead: "Regarding Combustion Experiments."[74]

Steadily increasing resources were another crucial factor in the technological progress made at Kummersdorf. As small and inexpensive as the rocket program was in 1933–34, it benefited from the Nazis' commitment to rearmament and from Becker's rising star in that process. Becker cultivated close contacts with Hitler and the Army leadership. During or after Hitler's visit to Kummersdorf in September 1933, the Führer promised Ordnance even more resources than it had hoped for. Schumann's research branch was expanded to a Section of Testing Division and pursued some rocket research of its own. One of Schumann's students, Kurt Wahmke, who had graduated before von Braun, was involved in experiments with hydrogen peroxide as an alternative oxidizer in the spring of 1934. During a careless experiment, Wahmke mixed the hydrogen peroxide with alcohol to see if he could produce a premixed single propellant. An explosion killed him and two assistants.[75]

At the end of 1934 the A-2s were finally ready. Two were shipped to the North Sea island of Borkum for launching; they were called "Max" and "Moritz" after the twins in the German version of the popular cartoon strip *The Katzenjammer Kids*. Secrecy, safety, or both must have motivated the choice of the island as a launch site. Von Braun, Riedel, and four others arrived on December 10, followed by Erich Schneider and by Leo Zanssen, who had joined Section 1 no later than mid-1933. Rudolph came with the liquid-oxygen tanker after finally clearing security checks and being hired. Dornberger was not able to attend because he had been rotated into the first solid-rocket artillery unit on October 1 and could not be released from duty. That was doubly unfortunate for him, because unguided 11-cm-diameter solid rockets were also to be tested on Borkum for their possible utility in anti-aircraft defense.[76]

On December 19 the 12-meter-high A-2 launch mast and all the measuring and photographic equipment were ready. Only the weather was uncooperative, with gale-force winds on a very cold North Sea day. Because the approaching Christmas holidays left little time for waiting, the first launch was carried out anyway. "Max" functioned perfectly. The engine burned for sixteen seconds, and the rocket reached about 1,700 meters in altitude before a wind gust produced precessions that tipped the vehicle right over. It was found buried in the beach sand 800 meters away. The next morning at dawn "Moritz" performed almost identically. It was a gratifying beginning.[77]

With the launch of the A-2s, the German Army liquid-fuel program had completed its first phase. From its uncertain beginnings as a minor activity on the margins of a powder rocket program, it had evolved into a successful research project that had produced flyable rockets more advanced than any so far built in Germany. As a result of the Nazi seizure of power, it had also been possible to impose the ultrasecrecy Army Ordnance had wanted, while capturing much of the talent that had grown up in the spaceflight movement and amateur rocket groups. Secrecy also laid the cornerstone for Army Ordnance's "everything under one roof" philosophy of rocket research and development. The future was clear, if vague in details: It was feasible to build a large liquid-fuel ballistic missile with a range of hundreds of kilometers. But for that to become a reality, the state would have to invest vastly increased resources.

Chapter 2

The Founding of Peenemünde

In March 1935 the National Socialist regime publicly repudiated the Versailles Treaty, instituted conscription, and unveiled the existence of the Luftwaffe, which had been forming inside the Air Ministry. Adolf Hitler's assumption that the Western powers were too indecisive to respond effectively was proved correct. Encouraged by that international climate and by the steady improvement of the German economy, Hitler began to accelerate rearmament in 1935–36. Army Ordnance's budget for weapons research expanded rapidly, it being the development and procurement arm of the largest armed service.

The rocket program almost inevitably profited from the Army's improving situation, but it also gained from the Luftwaffe's spectacular growth in the mid-1930s. The air force's commander-in-chief, Hermann Göring, had great political power as "the second man in the Reich," and the Luftwaffe had high priority because of the need to overcome Germany's backwardness in military aviation quickly. As a part of its search for the latest technology, the Luftwaffe began to take an interest in the rocket as a propulsion system for high-speed aircraft. Out of that interest arose an interservice rocket alliance and a revolutionary new center: Peenemünde.

If the Army Ordnance rocket group wished to justify large budget increases, however, it also had to produce results. The A-2 launches on Borkum were an important step toward a bigger and better-funded liquid-fuel rocket program. In mid-January 1935 the group presented films, slides, and lectures about the expedition for the benefit of Becker, von Horstig, Schumann, and other leaders in Ordnance, as well as representatives from the Navy and the Air Ministry. The reaction of at least one unnamed officer was so enthusiastic that Wernher von Braun had to respond to the idea of developing a scaled-up A-2 as a missile with a range of 50 kilometers.[1]

The leaders of the liquid-fuel program viewed that idea as a threat to their one overriding objective: the development, in absolute secrecy, of a large ballistic missile. In his response, von Braun admitted that a short-range A-2-type weapon might be produced quickly and could provide useful experience to the Ordnance group. But the accuracy of a missile stabilized by a massive gyro would be very poor—even assuming that the difficulties of stability and air resistance on a trajectory too low to escape from much of the earth's atmosphere could be overcome. (As the Paris Gun had shown, projectiles went much farther if they arced high enough to greatly reduce the friction produced by the air.) Poor performance might damage the case for a large ballistic missile within the Army and would disrupt the "unitary line of development" enjoyed by the Ordnance group so far. By revealing Germany's rocket development, an A-2-type weapon would also undermine "the effect of surprise" on foreign powers. As an example of what could be achieved by a ballistic missile, von Braun mentioned a payload of 1,500 kilograms over a range of 400 kilometers—each about one and a half times the later A-4's parameters. To bridge the gap in range between conventional artillery and the missile, von Braun suggested, it would be more cost-effective to pursue gun development or to fire the missile with a larger warhead over a shorter distance.[2]

Those arguments were effective in stopping the unwanted proposal, allowing the Ordnance rocket group to make new investments along the line of development it wanted. In early February the head of the ballistics and munitions section, von Horstig, outlined to Testing Division chief Becker a budget of nearly half a million marks to expand the offices and test facilities at Kummersdorf for solid-fuel and liquid-fuel rockets. The centerpiece was a new test stand that could accommodate larger liquid-fuel motors, such as the 1,500-kg-thrust one planned for the A-3. The test stand was to be surrounded by a square earthen blast wall with openings for a locomotive that would tow into position test rigs holding engines and whole rockets. The expanded facilities at Kummersdorf received, sometime in 1935, the label "Experimental

Center West"—a reference to their location in the artillery range. But those facilities were almost immediately inadequate because of new opportunities created by the Luftwaffe's interest in the rocket.[3]

ROCKET PLANES AND THE LUFTWAFFE ALLIANCE

Before 1935 the Air Ministry Technical Office and its predecessor, Section 8 (aviation) of Ordnance Testing Division, had dismissed the possibility of rocket propulsion for high-speed aircraft. In line with the Army's drive for a rocket monopoly, the Ministry had made "agreements" with Ordnance, leaving the latter in control of the technology. All that was to change soon after the beginning of the year.[4]

The key figure in forging the interservice rocket alliance was Major Wolfram Freiherr von Richthofen, a cousin and squadronmate of the Red Baron of World War I fame. He was an ace himself, having shot down eight enemy airplanes in 1917–18. Later a Field Marshal and one of the Luftwaffe's most successful operational commanders, von Richthofen in early 1935 was the enthusiastic new head of the Technical Office's Development Division. Von Braun remembers a visit by him to Kummersdorf in January, during which he showed a lively interest in their projects. In early February von Richthofen wrote to Section 1 about an accident at the Junkers Aircraft Company works in Dessau. An explosion there had injured a company official and had revealed the firm's sponsorship of Johannes Winkler's liquid-fuel rocket development. A week later Captain Leo Zanssen and Wernher von Braun went to Dessau to investigate and to impress upon the company Ordnance's obsession with secrecy. The two concluded that Winkler's work was backward compared to their own but that it was useful for him to explore alternative technological paths. Most important, the Junkers job satisfied Ordnance's interest in keeping Winkler out of the public eye. The results of the investigation and a company report, probably written by Winkler, were passed along to the Air Ministry, which had awaited them with interest.[5]

The cooperation between Section 1 and the Technical Office did not immediately foster an interservice alliance. But a delegation including Zanssen, von Braun, the important aerodynamicist Dr. Adolf Busemann, the aircraft designer Willi Messerschmitt, and a number of Air

Ministry officials went to Munich in March to observe the privately financed experiments of Paul Schmidt. Schmidt, an engineer and independent inventor, had been working with meager resources to realize the idea of a pulsejet, a form of air-breathing reaction propulsion with intermittent combustion. In heavily modified form, Schmidt's invention would become the engine of the Luftwaffe-developed V-1 cruise missile or "buzz bomb" launched by the thousands against London and the Belgian port of Antwerp in 1944–45. In 1935, however, the Luftwaffe was interested mostly in the pulsejet's possibilities for aircraft propulsion, which Schmidt had been pushing since 1930. Zanssen and von Braun represented Ordnance, because it was thought that the Army might wish to pursue an automatic "aerial torpedo," i.e., a cruise missile. At that time the concept was perceived as closer to an artillery projectile than to an unmanned airplane.[6]

It was not the first time the Army had received a cruise missile proposal. In October 1934 the inventor Hellmuth Walter had contacted Becker about the possibility of an "aerial torpedo" based on a ramjet. (A ramjet is essentially a tube that compresses air solely by the ram effect of the front opening at high speeds. The air is then burned with a fuel—Walter suggested oil—to produce thrust.) Since a ramjet has to be boosted to a high velocity to work, Walter had proposed burning the fuel in a rocket engine with highly concentrated hydrogen peroxide until supersonic cruise velocity was reached. Since 1930 he had been working with the Navy on hydrogen peroxide as a propellant for ship and U-boat turbines and torpedoes. He had also had discussions with the Air Ministry in 1934 on the use of the rocket–ramjet combination in "high-speed aircraft." The ramjet idea was as yet too technically difficult, although Walter did carry out some preliminary experiments a few years later, but Ordnance began to act as a consultant to Walter's hydrogen peroxide rocket development in late 1934, without investing any money. In the case of Schmidt's pulsejet, on the other hand, Ordnance contributed half the research funds to the Air Ministry, which supervised the work. Ordnance accepted this arrangement mostly so that it could keep its eye on the technology.[7]

Whatever the technological conservatism of of Ordnance's former aviation section, the Air Ministry's contacts with Walter and Schmidt show that it had become receptive to radical new technologies like the jet. As a service without an entrenched establishment, the Luftwaffe was unusually open to revolutionary ideas. German theoreticians were also the international leaders in supersonic aerodynamics and high-speed flight. Those aerodynamicists, who had close connections to the Air Ministry, were aware that the piston-engine, propeller-driven aircraft in a decade or less would reach the limits of its performance. Moreover, technological zeal combined easily with a nationalist or National Socialist zeal for rearmament. The Luftwaffe was imbued, as were the Army and Navy, with a desire to make Germany competitive with, or superior to, other powers as rapidly as possible.

In this expansive and aggressive milieu, von Richthofen promoted the rocket plane as an answer to the problem of high-speed flight. Because the idea of the turbojet engine was unknown outside two or three small groups in Britain and Germany in 1935, and because pulsejets and ramjets seemed as yet far from practical, the rocket was the only reaction-propulsion technology available. On May 10 von Richthofen met Zanssen to discuss the possibility of a Luftwaffe–Army–Junkers experimental rocket plane program. Zanssen mentioned the aviation section's earlier indifference to the rocket. Von Richthofen indicated that he was of quite another opinion. In the future, bombers could attack at high speeds and at altitudes of over 10,000 meters (33,000 feet). They would be above the ceiling of anti-aircraft fire, and it would be difficult for slow-climbing, propeller-driven fighters to intercept them. (Like almost everyone else at the time, he did not anticipate the invention of radar to increase warning time). A rapid reaction, high-speed interceptor would therefore become essential. It was basically the concept that would later appear as the world's only operational rocket fighter, the Messerschmitt Me 163 Comet.[8]

On May 22, Section 1 replied with a letter drafted by Zanssen and approved by the Chief of Army Ordnance. It endorsed the feasibility of a joint rocket aircraft program but expressed great reluctance about revealing anything at all to Junkers. Ordnance ruled out working with the Winkler group altogether for secrecy reasons, because the primary application of the rocket was the "liquid-fuel long-range missile," a revolutionary weapon that could achieve its greatest psychological impact through sudden deployment. "A considerable development lead over foreign countries has been achieved here, the loss of which is regarded as intolerable, above all because of the value to national defense of the moment of surprise."[9]

A little over a month later, the ballistics and munitions section and the Technical Office held a major meeting to work out the terms of the alliance. Professor Otto Mader, Junkers Engine Company's development chief, also attended. For that June 27 meeting at Kummersdorf, twenty-three-year-old Wernher von Braun wrote an official position paper that must be regarded as Peenemünde's birth certificate. Because a rocket engine differed little according to its application, he stated, it is "therefore advantageous that in the future as well, the development of the free-flying liquid-fuel rocket and the aircraft rocket engine could be carried out by the same center. Section 1 believes that this goal can be achieved through the future creation of an 'experimental rocket establishment'." This center should have some air force personnel, but they would be transferred to the employment of the Army or the center.[10]

Von Braun went on to raise a second reason why the rocket group was reluctant to become involved with private corporations:

> Because the previous development of liquid-fuel rocket propulsion has been financed by the state, Section 1 continues to place decisive importance on an agreement that drawings and design documentation of all sorts grounded in that experience not be made available to industry. Otherwise there is the danger that profit-making opportunities in industry would arise from development the state has carried out at a considerable expense.[11]

Not only military secrecy was at issue; working with an aircraft firm raised the specter of the commercial exploitation of rocket technology, for example, through the construction of takeoff-assist rockets for heavily loaded airplanes, one of the goals of Winkler's work at Junkers. But that would contradict the direction in which the Ordnance group was going: toward a large, secretive military laboratory in which corporations were only subcontractors.

Von Braun's misgivings about a corporate role in the rocket program not only reflected the attitudes of his superiors; it also accorded with the rather empty anticapitalist rhetoric of National Socialism. But von Braun did not join the Party until asked to do so in 1937, and although he and the officers in the program showed every sign of enthusiastically embracing Hitler's rearmament and "national regeneration," none shows any sign of having been a Nazi ideologue. Ordnance's construction of an empire of Army-owned munitions factories during the Third Reich drew less on National Socialist ideology than on centuries-old traditions of state ownership in Prussia and Germany. Ultimately, however, the obsession with secrecy and surprise and the distrust of the independent groups and inventors were the crucial factors in the rocket group's desire to restrict corporate access to the technology. Secrecy and von Braun's success at Kummersdorf had launched the Army firmly down the path of in-house development.[12]

As von Richthofen pointed out at the June 27 meeting, Ordnance's restrictive conditions would effectively obstruct cooperation with an aircraft firm like Junkers. Nor would the Luftwaffe accept junior-partner status in any joint "experimental rocket establishment." That statement did not, however, undermine the friendly tone of the meeting. Von Richthofen then sketched his rocket interceptor concept: an aircraft that could, after a forty-five-second boost, coast up to 15,000 meters (50,000 feet) and then glide or cruise at high altitude for some minutes. As a preliminary step, a small experimental rocket plane could be tested, perhaps by towing it into the air and igniting the engine. The projected contractor was Junkers, he announced; von Richthofen had earlier cleared this arrangement with the company representative, Professor Mader.[13]

During the summer of 1935 the Air Ministry brought a second large firm into the program, Ernst Heinkel Aircraft. Its owner and namesake was fascinated by high-speed flight. The Ministry may also have wished to mollify him after his company lost the single-engine fighter competition of that year. In September Ordnance, the Technical Office, Junkers, and Heinkel signed a joint agreement protecting the secrecy of Ordnance rocket development. Only five or six people at each firm were to receive access to the plans and documents, and they were to work on rocket aircraft in closed shops. Winkler's name was conspicuously absent from the list of Junkers employees inside the charmed circle. Sometime after late October the Kummersdorf group received a Junkers "Junior" single-engine light plane and experimented with the installation of a 300-kg-thrust A-2-type rocket in the tail. The funding and arrangements for the tests were made through the Research Division of the Technical Office in collaboration with the quasi-governmental German Experimental Establishment for Aviation in Berlin. The experiments aimed at developing takeoff-assist rockets for overloaded bombers as well as gaining experience in rocket-plane work. For unknown reasons the Junkers firm itself dropped out of the picture in the fall of 1935.[14]

Ernst Heinkel's firm thus became the sole aircraft contractor. On October 16, 1935, at Heinkel's plant near Rostock on the Baltic coast, Wernher von Braun, Walter Riedel, and two Air Ministry engineers met company managers, including the short, bald, bespectacled owner himself. They discussed how Ordnance's rocket technology might be accommodated in an aircraft.[15] The eventual decision was to proceed with an interim project before the construction of a pure rocket aircraft. A rocket engine would be installed in the tail of a Heinkel He 112 single-engine fighter, the loser to the soon-to-be-famous Messerschmitt Bf 109 in the 1935 competition. In December the firm specified an engine thrust of 1,000 kg, which became the Kummersdorf "4B" series of motors. (The variants of the 1,500-kg engine for the ongoing A-3 project formed the "3B" series.) During the same month von Braun requested 200,000 marks from the Air Ministry for "Project 112 R," noting that speed was of the essence since the work had already begun.[16]

By the end of 1935, then, both the Junkers Junior and the He 112 projects had been launched as part of the new Army–Luftwaffe alliance in rocketry. But the most important product of this alliance was yet to come. Shortly after New Year's Day 1936, von Braun's concept of an "experimental rocket establishment" would bear fruit.

PEENEMÜNDE AND THE A-4

A new test facility had been in the minds of the rocket group for some time before von Braun's June 1935 memorandum. According to Dornberger, "our area at Kummersdorf had long since become too small for us. Even at the firing of our powder rockets we were never quite at ease" because of the possibility that the devices might go astray. Liquid-fuel rockets were likely to be even more unreliable. For that reason the A-2s had been launched from Borkum, but that involved an elaborate and inconvenient expedition. Secrecy was another consideration; engine testing was very noisy. Finally, developing a large ballistic missile obviously required much more extensive facilities, something that became feasible with the free-spending Luftwaffe on the scene.[17]

Dornberger, who returned to head rocket development on March 1, 1936, recalls that von Braun had been searching for an appropriate firing range along the Baltic coast since mid-December 1935. The young engineer found an excellent location on the island of Rügen, but the German Labor Front, the mandatory Nazi union for all workers and employers, had claimed it as a beach resort. It was thus quite by chance that von Braun found the perfect site:

> Christmas 1935 I went home to my father's farm in Silesia. I told my parents about the new prospects, adding that we were on the lookout for a suitable site from which it was possible to fire rockets over several hundred miles; safety reasons required this site to be situated on the coast.
>
> "Why don't you take a look at Peenemünde" my mother suggested. "Your grandfather used to go duck-hunting up there."
>
> I followed her advice and it was love at first sight.

His mother, a baroness in her own right, had grown up on a family estate in the region near the town of Anklam.[18]

The new site was located on the northern tip of the island of Usedom, about 250 kilometers north of Berlin. It was a sylvan wilderness of dunes, marshes, and forests inhabited by deer and many kinds of wild birds. The relative inaccessibility of the site provided good security, and an offshore island, the Greifswalder Oie, was available as secluded and safe launch site. But Peenemünde was not totally remote. A number of popular summer beach resorts began just south of the site along the coast of Usedom.[19]

Once von Braun had found the site, things moved quickly. On January 6, 1936, von Horstig and von Richthofen, both now lieutenant colonels, met at the Air Ministry regarding the layout of the new joint center. Twelve days later Ordnance sketched out a division of territory roughly corresponding to the eventual Luftwaffe and Army facilities at Peenemünde-West and Peenemünde-East. Construction was assigned to the Luftwaffe. Dornberger says that decision was made because the rocket group were enamored of the Luftwaffe's architectural style, which ran toward Nazi neoclassicism. Of much greater importance, however, was the "entirely new, fantastic, unbureaucratic, fast-moving, decisive" character of Luftwaffe administration, to use the words of project engineer Arthur Rudolph. The Ordnance group had often been frustrated by the ponderous and penny-pinching Army bureaucracy in the early years of the program.[20]

Choosing a site was one thing; arranging for the money to pay for it was another. The Air Ministry started the ball rolling with a promise of 5 million marks. The remaining funds came from the Army in a display of interservice rivalry. Von Braun gives this account:

> General Becker . . . was wrathfully indignant at the impertinence of the Junior Service.
>
> "Just like that upstart Luftwaffe," he growled, "no sooner do we come up with a promising development than they try to pinch it! But they'll find that they're the junior partners in the rocket business!"
>
> "Do you mean," asked [Lieutenant] Colonel von Horstig in astonishment, "that you propose to spend more than five millions on rocketry?"
>
> "Exactly that," retorted Becker, "I intend to appropriate six millions on top of von Richthofen's five!"
>
> In this manner our modest effort[,] whose yearly budget had never exceeded 80,000 marks, emerged into what the Americans call the "big time."[21]

The Luftwaffe leadership had little problem spending such a large amount of money on the rocket, but the Army High Command needed further convincing. According to Dornberger, "Becker told me in January 1936, 'If you want more money, you have to prove that your rocket is of military value.'" In March—probably not long after Germany successfully remilitarized the Rhineland in defiance of the Western powers—the Commander-in-Chief of the Army, General Werner Freiherr von Fritsch, was brought to Kummersdorf for rocket motor firings and a detailed briefing. After hearing Ordnance's plans for the future, he posed the blunt question: "How much do you want?" The answer must have been even more than the 6 million marks promised by Becker. The estimated construction cost of the Army side of Peenemünde alone was 11 million for 1936 and 6 million more for 1937–39, both amounts to be shared equally with the Luftwaffe. A yearly operating budget of about 3.5 million marks was projected for the Army facility.[22]

Ultimately, to plan for the facilities and to justify this expense to the Army leadership, the ballistic missile had to become more concrete. In late March Walter Dornberger, Wernher von Braun, and Walter Riedel met to specify the characteristics of the A-4. The probable size of the engine was already known. According to Rudolph, the next engine-thrust goal had been set in late 1935 quite arbitrarily at 25 metric tons (55,000 pounds), nearly seventeen times more powerful than the 1,500-kg A-3 engine. Starting from that figure, estimates of engine efficiency, and the ratio of fueled to empty weight of the missile, it was possible to calculate combinations of range and payload. Dornberger cut through the discussions of his subordinates by laying down the following specifications:

> I am an old long-range artillerist. The most famous gun up to that time was the Paris gun. . . . This gun fired 22 pounds over a range of 78 miles, but possessed terrible weight in the firing position and a terrible dispersion.
>
> I wanted to eliminate this unhandy weight of the gun in the firing position by using a single-stage liquid fuel rocket to be launched vertically, and to be programmed later into an elevation of 45 degrees. The rocket should carry a hundred times the weight of the explosives of the Parisian gun [i.e., 1,000 kg] . . . over twice the range . . .

Furthermore, he wanted accuracy superior to conventional artillery: 50 percent of the missiles were to fall within a circle of two to three "mils"—artillery language for 0.2–0.3 percent of the total range. At the specified range of about 250 kilometers this accuracy was about half to three-quarters of a kilometer, far better than was feasible, it turned out. The missile's fins were also to be narrow enough to fit through a standard European railroad tunnel.[23]

Dornberger's specifications reveal the flawed thinking that lay behind the German missile program from the outset. The Paris Gun had been the greatest technical accomplishment of German artillerists up to that time, yet it had failed to have much effect on the French in 1918. The gun was a triumph of narrow technological thinking: the technical fascination of being able to break through traditional limits and fire over such unprecedented distances had overwhelmed any rigorous analysis of its likely impact on enemy morale. The interwar German artillery community completely failed to grasp that point, however. Those specialists, led by Becker, saw the gun only in terms of artillery reaching its technological limits in muzzle velocity and range.[24] Using the rocket as a ballistic missile certainly promised to eliminate the massive railroad-borne gun carriage and supporting equipment, to abolish all limits on range and to increase payload vastly. Yet the Army Ordnance missile enthusiasts must have understood that the investment required by the Reich would be huge, even if they underestimated the ultimate expenditure on Peenemünde and the A-4 by many times.

The most fundamental flaw in their thinking lay in the lack of any well-thought-out strategic concept of how the missile could actually affect the course of a war. Becker, Dornberger, and their associates counted on the psychological shock to the enemy of an unfamiliar and powerful weapon. Once that surprise had passed, they could only picture using the missile as a fairly accurate artillery shell against specific military and industrial targets. Interwar air power advocates, like the Italian General Guilio Douhet, had asserted that the strategic bombing of enemy cities would lead to the collapse of civilian morale, but the artillerists apparently ignored those theories. Dornberger, for example, did not contemplate using the A-4 as a terror weapon against whole cities until 1941. A comparison between the missile and the heavy bomber would have raised uncomfortable questions in any case. The development of that type of aircraft, unlike the ballistic missile, did not require a revolutionary leap in the technology of flight.[25]

Thus, in a fundamental sense the A-4 was another Paris Gun. It was the product of a narrow technological vision that obscured the strategic bankruptcy of the concept. The fact that Dornberger was also a spaceflight enthusiast, like his chief liquid-fuel rocket engineers—von Braun, Riedel, and Rudolph—only reinforced his tendency to substitute technological enthusiasm for careful strategic thought.

Given the flawed military logic of Ordnance's ballistic missile program, it is curious that the German Army leadership embraced it so readily. The blinkered strategic vision of German generals during the era of the two world wars is one likely explanation. The training and traditions of the Prussian Army officer corps after the Napoleonic era emphasized operational and tactical excellence at the expense of strategy and grand strategy. The political irresponsibility and strategic incompetence that tradition fostered were only furthered by a lack of effective civilian control over the military in the authoritarian systems of Prussia and the German empire. The result was the paradoxical combination of "battlefield brilliance" and strategic blundering that contributed so much to the "German catastrophe" of the twentieth century. In this context, it is easier to understand how Becker, Fritsch, and other generals might overestimate the effects of the A-4, or their predecessors the impact of the Paris Gun.[26]

The Army also embraced the ballistic missile because it was the pet project of the artillerists, a branch of the service very prominent in the Army leadership after that ultimate artillery war, World War I. Until Hitler himself took over command in December 1941, all the Army Commanders-in-Chief during the Third Reich came out of the artillery, as did Wilhelm Keitel, Hitler's chief of staff in the Armed Forces High Command (OKW) after 1938. Every Chief of Army Ordnance in this period, including Becker himself from 1938 to 1940, was an artillery man as well.[27]

Moreover, Becker and Dornberger probably argued that there was an international missile race in which the Germans had to stay ahead. Zanssen's 1935 letter to the Air Ministry had mentioned Germany's "considerable development lead" over other countries. In February and August 1936 Ordnance received news of Robert Goddard's activities in the United States, including, in all likelihood, his new Smithsonian report, which contained the first substantial public information about his rocket development since the early 1920s. Nothing in that report would have shaken the Ordnance group's confidence in its lead, nor did they glean any significant new technological concepts from Goddard. The report was not that specific, his patents were unavailable to the Germans, and almost everything he had done had been anticipated in the German-language literature or at Kummersdorf. But the American pioneer's advances sufficiently paralleled their own to provide an argument for a race with the United States. Dornberger would use such an argument during World War II in his battles for top priority.[28]

The expensive and militarily questionable ballistic missile program profited as well from the highly advantageous political and military context of the mid-1930s. Becker's great personal influence on armaments development kept growing; in early 1938 he became Chief of Army Ordnance. Although the Army's autonomy from Hitler and the Nazi leadership decreased rapidly over time, the senior service could still make some decisions independently, at least in technical questions. If the Army, supported by the politically influential Luftwaffe, wished to invest a lot of money in rocketry, no one in the Nazi hierarchy was likely to object, particularly as the Führer took little interest in what was still, for him, a small program. In any case, rearmament became an irresponsible free-for-all after 1935. Like children in a candy store, the services wanted everything they could order, and Hitler's demands for as rapid a buildup as possible, combined with weak coordination from the top, resulted in the showering of money upon politically favored programs until the system came up against shortages of skilled labor, foreign exchange, and raw materials in the late 1930s. It was a context in which the ballistic missile program and the Peenemünde rocket center could flourish.[29]

In the spring of 1936, with the basic configuration of the A-4 in hand and with a firm commitment from the Luftwaffe jointly to fund the Army half of the facility, it was necessary only to get final clearances from the top. On April 1 General Albert Kesselring, head of administration in the Luftwaffe, put his seal of approval on the construction plans for the Peenemünde project. The very same day an Air Ministry official was sent "in a high-powered car" to purchase the land. In Dornberger's words: "Here was action indeed!" Within weeks bulldozers began to cut into the pristine wilderness to build the facility that would further revolutionize the technology of the rocket.[30]

THE ALLIANCE DECLINES

The huge increases in spending brought the program increased bureaucratic stature within Army Ordnance. In the summer of 1936 the rocket group was elevated to an independent section under Dornberger. After a number of changes of name and the redesignation of Testing Division as Development and Testing Division in 1938, the rocket section was given the acronym *Wa Prüf 11* (Ordnance Test 11).

With his promotion to section head, Dornberger came into his own as the chief administrator of the solid-fuel and liquid-fuel programs. He had contributed a great deal to those programs in the early years; in 1935 he received an honorary doctorate, which Becker had arranged as Dean of the new Faculty of Military Technology at the Technical University of Berlin. But before he left for active duty from late 1934 to early 1936, Dornberger had always been under the command of von Horstig and Schneider, who now moved on to other positions outside the rocket program. The smiling, smooth-talking Dornberger proved himself to be both a master at salesmanship and bureaucratic maneuvering and a talented engineer in control of the complexities of his field. He became a respected leader among the core group of engineers at Kummersdorf and Peenemünde, acquiring the stereotypical Bavarian nickname "Seppl" for his tendency to wear Alpine lederhosen (leather shorts) on informal occasions. From his office in Berlin he energetically protected the rocket group from outside interference. Although extremely loyal to Becker and Ordnance, he was not above fudging paperwork or going outside the strict chain of command to get around some obstacle created by Army bureaucracy.[31]

Dornberger's counterpart among the engineers was Wernher von Braun, who possessed prodigious quantities of charm, tact, intellect, and leadership ability, not to mention social position and impeccable manners. The accounts of his subordinates and acquaintances show that his charismatic character quickly overcame doubts raised by his youth. Dornberger's only complaint was that von Braun tended to have too many ideas and enthusiasms and did not always stick to exactly what he should be doing. But that did not stop Ordnance from making the twenty-five-year-old aristocrat the head of the "East Works" of the joint "Peenemünde Experimental Center" when it opened in May 1937. He was provided with an initial staff of 123 white-collar and 226 blue-collar workers.[32]

Under Luftwaffe management the construction of Peenemünde had proceeded quickly enough to move most activities from Kummersdorf after only a year. The interservice character of the center was embodied in a Commandant's Office, headed by an Army general, that united the air force's West Works and the Army's East Works. Peenemünde-West took up about 10 square kilometers on the northwest corner of the island and featured a landing field for experimental aircraft plus hangars and administrative buildings. It would remain a moderately sized test station throughout its existence, because, unlike its Army counterpart, Peenemünde-West was not destined to become a research-and-development center. To its southwest, Peenemünde village contained a liquid oxygen plant and a new harbor. Peenemünde-East was a long strip of land stretching from the northern tip of the island down the Baltic coast to the small beach resort of Karlshagen, which was taken over for the center. Just north of Karlshagen, a "settlement" with pleasant steep-roofed rowhouses and apartments was created for the personnel of both services. It was linked to the other parts of the center by a local railway using worn-out old passenger cars. Somewhat farther to the north was the East Works' administrative, laboratory, and workshop area. Located here were the headquarters building ("House 4"); the manufacturing shops, headed by Arthur Rudolph; and the buildings of the other major subdivisions, including Walter Riedel's Design Office and the new Measurement Group built up under diploma engineer Gerhard Reisig from the fall of 1937. Everything was constructed in a comfortable, even lavish style, and as far as possible the trees were left in place for camouflage.[33]

For safety and secrecy reasons, the areas farther to the north were reserved for the test stands. The largest of them in the initial plan was Test Stand I for large rocket engines. As with the layout of the shop facilities, the rocket group thought big. This test stand was designed to take not merely the 25,000-kg thrust of the A-4 engine, but up to 100,000 kg (220,000 lb) of thrust, the projected next step for an even larger missile. Because the test stands would not be ready until 1938 and 1939, however, propulsion development had been left at Kummersdorf under the control of Dr. Walter Thiel. Thiel was a brilliant but often mercurial chemical engineer who had replaced Dr. Kurt Wahmke in Erich Schumann's Research Section after Wahmke had been killed in the 1934 accident at Kummersdorf. In the fall of 1936 Thiel was transferred to Dornberger's new rocket section to pursue development of the 25-ton engine. His group was to remain resident in Kummersdorf until the summer of 1940, although he did travel frequently to Peenemünde, especially after Test Stand I became operational for the big engine in the spring of 1939.[34]

While Thiel remained at Experimental Center West, the Test Stand Group was set up by Klaus Riedel, the Raketenflugplatz alumnus who had spent the intervening years at the large electrical engineering firm of Siemens in Berlin. He had come to Peenemünde in a unique way. In 1936 Rudolf Nebel had emerged once again to haunt Army Ordnance. He had launched yet another campaign picturing himself as a persecuted and ignored inventor and had succeeded in enlisting the support of Dr. Fritz Todt, the builder of the autobahns and head of the Nazi Party's engineering and technology organizations. About the same time Nebel received a belated patent with Klaus Riedel on Raketenflugplatz engine development. In order to neutralize him, the Army agreed to pay the two men 75,000 marks for the patent. As a part of the July 1937 agreement, Nebel was sworn to secrecy and excluded from the program. But not so Klaus Riedel, because von Braun wished to recruit his old friend for the expanded staff of Peenemünde. Riedel brought with him three other Raketenflugplatz veterans who had been at Siemens.[35]

While the joint facility at Peenemünde was being built and staffed, the cooperative rocket plane program had continued as the other principal pillar of the Luftwaffe–Army alliance; indeed, it was the air force's rationale for funding half the construction cost of the Army side of the facility. (Peenemünde-West was paid for solely out of Air Ministry funds.) Engine tests on the Junkers Junior began in early 1936. The objective was to gain further experience in installing and operating a liquid-fuel rocket engine in an aircraft, preparatory to putting one in an He 112 and then building an actual rocket fighter. In April von Braun wrote to the Air Ministry noting that a number of test firings had already been made, but the engine needed to be redesigned, and in any case it shifted the center of gravity of the airplane too far back.[36]

That fact would have been relevant only if the craft were to be flown. Walter Riedel remembers von Braun wanting to pilot it himself. With his dream of spaceflight and his family's wealth, it is not surprising that the young aristocrat had gone to glider school in 1931 and 1932 and had acquired a private pilot's license for powered aircraft in September 1933. At Kummersdorf the Army had supplied him with an airplane for business trips since about 1935. At some point it was even a Junkers Junior. To fulfill a military service obligation, from May to July

1936 he attended a Luftwaffe flying school, and he took further courses in 1938. As a result, he held the status of pilot in the air force reserve. But he was never allowed to fly the rocket-equipped Junior, because his superiors were understandably concerned for his safety. Plans to fly the aircraft were canceled anyway because of numerous technical problems with the redesigned 300-kg-thrust engine. After explosions and burnthroughs of the new lightweight design, many further changes had to be made, and the experiments lasted until at least August 1936.[37]

The Junkers Junior ground tests became primarily a pathfinder program for the bigger Heinkel project, which was funded by the Development Division of the Air Ministry Technical Office. During 1936 Kummersdorf designed, ordered the construction of, and test-fired the new "4B" series of 1,000-kg-thrust engines (later scaled down to 725 kg). An additional spectacular set of ground tests evolved from cooling and burnthrough problems in the Junior experiments. For the first time Kummersdorf's rocket engines were fired horizontally rather than vertically, changing the fuel flow through the cooling jacket around the engine nozzle. The Heinkel works was concerned that the forces of acceleration acting on a turning rocket aircraft would disturb the flow of the liquid propellants around the engine, creating further overheating and burnthrough problems. Sometime in 1936 the von Braun group built an iron "carousel" or centrifuge at Kummersdorf, 10 to 12 meters across. The engine at the end of one arm was counterbalanced by propellant tanks and gas pressurization bottles at the other. Riding on the pivot in the middle was an armored control booth. Von Braun operated the engine controls and a brake while being spun around with the device at ten to twenty revolutions per minute. The feared effect of centrifugal force on engine cooling was not found.[38]

Around the end of 1936 a "4B" motor was installed in an He 112 fuselage, but there were still distressing explosions and failures, usually caused by delayed ignition. Modifications were imperative. A sort of pilot light, or small flame, became the new igniter. To fit a rocket to a manned aircraft meant that the thrust had to be throttlable and the controls simple. The aim was to provide the pilot with a single lever that controlled engine thrust by controlling the release of the nitrogen gas that forced the propellants out of the tanks. But the Ordnance group did not have high confidence in this system. When the Luftwaffe test pilot assigned to fly the rocket-equipped He 112, Captain Erich Warsitz, first came to Kummersdorf, he stood beside the aircraft and watched as von Braun started the engine from the cockpit. The noise was ear-splitting. Later that night von Braun told him in a Berlin bar that he had witnessed the first time ignition had been done from inside the aircraft. Usually the engine was controlled from a concrete bunker many meters away. But von Braun and one of Heinkel's designers had feared that Warsitz would never get in the cockpit if he observed the engine test that way![39]

Eventually the He 112 was flight-ready. On June 3, 1937, Warsitz attempted to ignite the rocket engine in the air for the first time after taking off under normal power. The experiment was carried out at an out-of-the-way airfield at Neuhardenberg, north of Berlin, because Peenemünde-West was unfinished. Warsitz started the ignition flame and then attempted to turn it off. Since it would not go out, he ignited the engine at half-power to prevent overheating. The acceleration was mild, and after ten seconds he stopped it again. The official report states:

> After a short gliding flight, the pilot noticed a strong acrid odor of burning rubber and paint and clearly perceptible hot gases flowed under the pilot's seat. The pilot looked to the rear and noticed a strong flickering in the tail area. The airplane at this time was still at an altitude of about a hundred meters. Because the pilot had to fear that the mobility of the control surfaces would be compromised by the fire in the tail section, and because the nitrogen for fire extinguishing was completely exhausted, he decided on an immediate landing. Sufficient altitude to extend the landing gear was no longer available. The aircraft landed with fully extended flaps on its belly and skidded about 45m along the ground.

Damage was significant. An unanticipated region of low aerodynamic pressure around the tail had caused alcohol fumes to be sucked back into the fuselage, where they were ignited by heating or the ignition flame.[40]

The aircraft was repaired and flown at least a few more times over the summer by Warsitz. But the safety of the system was doubtful, so a redesign was in order. Ordnance, Heinkel, and the Air Ministry decided to use turbopumps instead of gas pressurization as the means of forcing the propellants into the combustion chamber and to employ an electric glowplug instead of an ignition flame. The development of turbopumps had begun as far back as mid-1935 because of the anticipated need for them in large rocket engines. Employing them in an aircraft had been discussed in early 1937, either for the He 112 or the pure rocket aircraft project, "P 1033," which received the official Air Ministry designation Heinkel He 176 in December 1937. From the fall of that year, Ordnance engine development for the Luftwaffe followed two parallel tracks: similar but not identical turbopump-driven motors and tankage systems for the He 112 and 176. Innumerable design problems, however, caused both aircraft to lag farther and farther behind schedule. The He 112 equipped with an Army alcohol/liquid-oxygen engine was not to fly again until the autumn of 1939. The tiny He 176 never flew with such an engine at all.[41]

Through 1938 and the first half of 1939, interservice relations between individuals in the rocket plane program continued to be cordial. A new project was even added. In August 1938 the Air Ministry asked the Army to design an alcohol/liquid-oxygen takeoff-assist system for heavily loaded aircraft. Two teardrop-shaped pods of 1,000-kg thrust each would be strapped under the wings of an airplane and then jettisoned and parachuted to the ground after use.[42]

But the Luftwaffe strove, from 1936 on, to acquire independence from the Army in the rocket field, and the alliance declined after 1937. From late 1935 on, the Air Ministry became more and more deeply involved in Hellmuth Walter's hydrogen peroxide rocket development in the port city of Kiel. In March 1936 Walter, who had recently set up an engineering company with the assistance of the Navy, notified Ordnance that he no longer needed consultation in rocketry because he had received Air Ministry contracts for takeoff-assist rockets and engines for "aerial torpedoes" and aircraft. Parallel to the cooperative program with the Army, the air force had Heinkel build He 112 and He 176 versions with the less efficient but more practical hydrogen peroxide rocket. Other aircraft were experimentally fitted with Walter motors as early as January 1937.[43]

For military use in the field, peroxide had a number of advantages over liquid oxygen, which had a tendency to freeze valves and boil away. Hydrogen peroxide (H_2O_2) in high concentrations (80 percent or more) was not easy to deal with either, because of its tendency to explode, but Walter was able to develop a system for handling it. He could also offer two different versions of his rocket, "hot" and "cold." In the "cold" version, the inherently unstable peroxide was run over a catalyst and decomposed into superheated steam and oxygen; in the "hot" version catalyzed peroxide was burned with a hydrocarbon fuel, producing more thrust. Because of their adaptability, Walter's rockets had useful applications in a number of Luftwaffe projects, despite their relatively poor efficiency compared with Ordnance's engines. As time went on, the air service committed itself more and more to this technology. Around the turn of 1938–39, it initiated the Me 163 rocket aircraft project at Messerschmitt by combining Walter motors with the radical tailless, delta-wing glider designs of Alexander Lippisch. The main line of takeoff-assist systems came to be designed by the Kiel company as well. Apparently the Air Ministry had asked Ordnance to design a liquid-oxygen/alcohol version in 1938 only because of fears that hydrogen peroxide would be in short supply.[44]

If the Luftwaffe's cooperation with Walter suggests a gradual evolution away from dependence on the Army rocket program, the strange story of Eugen Sänger indicates that at least some people in the Technical Office wanted to lay the groundwork for independence even at the height of the alliance. Sänger (1905–64) was an Austrian spaceflight enthusiast and engineer who had pursued rocket research at the Technical University in Vienna beginning around 1929. He put in a proposal to the Germans in 1934 after its rejection by the Austrian military. Although the Ordnance rocket group was aware of his useful publications, it was not keenly interested, but it did ask for an investigation of his political views by the SA, which no longer posed a threat to the Army after the bloody purge of mid-1934. (Ordnance was unaware that Sänger had briefly been a member of the Austrian Nazi Party and SS in 1933.) Because the SA never responded, and because Sänger was not a German citizen and his work apparently was no more advanced than von Braun's, Ordnance did not take up his proposal. Zanssen suggested that the Air Ministry might be more interested, because the Austrian had made theoretical investigations into rocket aircraft. But in October 1935, after the founding of the alliance with the Luftwaffe, von Braun recommended that the Air Ministry not hire Sänger since his efforts would be duplicative. A 1937 document suggests that the aristocratic young engineer may well have perceived the Austrian as a rival.[45]

In February 1936 the Research Division of the Technical Office hired Sänger anyway for a projected massive aeronautical research establishment near Braunschweig in north-central Germany. Shortly thereafter, the Air Ministry gave him the funds to create a rocket research institute at Trauen, some distance from the main establishment. That institute received a cover name, the Aircraft Test Center, with the apparent intent of obscuring its existence as much from the Army as from foreign intelligence services. The Air Ministry allegedly even asked Sänger to change his name. He refused but signed documents with only the initial "S." His group began to test a 1,000-kg-thrust liquid oxygen/diesel oil rocket motor in 1939 and drew up a design for a 100-ton-thrust engine, but in 1942 the Air Ministry ended its duplicitous attempt to set up a large rocket-engine program parallel to the Army. There was skepticism about its value to the Luftwaffe. Besides, intense disagreements had arisen between Sänger and the director of the Braunschweig establishment. In any case, the Austrian engineer had never received funds adequate to challenge Peenemünde-East in the high-stakes rocket business.[46]

The decision to hire Sänger and pursue an independent course may have come about because of the decline of von Richthofen's influence. He left the Development Division in early 1937, six months after Göring had appointed Ernst Udet, a renowned World War I fighter ace, to head the Technical Office. In the long run Udet was a disastrous choice, being such a poor administrator, but had the virtue of subservience to Göring's wishes. The expenditure on Sänger's facility at Trauen could not have been made without the approval of Göring or Udet, and it would have been consistent with their desire to assert independence from the Army.

A final indication of the decline of the Luftwaffe–Army alliance is the breakup of the joint facility at Peenemünde less than a year after it was opened. The cause primarily lay elsewhere than in Luftwaffe policy. A certain Brigadier General Schneider (no relation to the former rocket group member Erich Schneider) was appointed as Commandant of the Peenemünde Experimental Center. As an old-fashioned officer of the combat engineers (Pioneers), he proved to be a poor choice. He was very bureaucratic and even threatened Rudolph with legal action for ordering large quantities of materials in advance. Rudolph's policy ensured fast progress in development work at the cost of some wastage, but it ran against the tradition of ordering what was needed according to the lowest bid and waiting months for it to show up. Schneider had another run-in with Peenemünde-East over who was to control the receipt of shipments, and he wanted to meddle in technical correspondence between von Braun and Dornberger's office in Berlin.[47]

The Peenemünde organization simply did not work. The two facilities were under the authority of the Commandant for some functions and under their respective service administrations for others. The demise of that unwieldy arrangement was hastened by Schneider's obstructiveness. By the fall of 1937 he was on bad terms with Dornberger and everyone else. According to an October 28 memorandum from Dornberger to Schneider: "The constant small conflicts with the Construction Office and the RLM [Reich Air Ministry] have led to RLM's desire to separate itself from the Commandant's Office. The basis for maintaining a general as Commandant has therefore been eliminated." The Air Ministry officially separated Peenemünde-West from the joint command on April 1, 1938, after notifying the Army that the 1937–38 payment of 1 million marks into the development and test budget of Peenemünde-East would not be continued. In the future the air force would return to paying the actual costs of any work done. As a result of the separation, what remained of the facility became the Peenemünde Army Experimental Center, and Schneider was retired. In the summer of 1938 Leo Zanssen returned to the program as Commandant after serving more than two years as a battery commander and administrator in the solid rocket/chemical warfare units (the *Nebeltruppen*).[48]

The joint character of Peenemünde had thus collapsed in less than a year, although daily cooperation and coordination were still a necessity. While the separation was neither a direct product of interservice rivalry nor of a Luftwaffe policy to assert its independence, the net effect was to loosen the rocket alliance further. It is symbolically appropriate that Peenemünde-West eventually put up a fence around its perimeter.

SUCCESSFUL FAILURE: THE A-3

The momentous events of the years after 1935—the rise and decline of the Luftwaffe alliance, the experiments with rocket aircraft, the founding of Peenemünde, and the birth of the A-4—tend to overshadow the primary job of the engineers at Kummersdorf and Peenemünde, which was to carry forward the development of the Ordnance rocket series and its associated technologies. Following the A-2 success in December 1934, the Army planned an ambitious new step. Not only would the A-3 be much bigger, necessitating greatly increased thrust, but above all it would require an active guidance system to replace the crude stopgap measure of a single massive gyroscope.

Developing the new "3B" series of 1,500-kg-thrust engines was the most straightforward part of the job. The basic concept of the A-2 engine was just scaled up: Welded inside the alcohol tank was a long cylindrical combustion chamber. The length of the chamber was intended to give the propellants more time to burn completely. A double wall allowed regenerative cooling by the circulation of the watered alcohol before injection. The injection system was, however, changed under Walter Riedel's influence to one similar to the Heylandt systems. Whereas the "2B" engines, derived from Raketenflugplatz designs, had only fuel and oxidizer jets pointed at each other, the "3B" engines had a mushroom-shaped injector sticking down from the top of the engine. From the underside of the cap, alchohol jets sprayed upward against liquid oxygen jets coming down from a number of small injectors at the top of the combustion chamber. This innovation increased exhaust velocity from 1,600 m/sec to more than 1,700 because of more efficient combustion, with a resulting increase in performance. But this necessarily worsened the cooling problem because of the increase in the temperature of the burning gases.[49]

The solution was an endless series of experiments with different aluminum alloys and variations on the basic engine concept. After successfully testing a steel configuration of the 1,500-kg engine in the summer and fall of 1935, the Kummersdorf group went on to test aluminum alloy engines and tankage built by Zarges in Stuttgart and a few other firms that had been let in on the secret. Secrecy was such an obsession that in early 1935 manufacturers were asked to send shipments to a shadow firm under Rudolph's name in a town next to Kummersdorf, rather than use a military address. The inconvenience of shipping the highly secret components across the country to Kummersdorf and, not infrequently, back to the firms for repairs was a factor in the decision to concentrate manufacturing capability in Rudolph's workshops when Peenemünde was planned and built. Another factor was growing dissatisfaction with the work of the primary contractor, Zarges, whose small company was based in Stuttgart. Zarges lacked the highly skilled welders necessary to carry out precision work on difficult alloys, and it was not easy to find manufacturing capacity elsewhere. Those experiences reinforced the Ordnance group's preference for an Army-run facility with "everything under one roof."[50]

By contrast, designing and building a three-axis, gyroscopic guidance and control system for a flying rocket was beyond the capability of Army Ordnance. In this case, contracting the whole problem to a company was unavoidable. In 1933 or early 1934 the Navy recommended that the rocket group contact Aerogeodetic, a firm primarily based in Berlin. The Navy had surreptitously bought the Dutch company in 1926 and had used it as a cover for secret work, mostly in heavy ship-based gyroscopic navigation and fire-control systems. A year or two after the Nazi seizure of power the company changed its name to Kreiselgeräte GmbH (Gyro Devices, Ltd.) and gave up the headquarters in the Netherlands that served as a front.[51]

The heart and soul of Kreiselgeräte was its technical director, Johannes Maria Boykow (1879–1935). Von Braun gives a striking description of the man:

> Boykow was one of the strangest and most charming characters I have ever met. A former naval officer of the Imperial Austrian Navy, he had seen the whole world and knew how to spin a yarn. Before the First World War, he had quit the services to become a dramatic actor. Drafted back when the war broke out, he became a destroyer captain, a naval aviator and finally got in touch with torpedo development. And it was here that he ran into the problems of the gyroscope which were to concern him for the rest of his life. He acquired hundreds of patents and

> gradually became the [German] Navy's No. 1 expert in gyro compasses and . . . fire control equipment. He was a true genius, but . . . he did not bother much about the mundane engineering phases of his inventions. His company's design office often found it necessary to deviate considerably from his original ideas, and therefore the end products but vaguely resembled his initial proposals. Unfortunately, I found this out only after severe setbacks. When I first met Boykow, I was left spellbound by his analytic sharpness and imagination and, being a novice in the gyro field myself, I took everything he said for granted.

By October 1934 Boykow had begun designing what would become the A-3 guidance and control system. For the task he could draw on experiments with aircraft autopilots he had made independently of Kreiselgeräte. But he died not much more than a year later, leaving it to the company to complete.[52]

After preliminary laboratory experiments with stabilization in one axis, Kreiselgeräte assembled the first version of what it called the "Sg 33" in mid-1936. Its final form for the A-3 is illustrated in Figure 2.1. The Sg 33 had the function of simply holding the rocket to a vertical course, yet it was, in the end, too complicated for the technology of the time. Two gyros were to hold a stabilized platform horizontal. When the rocket tipped in pitch (nose backward or forward) or yaw (side to side) the corresponding gyro wheel, spinning at 20,000 rpm, would move ("precess") at right angles, as the laws of physics dictate. This movement was sensed by electrical contacts, which in turn released nitrogen gas through small nozzles to push the platform back into place. (Unlike succeeding systems, the platform gyros had no direct influence on the attitude of the rocket.) Located on top of the platform were two devices to measure the movement of the rocket in a horizontal direction away from the initial vertical trajectory. The primitive accelerometers used little wagons on tracks to convert horizontal acceleration into a measurement of horizontal speed, which was then sent to the control system of the rocket. Under the platform were three "rate gyros." Their function was to measure the rate at which the rocket was turning away from its specified direction, whether in pitch, yaw, or roll (turning around the longitudinal axis). The signals from the rate gyros were used to push the rocket back into its initial vertical attitude.[53]

FIGURE 2.1
Kreiselgeräte's Sg 33 Stable Platform for the A-3

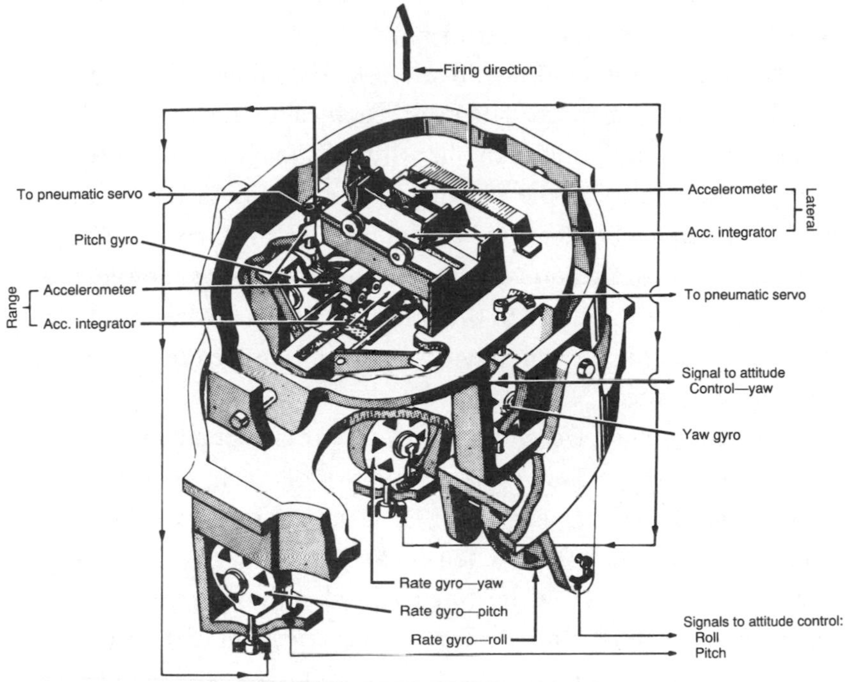

The control forces commanded by the wagons and rate gyros were sent to "jet vanes" in the rocket exhaust, which deflected the direction of thrust—an idea anticipated by Oberth and other pioneers. (Goddard had already experimented in New Mexico with jet vanes and a less ambitious gyro system as early as 1932.) But it was no easy task finding materials that would withstand the fiery temperatures and erosion of a rocket exhaust. The Kummersdorf group were finally able to develop, in conjunction with a contractor, molybdenum and tungsten vanes that were at least adequate to the task, but only after hundreds of test failures. Those vanes were rotated by rods that came down from electrical servomotors in the guidance system at the top of the A-3.[54]

Also guiding the rocket were the long, narrow fins that gave it longitudinal stability or, to use the more picturesque German term, "arrow stability." They ensured that, when the vehicle pitched or yawed around its center of gravity, the lift forces generated by the fins would tend to force the vehicle back to its original position—nose-on into the airflow—so it would have an inherent aerodynamic stability like an arrow. (In technical terms, the fins ensured that the rocket's center of pressure was behind its center of gravity.) Finding the appropriate shape for the fins was another difficult task. The Luftwaffe alliance was helpful here, because in late 1935 the Technical Office was able to introduce von Braun to one of the handful of supersonic wind tunnel groups in the country, at the Technical University in Aachen, near the Dutch and Belgian borders. An assistant professor there, Dr. Rudolf Hermann, made the preliminary drag measurements that allowed a calculation of the performance of the rocket. He then worked on the fin form so that stability through the whole range from zero velocity to supersonic was assured.[55]

At the beginning of December 1937, a year later than von Braun's 1935 estimate, four A-3s were finally ready for launch. They were not small: 6.5 m (22 ft) long and 0.7 m (2.3 ft) in diameter, with a fueled weight of 750 kg (1,650 lb). Each rocket carried registering instruments to measure either the heating of the skin through friction or atmospheric temperature and pressure during a parachute descent from a peak altitude of 20 km. The launch site was the Greifswalder Oie, the small island with high cliffs a few kilometers offshore from Peenemünde. Ironically, it was the same location Oberth had requested for the launching of his *Frau im Mond* rocket in 1929, only to be refused by the Prussian authorities because he might endanger the lighthouse there. In the Third Reich, however, a request from the military was not likely to be turned down.[56]

Converting the island proved to be a major task and expense for the Army, because about the only thing on the island, except for the lighthouse, was the combination farmhouse–guesthouse run by the island's lessee. It was fortunate that a small-gauge railway had been left in place from the erection of the lighthouse, because there were no roads. When a liquid oxygen tanker truck was brought over to the island, the launch crew spent hours trying to pull it out of the mud. The Army New Construction Office built a dock, a launch bunker, a generator building, and temporary barracks. Telephone lines were strung to link the buildings. A large tent was put up in a wooded area for workspace. An ancient ferry was leased to haul the equipment from the mainland.[57]

Toward the end of November a select crew of about 120 individuals from Peenemünde and Berlin, headed by Dornberger, Zanssen, and von Braun, assembled on the island for "Operation Beacon." Most were new to the launch business, as the rocket program had grown so much since the A-2s. Enthusiasm ran high, which was fortunate, because conditions were trying. The weather became miserable: It rained for days, which delayed the launches, and then it was bitterly cold. The wind threatened to tear the tent pegs right out of the ground. An "extraordinary plague of mice and rats" emerged to gnaw on the tar paper of the bunkers, so that constant tearing sounds could be heard, and rain seeped through after ten days. More ominously, the field mice showed a taste for cable insulation, causing short circuits. Technical delays in the launching tried the patience of the many high-level visitors and caused problems with the launch organization, because so much was on loan from other organizations—airplanes from the Luftwaffe, boats from the Navy, photo and measuring equipment from other branches of Ordnance.[58]

Finally, about 10 A.M. on December 4 the crew managed to launch the first A-3, patriotically named "Deutschland." For the first three seconds the rocket ascended vertically, then suddenly the parachute popped out of the side, trailed behind the still accelerating vehicle, and was incinerated. The rocket turned into the wind, and the engine shut off automatically when it tipped over too far. After about twenty seconds it crashed back onto the island only about 300 meters from the launch site, exploding violently on impact. According to Dornberger: "Eyewitness accounts were wildly contradictory. Everyone claimed to have seen something different. We decided to venture on a second launching." When the second A-3 was sent on its way two days later, virtually the same thing happened, with the vehicle crashing only 5 meters offshore. Now that it was clear that the parachute had been deployed, it was only natural to blame the powder charge that pushed it out. The parachute was omitted for the third launch on December 8, and a signal flare was put in its place. The wind was stronger than on earlier attempts, and the rocket turned quickly into it, ejecting the flare after four seconds. Again the engine cut out automatically and the rocket crashed 2 kilometers out to sea. The last attempt on December 11 was almost identical.[59]

Those results were shocking and discouraging, but already during the many interminable delays on the Oie, Dornberger, von Braun, and the chief engineers threw themselves energetically into explaining the failures. Attention focused initially on the possibility of a static electricity buildup on the skin of the rocket, setting off the parachute charge. But ground tests conducted later in December indicated that that was definitely not the case. The fact that the A-3 tended to turn into the wind rather than stay on a vertical course also implied that the control system was too weak. The rocket appeared to be excessively stable; the fins had apparently moved the center of pressure so far back that the jet vanes lacked the power to fight back against the aerodynamic forces. The servomotors that moved the vanes also seemed to lack sufficient power, a consequence of the undeveloped state of this technology.[60]

While it was in fact true that the control system was too weak and the rocket too stable, those problems did not explain the ejection of the parachute or flare every time. Only after review of the launch films, repeated ground tests, and meetings with Kreiselgeräte did it become clear in January 1938 what had gone wrong. The Achilles heel of the Sg 33 guidance and control system was its inability to stop a rapid rolling of the A-3. For reasons of simplification, the stable platform had no ability to turn around the vertical axis and no roll gyro to sense whether the rocket was moving in that axis. If the vehicle rolled at a rate of more than six degrees per second, the forces acting on the platform gyros would quickly overwhelm their ability to compensate for the precession induced by the rolling. When one of the gyros hit the end of its allowed range of motion (30 degrees), it would lurch back, and the platform would tumble over, losing its ability to control the vehicle. The circuitry for letting out the parachute had been linked to the platform on the assumption that at the peak of the trajectory the rocket would turn over, upsetting the platform. But the fundamental flaw was that the control forces exerted by the jet vanes, on command of the rate gyro for the roll axis, were far too weak. Assymmetries in the fins, in conjunction with wind, would be enough to start a roll that overpowered the control system. In every case this had happened so fast that the platform toppled in the first three or four seconds.[61]

Because of a lack of experience, no one in Ordnance or Kreiselgeräte had seen this coming. Kreiselgeräte specialized in heavy naval systems, and the engineers in Army Ordnance had been completely dependent on the company and on Boykow's original design. Thus von Braun came to rue his uncritical enthusiasm for the late inventor. It was clear that much more effort and resources had to be put into guidance and control and that competing companies had to be pulled into the program. Dornberger and his subordinates saw as well that it was a mistake after the A-2s to conclude that frequent launches were unnecessary. A new vehicle, called the A-5 (since the A-4 designation had already been assigned), would have to be built to test guidance systems systematically in the air, rather than only with the burning rocket on a test stand, as had been done at Kummersdorf. The excessive stability of the A-3 also confirmed earlier impressions that the wind tunnel testing at Aachen had been far too limited to give an adequate understanding of the forces acting on a flying rocket. Only the engine system of the A-3 had worked without a hitch. But the 25-ton-thrust A-4 engine was a huge step that required a massive infusion of resources and more systematic work. That at least had begun with Thiel's transfer from Schumann's research section and the construction of Peenemünde.[62]

The failure of the A-3s thus confirmed and strengthened the trend that had begun in 1936. If breakthroughs in the key technologies were needed to build something as revolutionary as a ballistic missile, the rocket program would have to spend much more money and build much more in-house expertise. But the A-3s were the epitome of what von Braun later called "successful failures" in the rocket business. So much had been learned from this experience that, given the highly favorable political and budget climate of the late 1930s, the technical obstacles could almost certainly be overcome. In the years between 1936 and 1941 the Army Ordnance group would do precisely that.

Chapter 3

Breakthrough in Key Technologies

Notwithstanding the important advances the Army group had made in the A-3 and rocket aircraft programs, the technological challenge of the A-4 remained gigantic. The engine would have to be seventeen times more powerful than the largest rocket motor so far constructed; the missile would have to fly at nearly five times the speed of sound when no Ordnance rocket had even approached the sound barrier; and the vehicle would have to be guided to targets nearly 300 kilometers away, when no liquid-fuel rocket built by the Germans had ever traveled more than a few thousand meters vertically. The A-3 failures only underlined how far away the engineers were from solving the guidance and control problem in particular. Yet by late 1941 Peenemünde had in its possession the technologies essential to the success of the A-4, and the first versions of that rocket were on the test stand.

The foundation for that remarkable technological achievement was Ordnance's ability to mobilize money, manpower, and matériel for the ballistic missile project—something it was able to do because of the high priority placed on rocketry by the Army High Command. Access to resources alone, however, did not automatically lead to the dramatic breakthroughs necessary for the A-4. Under the leadership of Becker, Dornberger, and von Braun, the liquid-fuel program had to expand its engineering staff greatly, put innovative leaders at the head of critical projects, and gain control over additional research capability in universities and corporations. The research process itself had to be altered so that trial-and-error testing was replaced, where possible, with a more scientific and theoretical approach, although that became apparent only over time. The result, especially after the A-3 guidance failures, was to accelerate further the growth of the large government laboratory at the heart of Peenemünde-East. At the beginning of 1938 the facility had 411 employees. By September 1939 that number had tripled.[1] Although no figures are available for late 1941, the number of people in development (as opposed to the new A-4 Production Plant) must have nearly tripled again to at least three thousand engineers, craftsmen, and office workers. With that vastly expanded staff came a corresponding increase in the facilities and materials available for research and testing.

While access to additional university and corporate laboratories was essential to the project, the massive buildup of in-house research and development capability was a critical factor in Peenemünde's success. It was not enough to attract highly talented engineers who could produce fundamentally new ideas, nor did it suffice to have those individuals led by excellent managers like von Braun and Dornberger. Only the possession of a lavishly funded and staffed organization allowed the rocket group to create working technology in a very short time. Dornberger's in-house or "everything-under-one-roof" philosophy made a further contribution by fostering internal communication and increasing efficiency. In combination, these assets and strengths gave Peenemünde mastery, in only five years, of the three technologies key to the A-4's success: large liquid-fuel rocket engines, supersonic aerodynamics, and guidance and control.

THIEL AND THE BIG ENGINE

Walter Thiel's transfer to the rocket section toward the end of 1936 was a milestone on the road to the ballistic missile. Within months his analytical and scientific approach would result in a reconsideration of the entire direction in which engine design had been proceeding under Walter Riedel and Wernher von Braun. Their 1,500-kg-thrust motor, the one that powered the A-3 and the A-5, was a big step forward in size and efficiency, but it was taking the Ordnance group down a dead-end road. Based on practical experience and the limited theoretical calculations in von Braun's 1934 dissertation, Kummersdorf's engines had become longer and longer. That had been done to give fuel and oxidizer droplets enough time to evaporate, mix, and burn properly. But the 25-ton engine threatened to become completely unwieldy, and the efficiency of combustion in the 1,500-kg motor still left something to be desired. It was significantly below the target performance—an exhaust velocity of about 2,000 m/sec—that would be needed to get the most out of the chosen combination of alcohol and liquid oxygen at a combustion chamber pressure of 10 atmospheres.[2]

Thiel, a pale, dark-haired, intense individual in horn-rimmed glasses, fitted one of the stereotypes of the German scientist of the Nazi period. He was loyal to the regime but too focused on his work to be very political. As far as is known, he never joined the Party. In the style of the German university professor, he could be authoritarian and arrogant to his subordinates. He was also high-strung and subject to episodes of depression when under stress; Dornberger and von Braun had to smooth over many conflicts. But Thiel brought to the rocket group a doctorate in chemical engineering, keen theoretical insight, tremendous ambition, and an imaginative mind. As Wahmke's replacement in the research section, he had been a consultant to Hellmuth Walter, had experimented with hydrogen peroxide engines in the laboratory himself, and had supervised a graduate student working on the fundamental processes of combustion in a Heylandt 20-kg-thrust motor.[3]

Thiel's initial program in early 1937 continued to focus on basic research into all areas of rocket propulsion, including exotic propellants like liquid hydrogen. He also outlined ambitious plans for cooperation with academic institutes in developing more heat resistant metal alloys, a better theory of combustion, and more thorough temperature and composition measurements of burning exhaust jets. Thiel was forced, however, to depend on Ordnance's own resources at Kummersdorf and Peenemünde. Although von Braun's group had been working with two or three academic institutes in aerodynamics and measuring techniques since 1935–36, the Army's obsession with security kept contacts with research institutions to a minimum before the outbreak of World War II.[4] Ordnance's goal was to develop and produce a ballistic missile in the deepest secrecy and then to use it without warning during a war. For any hint of the German rocket program to reach the outside world not only would ruin the effect of surprise but might also encourage other powers to pursue the technology more intensely. Virtually all proposals for contracting research outside Ordnance were therefore rejected to minimize the danger of security leaks.

Despite that handicap, in 1937–38 Thiel came quickly to four of the innovations that would make an efficient 25-ton-thrust motor possible. The first was an injection system that greatly improved the atomization and mixing of the two propellants. The 1,500-kg engine had used a modification of the old Heylandt system, with a mushroom-shaped injector extending down from the top of the motor, spraying watered alcohol upward toward the liquid-oxygen injectors. Dornberger claims the credit for having suggested small "centrifugal" nozzles that tended to atomize propellant droplets more completely, while spraying them outward in a rotational motion that produced better mixing. Thiel promptly began working with the Schlick firm, which produced the nozzles. By July 1937 he had demonstrated that fitting an injector with centrifugal nozzle holes to a 1,500-kg motor produced an immediate increase in exhaust velocity from 1,700 to 1,900 m/sec. A higher exhaust velocity meant a more efficient use of propellants and also improved the steering forces of the jet vanes by up to 20 percent.[5]

A further improvement in performance was promised by mid-1937 experiments with the "pre-chamber system." This second innovation placed the injector holes for both fuel and oxidizer in their own small chamber on top of the combustion chamber, producing better mixing before burning. Moreover, it helped to prevent heat damage and burn-throughs by keeping the flame front farther from the nozzles. Figure 3.1 shows the later configuration of the A-4 25-ton motor with eighteen of these small injection chambers.[6]

Two other Thiel innovations fundamental to the success of the A-4 took longer to emerge but can be glimpsed in the thorough and scientific research program that he laid out in the summer of 1937. The first was shortening the combustion chamber. Throughout 1937 Thiel and his assistants at Kummersdorf carried out a number of experiments of the most varied types. They included designing a small 100-kg motor for basic research and using gasoline and compressed air for ease of handling in repetitive testing. He quickly came to the conclusion that the volume of the combustion chamber was crucial to efficient burning, not the length. Further experiments in 1937–38 proved that it

FIGURE 3.1

The "Eighteen-Pot," 25-Ton-Thrust A-4 Engine

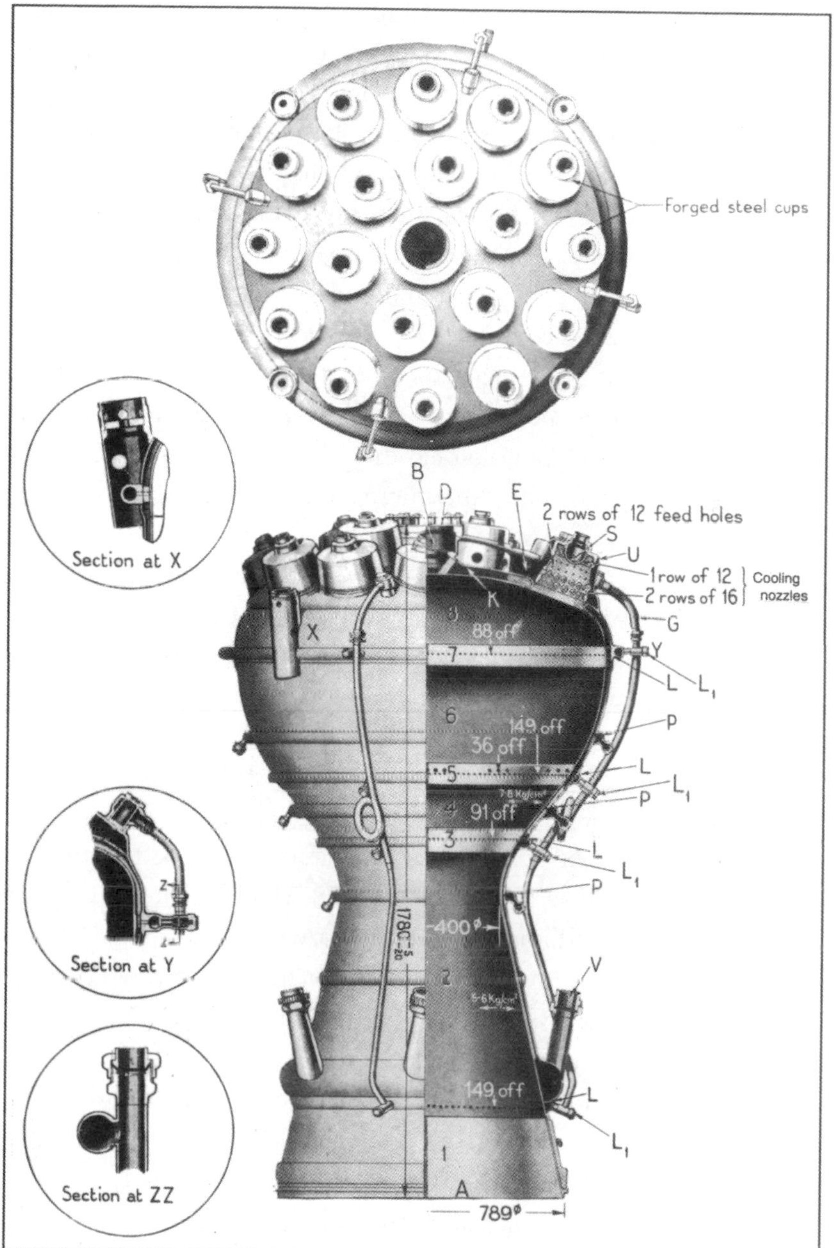

British War Office illustration, 1946

would be possible to reduce the length of motors by enlarging their cross-section. Better injection systems also contributed to more complete combustion, lessening the need to give the propellant droplets a relatively long time to remain in the chamber. As shown in experiments on a new 1,500-kg motor, it was therefore possible to reduce the volume of the combustion chamber to a fraction of that of the old engine. A short chamber with a nearly spherical shape also lessened other problems inherent in long engines, including pressure fluctuations and poor mixing. The 25-ton engine, Thiel decided by August 1938, could be dramatically shorter than earlier planned, thus making it much easier to manufacture and incorporate into the rocket.[7]

In the summer of 1937 Thiel realized that the question of the length and shape of the engine nozzle needed to be reexamined too. He created a systematic test program that led directly to his fourth innovation. Although the compressed-air experiments at Heylandt in 1931–32 had shown that a long, very narrow nozzle was inefficient, von Braun's group had not carried that fundamental research further. Practical experience and the existing engineering knowledge of nozzles seemed to confirm the accepted norm after 1932: a 10–12-degree angle of opening between the sides of the cone. Thiel was able to show in 1937–38 that an angle of 30 degrees reduced friction between the exhaust gases and the wall. Nozzles could also be shortened, because excessive length caused turbulence at the end of the expansion nozzle. The net effect was to increase exhaust velocity and shorten the length of the 25-ton-thrust engine further. By the end of 1938, Thiel's team was close to achieving the target performance of 2,000 m/sec.[8]

The cumulative effect of those four innovations—improved injection, the "pre-chamber system," a spherical combustion chamber, and a widened nozzle—was nothing short of revolutionary. It showed what could be accomplished when increased money and manpower were combined with a systematic attack on a technology ripe for transformation. Thiel's penetrating mind, practical talent, and capacity for hard work had inspired the propulsion group and had driven it forward, notwithstanding his sometimes difficult personality. Von Braun's leadership role was less important, since he had turned his attention to guidance-and-control problems once he saw that Thiel had engine development firmly in hand. Yet there is no doubt that von Braun's charismatic and visionary personality, combined with his excellent management skills, further inspired Thiel and his group to push the frontiers of rocket propulsion.

The theoretical feasibility of building a powerful yet relatively compact rocket engine did not, however, mean that the practical problems were solved in 1938. Thiel's group faced perplexing difficulties in two areas: scaling up the injection system to 25,000 kg of thrust and cooling an engine that produced considerably more heat per unit area because of its efficiency and reduced size.

Faced with increasing the injection of the propellants by a factor of seventeen, Thiel took the logical yet imaginative step of building a motor of intermediate size using a cluster of injection chambers or "pots." Since Peenemünde's Test Stand I would not be ready for 25-ton engine tests until April 1939, Thiel designed a motor of a size that could be accommodated in Kummersdorf: a 4.2-ton thrust one that grouped three injectors of 1,400-kg thrust each from the experiments of 1937–38. Those experiments showed that clustering raised combustion efficiency slightly by improving mixing. There was no straight line, however, from the three-pot, 4.2-ton engine to the eighteen-pot, 25-ton engine depicted in Figure 3.1. In 1939 Thiel and his associates favored the "star" configuration of six or eight larger injectors arrayed around the sides of the combustion chamber. Those configurations were actually tested in Peenemünde but must have failed because of cooling and burnthrough problems. By mid-1940 efforts concentrated on refining the eighteen-pot configuration of smaller injectors, which promised to deliver a working A-4 engine more quickly—something Army Ordnance demanded even more urgently from Thiel after World War II started in September 1939.[9]

The eighteen-pot concept may have been promising, but the engine failures did not cease. One difficulty had plagued liquid-fuel rocket engineering from the outset: The more efficient the engine, the more urgent the cooling problem. At 2,400° C, the temperature of the combustion chamber gases in the A-4 engine greatly exceeded the melting point of the metal shell that held them. Regenerative cooling (circulation of the fuel through the jacket around the engine) was only part of the answer. Improved alloys helped, but metallurgical research for the rocket program was limited before the war by secrecy restrictions. After the war started, an increasingly severe shortage of aluminum, magnesium, and other metals forced the Ordnance rocket program to go back to steel as much as possible, and not always steel of the best quality. That was ironic, considering the great amount of effort expended in the 1930s on using aluminum alloys to reduce weight.

The answer to the burnthrough problems proved to be another important innovation: film cooling. Diploma engineer Moritz Pöhlmann, who came to head the propulsion design office at Kummersdorf in August 1939, almost immediately suggested that a film of alcohol fuel along the wall of the combustion chamber and nozzle would provide an insulating layer against the massive heat flux from the burning gases. Tests with smaller engines immediately proved the validity of the concept. The final configuration is shown in Figure 3.1. Four rings of small holes seeped alcohol into the chamber, taking up 70 percent of the total heat flux. The fuel circulating through the cooling jacket absorbed the rest.[10]

It was by no means a simple task, however, to make this idea work, any more than choosing the form of the injection system ended all problems in that area. It actually required two years of repetitive trial-and-error experiments with the film cooling orifices and with the configuration of holes and nozzles in the injectors to make the eighteen-pot engine work consistently. Test stand failures often resulted in spectacular burnthroughs or explosions that ruined the engines. Metal fatigue would wear out even successful combustion chambers after limited use.[11] Only large resources of material and manpower, plus extensive manufacturing and test facilities at Peenemünde and Kummersdorf, allowed the innovations of Thiel and his assistants to be quickly transformed into working technology.

In addition to having primary responsibility for developing the 25-ton engine, the propulsion group had to develop related devices in conjunction with Walter Riedel's design bureau. Problems with the jet vanes were largely solved in mid-1938, when a new draftsman at Kummersdorf suggested that graphite replace the expensive and unsatisfactory tungsten-molybdenum alloys employed in the A-3 vanes. Thiel's group also inherited the development of the turbopumps that would be needed to move the large volumes of propellants. Von Braun had initiated that project in mid-1935 at the southwest German firm of

Klein, Schanzlin & Becker. Apparently the company had experience with large firefighting water pumps, but extreme temperatures, difficulties lubricating the liquid-oxygen side, and the need for lightness and compactness posed fundamental problems that would require years to overcome. The demands of liquid-fuel rocket engineering pushed turbopump development to the absolute limits of the technology.[12]

The Kummersdorf engineers also had to find a way to drive the pump's central turbine that would provide the energy to suck liquid oxygen and alcohol from the tanks. Ordnance patented various schemes in the spring of 1936, including drawing hot gases from the combustion chamber to push the turbine blades. But the eventual solution actually derived from collaboration with Hellmuth Walter's hydrogen peroxide rocket work. In March 1936 von Braun asked Walter to design what would soon be called the "steam generator." When peroxide was mixed with a catalyst, it produced superheated steam that could be used to drive a turbine. The first design, based on a torpedo system, did not work well but promised a more practical solution than employing the fiery exhaust gases of the rocket engine. After 1937 the smaller turbopump/steam-generator system for the rocket aircraft projects provided further development experience (and funds from the Luftwaffe) to help solve the problem. The first turbopump-fed Ordnance rocket engines were actually the ones developed for the second version of the He 112 flown in 1939–40. By 1941 Thiel's propulsion group, aided by Riedel's design bureau, had the basic configuration of the A-4 turbopumps and steam generator in hand, but numerous development and manufacturing problems continued.[13]

Only in the case of these auxiliary devices did corporations play an important role in propulsion design and development. Otherwise the civil service engineers at Kummersdorf and Peenemünde dominated the process. After the war began, Army Ordnance began to draw the universities into rocket research as a third force. Although dissertation work had been important in propulsion before 1937, it took the war to force the partial lifting of the secrecy restrictions. The catalyzing event was the Army Commander-in-Chief's acceleration of the A-4 project immediately after the outbreak of the war. In order to solve the missile's perplexing technical problems in time for its projected military deployment in September 1941, Army Ordnance decided to bring academic research institutions into the program on a large scale. General Becker, chief of Ordnance since March 1938, probably suggested and certainly approved this action.[14]

In addition to accelerating the A-4 schedule, Becker had a second motivation for bringing Peenemünde and the universities together: securing the Army's place in wartime academic research. In the intervening years he had become one of the leading science policymakers of the Third Reich. As Dean of the Faculty of Military Technology at the Technical University of Berlin, Becker had played an important role in the increasing militarization and state domination of German scientific institutions. Although his faculty never came into effective operation, and Becker delegated most of his duties to the Associate Dean, his academic position helped to further his national prominence and his relations with the Nazi elite. In 1937 Reich Education Minister Bernhard Rust named him President of a newly created Reich Research Council. Like so many leaders in the National Socialist regime, Becker acquired at least the illusion of power by accumulating many different hats. The Research Council turned out, however, to be stillborn. Rivalry among the various ministries and armed services made a coherent science policy impossible to achieve, especially because Rust was a nonentity in the ruthless behind-the-scenes struggle that was the Third Reich.[15]

When a general war rather unexpectedly erupted in 1939, chaos reigned in German education and science policy. No plan for the mobilization of science and engineering existed, and massive callups drained the research institutes of students and younger scholars. There was even talk of closing the universities for the duration of the war. For both patriotic and selfish reasons, leaders of academic institutions eagerly sought military projects that would provide funding and draft exemptions. Self-interest also compelled the armed services to grab what institutes they could: Not only did they need more research capacity because of the demands of war, but they also wished to prevent other services or bodies from taking over institutes they wanted. By its very nature, the Third Reich was a collection of competing bureaucratic empires. Hitler had further exacerbated interservice rivalry in the late 1930s by failing to provide effective tri-service coordination or clear priorities for rearmament. In September 1939 Becker thus had to move quickly to secure the Army's role in academia, which could scarcely be

General Karl Emil Becker initiated the German Army rocket program in the early 1930s. He was convinced that liquid-fuel rocketry provided the key to a devastating new secret weapon: the long-range, ballistic missile. *(Smithsonian Institution, SI neg. no. 80-14963)*

On July 23, 1930, the pioneer of the Weimar spaceflight movement, Hermann Oberth (center, in profile), demonstrated his rocket engine in Berlin. At far left is Rudolf Nebel, who would soon found the Raketenflugplatz, the most important amateur rocket group. Klaus Riedel, the chief designer of that group, holds what may be a traditional solid rocket. Immediately behind him to the right is eighteen-year-old Wernher von Braun. *(SI neg. no. A5347-H)*

Max Valier tests his new liquid-fuel rocket-car motor at the Heylandt Company in Berlin, March/April 1930. He was killed in a laboratory experiment soon afterward, due in part to his cavalier attitude toward safety. *(Imperial War Museum)*

In 1931, Heylandt built a new rocket car under the direction of three engineers. Arthur Rudolph (left) and Walter Riedel (right) had worked with Valier and would become prominent in the Army rocket program. The man in the center is almost certainly their foreman, Alfons Pietsch. *(Imperial War Museum)*

In late 1931, the Army began its investigation into liquid-fuel rocketry with a contract to Heylandt, which conducted compressed-air experiments on the shape of nozzles. In this experiment, the nozzle, which is pointing upward, has eight pressure measurement gauges along its lower half. *(Imperial War Museum)*

This April 1931 photo of the Raketenflugplatz shows, from left to right, Rudolf Nebel, space popularizer Willy Ley, and Klaus Riedel. Behind them is an engine test stand made out of the launch rail for the ill-fated Oberth *Frau im Mond* rocket of 1929. *(SI neg. no. 82-4628)*

A soldier holds the rocket the Raketenflugplatz launched for the Army at the Kummersdorf test range on June 22, 1932. This demonstration's failure confirmed General Becker's decision to concentrate Army efforts on in-house development. *(Imperial War Museum)*

An A-3, missing some of its exterior skin, undergoes guidance testing at Kummersdorf, 1936–37. Ground tests alone failed to show the weaknesses of the guidance system built by an outside contractor. *(SI neg. no. 77-14790)*

An A-3 is prepared for launch on the Greifswalder Oie, an island near Peenemünde, in December 1937. Standing to the right, with his hand in his vest, is Dr. Wernher von Braun, the technical director of the Army side of Peenemünde at age twenty-five. *(Deutsches Museum Munich)*

Test Stand I at Peenemünde, designed to accommodate engines or missiles with up to 100 metric tons of thrust, was finished in spring 1939. The Army had made a massive investment in ballistic-missile technology since the mid-1930s. *(SI neg. no. 79-12318)*

The Heinkel He 176 rocket plane made limited flights in June-July 1939. It was a product of the joint Army-Luftwaffe rocket program that led to the construction of Peenemünde. (*Deutsches Museum Munich*)

Walter Dornberger signing papers in his office in Berlin or Peenemünde sometime after his promotion to general in June 1943. He was the Army rocket program's energetic leader and administrator. *(SI neg. no. 90-2937)*

Between 1938 and 1943, regular launches of the A-5, which looked like a scaled-down A-4 (V-2) ballistic missile, proved essential to guidance-and-control development. *(SI neg. no. 76-15523)*

A key technical innovation for the A-4 was the creation of the *Vertikant* guidance system. The version shown used three gyroscopes: two to control the orientation of the missile in space, and a third (at top) to shut off the engine when the correct velocity was reached. *(SI neg. no. 86-1067)*

After the outbreak of war, the rocket program was drawn into political battles. Armaments Minister Fritz Todt (center left, in profile) visited Peenemünde in early October 1940. At far left is General Emil Leeb, Chief of Army Ordnance after Becker's suicide in April. Third from right, in the front row, is Colonel Dornberger. Third from left, in the background, is Heinrich Lübke, an official in Albert Speer's Construction Group Schlempp who was later president of West Germany. *(Deutsches Museum Munich)*

accomplished through the Reich Research Council. The fact that the Ordnance chief himself had little use for that body, although he headed it, is shown by Ordnance's seizure of the nuclear project from the council in that same month. Erich Schumann's research division was given supervision over this small effort to investigate the feasibility of atomic bombs and reactors, which had arisen from the discovery of uranium fission by German physicists at the end of 1938.[16]

With the encouragement of Becker, Dornberger's rocket section (Wa Prüf 11) immediately began to establish direct contacts with academic institutes, mostly at the technical (engineering) universities. The first known meeting took place on September 14 at the Technical University of Dresden. Precisely two weeks later, Dornberger, Thiel, and Walter Riedel headed a conference at Kummersdorf on propulsion research attended by representatives of four universities. Around that time or shortly thereafter, three dozen professors visited Peenemünde. That occasion soon acquired the ironic nickname "The Day of Wisdom"—the day on which so many great minds were at the center. It went into Peenemünde mythology as the day on which the connection between the rocket program and the universities was born, but the academics would never have been admitted to the ultrasecret center unless they had already agreed to cooperate. Only the details as to which institutes would take which projects remained to be settled.[17]

In forging the alliance with academia, Dornberger followed Becker's lead and ignored the Research Council completely. Even the Education Ministry was treated dismissively. On September 19 the chief of Wa Prüf 11 merely asked Schumann, as the liaison person, to notify Rust's Ministry that certain institutes would be working for the Army and more were expected to do so. The leader of the rocket section gave this revealing excuse for violating channels: "The accelerated execution of agreements was necessary because, insofar as the relevant persons had not already been drafted, they were going to be [contractually] obligated to the RLM [Air Ministry]."[18] Notwithstanding the close Luftwaffe connections that still prevailed in some aspects of Army rocket development, interservice relations continued to erode. That situation did not bode well for the German war effort. Nor, for that matter, was the commitment of so many resources to the militarily dubious A-4 project a particularly good sign.

If the integration of university institutes into the rocket program was doubtful for the country, it was certainly beneficial to Wa Prüf 11's technology development, including Thiel's efforts in propulsion. While the fundamental design of the 25-ton motor was not changed by academic research, Professor Wewerka of the Technical University of Stuttgart made significant suggestions for overcoming turbopump design problems. In the realm of theory, Wewerka verified Thiel's findings regarding the best angle of opening for expansion nozzles. To choose only one other important example, Professor Beck of the Technical University of Dresden (later Berlin) was instrumental in proposing and testing alternative designs for the A-4 injection system. A new design was certainly needed. The eighteen-pot motor could be made to work, but it was a "monstrosity and a plumber's nightmare," to use the words of Thiel's later replacement, Martin Schilling. The eighteen small injection chambers required eighteen separate liquid oxygen lines. The piping for fuel flow to the cooling jacket and film cooling outlets was equally complex. After the propulsion group withdrew completely to Peenemünde in August 1940, one of the test stands at Kummersdorf was handed over to Beck's institute full time for experiments. The alternate injection systems, called "mixing nozzles," used a series of concentric ring slots, or a plate covered with injection holes. But the hoped-for quick success never materialized.[19]

The energetic and imaginative Thiel also pursued alternatives to the basic 25-ton engine. Since 1937 he had experimented with higher chamber pressures, which promoted more efficient burning. He put some effort into developing a 25-atmosphere, 725-kg-thrust engine for the He 176 rocket plane. The existing technology did not, however, seem to allow much success in this area. Ultimately he had to settle for raising the A-4 motor's pressure to thirteen atmospheres, which could be done without major changes to the design. He also looked anew at alternative propellants in 1941, because super-cold liquid oxygen was extremely inconvenient to use and in potentially short supply. Thus, when Thiel declared the basic eighteen-pot A-4 motor officially finished on September 15, 1941, it was far from clear that it was satisfactory for mass production.[20] Nonetheless, in the five years since he had joined Dornberger's rocket section, he and his assistants had created a revolution in rocketry. It was now possible to lift a missile with a takeoff weight of more than 12 tons and hurl it a couple of hundred kilometers. It was truly a remarkable accomplishment.

KURZWEG, HERMANN, AND SUPERSONIC AERODYNAMICS

Possessing the raw power to propel a missile over such distances was one thing; stabilizing and guiding it through velocities exceeding Mach 4.5 (four and a half times the speed of sound) was quite another. When Wernher von Braun began to investigate the best form for the A-3 in 1935, he knew essentially nothing about the supersonic aerodynamics of fin-stabilized bodies. The ballisticians, led by Becker and his mentor, Professor Carl Cranz, had experimented with spinning rifle bullets and artillery shells moving at those velocities; indeed, the fuselage of the A-3 was based on the infantry "S" bullet, because it was known that its shape worked at higher Mach numbers. But some artillery specialists told the rocket group that stabilizing a supersonic body with fins was impossible.[21]

That Army ballisticians could have made this statement indicates how little contact they had with the aerodynamics community, which was funded by the Transportation Ministry before 1933 and the Air Ministry thereafter. The areas of interest of the two communities were in any case far apart, since the aerodynamicists concentrated on airfoils and aircraft moving at the low speeds typical of the day. But from the late 1920s on the theory of winged bodies moving at supersonic and high subsonic velocities, where air begins to compress significantly, had made large advances, especially under the direction of the grand old man of aerodynamics, Professor Ludwig Prandtl of Göttingen. Practical experiments had also begun in small supersonic wind tunnels at this time. Although the first was built in Switzerland, Germany had a dominant place in this area too. Based on testing done in tunnels at Göttingen and Dresden, one of Prandtl's rising stars, Dr. Adolf Busemann, first revealed in October 1935 that swept-back wings worked much better than straight wings at velocities approaching and exceeding Mach 1 (the so-called sound barrier). Swept wings delayed the onset of turbulence near Mach 1 and had much better lift and drag characteristics at high speeds. That discovery interested few aerodynamicists at the time, as it did not seem to have much practical application.[22]

In the meantime, the rise of the rocket aircraft program had put von Braun in touch with Busemann and the aerodynamics community. With the help of the Air Ministry, a few A-3 tunnel tests were made in 1935, but the real collaboration began with von Braun's visit to the Technical University of Aachen on January 8, 1936. There he encountered Dr. Rudolf Hermann, an assistant professor who had constructed a Luftwaffe-financed supersonic wind tunnel. The test section of this tunnel, where the aircraft or missile models were placed, was square and measured only 10 centimeters (4 inches) on a side. During 1936 and early 1937 Hermann made basic A-3 measurements up to the tunnel's maximum velocity, Mach 3.3. But the models had to be very small, and his task was only to modify a design already chosen. By enlarging the A-3's highly swept-back fins, he was able to create an "arrow-stable" vehicle, but its form was far from ideal. If the A-3 had made it to high altitudes, reduced atmospheric pressure would have allowed the engine exhaust jet to expand, burning off the fins or the antenna ring around the ends of the fins. Actual flight testing also indicated that the A-3 was *too* stable, making it difficult for the guidance-and-control to push the vehicle back to the desired attitude when the aerodynamic forces were so overpowering.[23]

Long before the unfortunate outcome of those launches became known in December 1937, however, von Braun had decided that the rocket program needed its own supersonic wind tunnel complex if it was to take on the challenge of the A-4. It was not only inconvenient to go to Aachen, but the tunnel was too small, and ready access was not assured. The Air Ministry controlled the aeronautical research establishment; more important, the existing tunnels were greatly overbooked until the Ministry's massive investments in new research facilities bore fruit in the early war years.[24]

Dornberger agreed with his able assistant that a supersonic tunnel would be a good idea. It certainly fitted his "everything-under-one-roof" concept for the new Peenemünde center, but "the cost frightened me; the estimate was 300,000 marks. I had enough experience with building to know that there wasn't the least chance of remaining at that figure, especially with von Braun about. The supersonic tunnel was more likely to cost a million marks." Dornberger decided to move, however, once Hermann had demonstrated in the autumn of 1936 that he could make the A-3 stable at supersonic velocities. He went to Becker, then chief of Testing Division, and asked for the tunnel. His boss agreed, but only on the condition that at least one other section make use of it. Curiously, Dornberger was not able to secure the support of Section 1 (ballistics and munitions), which his rocket group had so recently left, but he did convince the head of anti-aircraft artillery that the shape of shells could be refined in the tunnel. Becker issued the order on November 30, 1936, and in April 1937 Rudolf Hermann joined the staff of Peenemünde.[25]

The thirty-two-year-old aerodynamicist, described by Dornberger as "slender . . . with a lofty brow and light-brown, wavy hair brushed straight back," was a talented and energetic engineering scientist. According to American records, he was also an "extremely ardent Nazi" in the later war years, although he did not join the Party until membership was reopened in mid-1937. When Dornberger and von Braun recruited Hermann away from academia, they doubtless appealed to his patriotism, but above all they stroked his ambition: He would have a chance to build the world's most advanced supersonic wind tunnel. Hermann immediately began to recruit a staff and to plan for the facility. The heart of the "Aerodynamic Institute," to be constructed in the middle of the laboratory and shop area of Peenemünde-East, would be a scaled-up version of the 10-by-10-centimeter tunnel built at Aachen. It would have a test section measuring 40 centimeters (about 16 inches) to a side and a maximum running speed of Mach 4.4, a world's record equaled but not exceeded before the end of the war.[26]

The configuration and principles of the main Peenemünde tunnel are shown in Figure 3.2. It was of the open or "blow-down" type. A spherical vacuum reservoir with a diameter of about 12.5 meters (40 feet!) was emptied out by six pumps exerting 1,100 horsepower. When the quick-acting valve was opened, air rushed in from outside and filled the vacuum chamber in about 20 seconds, the maximum running time for an experiment. The velocity of the air through the test section was determined by the shape and size of the opening in the "Laval nozzle" through which the inrushing molecules must first pass.

A "three-component balance" measured the lift and drag forces on the model; the airflow patterns around it could be photographed or measured with elaborate optical equipment. A smaller 18-by-18-centimeter tunnel was also built and connected to the same reservoir. By pumping continuously, this tunnel could be operated without interruption, although only for lower Mach numbers.[27]

The planning and construction of this expensive state-of-the-art facility took a long time, which must have resulted in considerable pressure on Hermann and his staff from the Ordnance Office. The aerodynamicists were not able to put the big tunnel into operation until about May 1939, and the small one came even later. Even then there were numerous startup difficulties. The design of the Laval nozzles that determined the Mach number was a problem of great theoretical complexity and strenuous trial-and-error correction. Even at the beginning of 1941 the nozzle for Mach 3.1 was still being refined, and the highest working speed was Mach 2.5. The nozzle for Mach 4.4 was not ready until 1942 or 1943. Another problem was condensation clouds that formed because the rapidly expanding air cooled dramatically after going though the throat of the nozzle. To get accurate results it was necessary to place a special air-drying silica-gel honeycomb across the mouth of the opening (see Figure 3.2). Until that system was finished in the spring of 1940, the effectiveness of Hermann's facility was diminished by measurements of questionable reliability.[28]

The construction of this world-class aerodynamics institute at Peenemünde, with a staff of sixty in mid-1939 and two hundred in 1943, was another large stride on the road to massive in-house research and development capability. But the delay in putting the tunnels into operation meant that the critical aerodynamic innovations for the A-4 were largely obtained through educated guesses and improvised experiments. When the decision was made in January 1938 to build a new vehicle (the A-5) to test guidance systems, the question of the external form for a rocket again became pressing. Two problems in particular forced Hermann and his staff to improvise: fin design and the stability of rocket bodies as they passed through the sound barrier.

In designing the A-5, the body form was not the problem. Since the A-3 fuselage was considered adequate, and because extensive wind-tunnel testing was still unavailable, the body was made only slightly

FIGURE 3.2

The Supersonic Wind Tunnel at Peenemünde

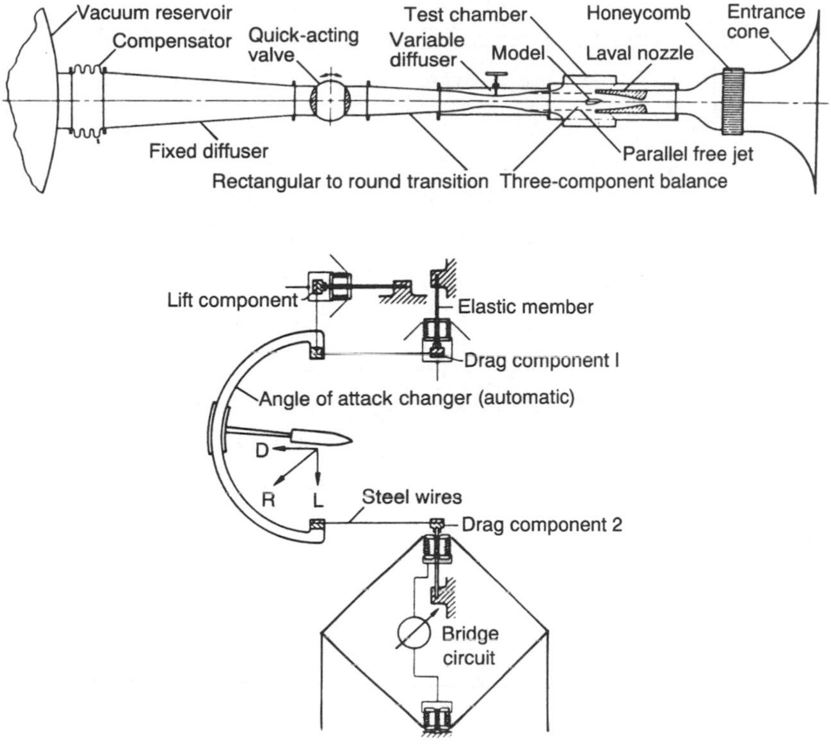

fatter on the A–5, which then became the model for the A-4. By contrast, no one knew how to design fins for supersonic flight. In 1937 Hermann had lured to Peenemünde as his chief assistant and head of research Dr. Hermann Kurzweg, an acquaintance from the University of Leipzig then working at the eminent optical company Zeiss. Kurzweg, an inactive but contributing member of the SS since 1934, tackled the A-5 fin problem in early 1938 with only the most limited information: a knowledge of supersonic aerodynamics, a rough estimate of the expected pressure distribution over the rocket body, and wind tunnel information on the characteristics of flat plates. (Another consideration must have been Dornberger's injunction that the fins, when scaled to A-4 dimensions, be able to pass through a standard railroad tunnel.) To allow for the expansion of the exhaust jet at high altitude, where air pressure is low, Kurzweg made the rear of the A-5's fins open at a much wider angle than those of the A-3. For good supersonic characteristics he swept the forward edge more strongly, kept the fins relatively thin, although thicker than on the A-3, and not quite as wide as they were long. The result was the first critical aerodynamic innovation for the A-4: a broad fin shape familiar from later pictures of that missile.[29]

Lacking any way to get quick answers about the appropriateness of his design, Kurzweg resorted to homemade improvisations. One weekend he carved a rocket body out of a Peenemünde pine branch, inserted weights into holes to get the proper balance, and made hard rubber fins in the proposed shape, but in three different sizes. To obtain low-speed stability information he tried throwing the model off the roof of his house. When that proved unsatisfactory, the model was mounted on a wire through its center of gravity, and Kurzweg drove down the Berlin-Anklam highway at a speed of 100 km/h (about 60 mph). The largest and second-largest versions of the fins seemed to be stable, but not the smallest. Since it now seemed that the aerodynamics group had a workable A-5 concept in hand, wind tunnel testing was done in 1938 and 1939 at Aachen and in the subsonic installation at the Zeppelin Airship Construction Company in Friedrichshafen on Lake Constance. Kurzweg's fin design came to be embodied in the first unguided A-5s launched from the Greifswalder Oie in October 1938. With small modifications, this shape was also used on the A-4.[30]

But before the aerodynamics group could make a definite decision, a more systematic attack on fin form was necessary. In conjunction with the 1938 A-5 launches, which aimed to exercise the launch organization and demonstrate the aerodynamic stability of the rocket, Peenemünde-East sent aloft a number of models equipped with various fin shapes and propelled by small Walter hydrogen-peroxide motors. The exact purpose of those model experiments is unclear, but the most likely one was testing for an unguided anti-aircraft missile proposed by Wa Prüf 11. Inspired by that effort, the aerodynamics group then fired off forty subscale A-5 models using Walter motors in March 1939. Eight different fin shapes were given to the 1.6-meter-(5-foot)-high models, and the launches were systematically photographed. The results confirmed that, of the designs tested, Kurzweg's original had the best subsonic stability characteristics. Later, extensive wind tunnel work at

Peenemünde and Zeppelin substantiated this research for the whole velocity range and refined the shape for the A-4. It is a tribute to Kurzweg's ability that he had so successfully defined the solution from the outset.[31]

A second critical innovation was finding a method to determine whether the A-5 and A-4 would be stable as they passed through the speed of sound. No wind tunnel before the late 1940s could successfully produce velocities between Mach 0.85 and 1.2, and theory regarding this "transonic" region was in an equally primitive state. An unreliable Mach 0.95 test run (at Aachen?) suggested to Hermann that the A-5's center of pressure would move far in front of its center of gravity, making the rocket unstable—it would tumble out of control. To gain some data on the problem, the head of the aerodynamics group proposed in July 1938 that an iron model could be dropped from an airplane at an altitude of 7,000 meters (about 23,000 feet). It would pass through the speed of sound at 1,000 meters while being carefully photographed from the ground and observed from the air.[32]

Those tests were started no later than the autumn of 1939, using a Luftwaffe He 111 bomber from Peenemünde-West. A new recruit to the guidance group, Dr. Walter Haeussermann, witnessed one of the drops at the end of 1939 from a second plane piloted by Wernher von Braun. After the release from the He 111, von Braun dove his airplane after the model to observe its behavior and to radio to recovery crews where it hit the water. In the end, the drop tests laid aside concerns about serious transonic instability problems for the A-5/A-4 design but were not exacting enough to show that no problem existed. In fact, repeated launches of the A-5 from 1939 to 1942 demonstrated that the rocket was marginally unstable in the transonic region. Its nose would wobble in a circular motion of a few degrees radius, creating enough drag to prevent it from ever passing through Mach 1. Thus, when the first A-4 launches were attempted in 1942, the aerodynamics group, along with everyone else, had to cross their fingers and hope that, in the quick transition through the sound barrier, the control system would prevent any dangerous movement of the missile.[33]

It is thus clear that the Peenemünde wind tunnel was not essential to the fundamental shape of the A-4, but assembling a highly competent aerodynamics staff was. Even so, the systematic work that began in the tunnel in 1940, along with the subsonic experiments of Dr. Max Schirmer at Zeppelin, was highly important for the refinement of the rocket's design and the elimination of uncertainties in many areas. When the rocket group began working on the A-3 and the A-5, virtually nothing was known about the heating caused by friction with the air. For the A-4 the problem was especially urgent because of much higher velocities and the need to survive reentry into the earth's atmosphere after passing the 80-kilometer (50-mile) peak of the trajectory. Peenemünde aerodynamicists undertook fundamental heat-transfer research in their tunnels using simple shapes and rocket models equipped with temperature sensors. Their data helped verify theoretical calculations and provided guidance to the designers as to the steels that would have to be used. Elaborate pressure measurements were also taken on the surface of wind tunnel models cut in half lengthwise, allowing many more sensors to be installed. That procedure helped verify the predicted loads on the vehicle—information crucial to the structural designers. Another area where Peenemünde broke new ground was in assessing the impact of engine exhaust jets on the aerodynamic characteristics of missiles. Using compressed air jets in the tails of models, the group found that drag increased in the subsonic range but decreased in the supersonic. Without that kind of information, launching the A-4 would have been very much a shot in the dark.[34]

As compared with propulsion and, especially, with guidance and control, the contributions of university research after the start of the war were less important, in part because virtually all aerodynamicists were already working for the Air Ministry. But academic contractors to Peenemünde made a couple of valuable theoretical innovations, and they built or refined measurement technology for the tunnels. Corporate research, in the form of Schirmer's work at Zeppelin, was of more significance, although still modest. The end of the airship era, resulting from the *Hindenburg* accident and Hitler's and Göring's dislike for those fragile vehicles, must have created an opening for the rocket program in the company's aerodynamic facility.[35]

Among the most important contributions of Zeppelin—in conjunction with supersonic work at Peenemünde—was new research into winged missiles. In June 1939 a member of Riedel's design bureau, Kurt Patt, proposed that the high energy of an A-4-type missile as it approached impact be employed instead to generate aerodynamic lift. Through the use of wings, the A-4's range could thereby be doubled to 550 kilometers. That idea—later called the A-9—was taken up enthusiastically by von Braun's group because it seemed a relatively cheap way to extract extra range from a missile powered by the 25-ton engine. Patt's design, essentially a fuselage-less flying wing, was too radical, but affixing aircraft-type wings to an A-4 appeared to be a feasible alternative. That idea would require basic research into the configuration of a supersonic airplane that could ascend as a missile and descend as a glider. By 1941 a workable design emerged: a simple sweptback wing. More radical airfoil shapes were also tried, but producing a compromise that worked at all velocities proved to be anything but simple. A-9 research continued into 1942 before it was halted altogether by higher-priority projects. In the last months of the war it was revived and given the designation A-4b.[36]

A prominent feature of the glider missile project, as in the case of the A-5/A-4, was the repetitive, systematic wind-tunnel work required to evaluate different designs at different Mach numbers and different angles of attack (the angle of the missile's nose to the direction of airflow). Doing comprehensive pressure measurements on a half-model, to take an extreme example, entailed more than 100,000 gauge readings recorded by hand by twenty people for two weeks in two shifts. As the war went on at Peenemünde, the inherent character of aerodynamic research, combined with intense political pressure to finish the A-4, made the work process increasingly stressful, routinized, and factory-like. A second 40-by-40-centimeter test section was built so that one would be available while changes were being made on the other. The aerodynamics group went to multishift operation nearly around the clock. In addition to all the missile work, extensive research was also done for Army Ordnance on artillery shells, including the fin-stabilized "Peenemünde Arrow Projectiles" designed in-house.[37]

This mass production of research reinforces a basic point about the revolutionary breakthroughs in key technologies that Peenemünde achieved between 1936 and 1941. In aerodynamics as in propulsion, brilliant ideas and excellent management were not by themselves sufficient. Only the existence of a massive and well-funded organization allowed the rocket group to create working technology in a short period of time. The case of guidance and control—the most difficult challenge in the whole A-4 project—illustrates this point even more clearly.

VON BRAUN, STEINHOFF, AND MISSILE GUIDANCE

Unlike the other two key technologies, the transformation of research in guidance and control did not begin around the turn of the year 1936–37. Until just before the disturbing failures of the A-3s in December 1937, Ordnance remained dependent on Kreiselgeräte as its sole contractor in this area. Those failures then greatly accelerated a twofold shift in philosophy: toward competitive development by a number of firms in the gyroscope and autopilot sector, and toward a buildup of in-house expertise at Peenemünde. Although those two processes overlapped, it was not until 1939 that Wernher von Braun began in earnest to put together a large guidance laboratory. He did so because he was increasingly dissatisfied with the corporations. The extreme and specialized demands of ballistic missile guidance, combined with Ordnance's pressure to produce results quickly, strained the research and development capability of contractors already overburdened with Luftwaffe and Navy work. Rather than accept delays, von Braun used his rapidly expanding budget to construct a new laboratory at the center.

Competition for Kreiselgeräte was first formally discussed on November 9, 1937, when Dornberger and von Braun went to a meeting at the aviation instruments division of Siemens, the giant electrical concern. Among those attending was Siemens's Dr. Karl Fieber, who remembers the "unusual declarations of secrecy with threats of the death penalty." Also present were Karl Otto Altvater, director of the aviation instruments division, and Klaus Riedel, who had worked for the company for three or four years after the collapse of the Raketenflugplatz. Altvater, a retired U-boat captain and Naval Ordnance section chief in the 1920s, was Riedel's uncle. He had given his nephew and two or three other rocket enthusiasts jobs when the amateur group fell apart. After the settlement of the Nebel–Riedel patent in July 1937, Klaus Riedel went with his friends to Peenemünde, where he in all likelihood suggested Siemens as an alternative source of gyroscope expertise. The firm had become increasingly involved in aircraft autopilots and navigation instruments since the 1920s and had purchased Boykow's patents in this area in 1931.[38]

During the November 9 meeting von Braun lectured on the A-3 guidance system, giving no indication that the rocket group had anything but confidence in it, and the Siemens people made no comment. Nevertheless, Dornberger indicated that Ordnance had "the conscious intent . . . to create a competitive line of development to Kreisegeräte Ltd." Nothing came of the discussions right away. For the next two months the preparations for the launches on the Oie and the postmortems on the failures completely absorbed the attention of Dornberger and his assistants. Only toward the end of January 1938 was it possible to meet again. Now the tone was completely different. With hindsight provided by the A-3 launches, the Siemens representatives expressed their reservations about the Sg 33's inadequate control forces and its inability to stop a rapid rolling of the vehicle.[39]

Meanwhile, the discussions between the rocket group and Kreiselgeräte had produced some basic decisions. There was a need for systematic launch testing, a dramatic increase in the forces exerted by the jet vanes, and a refinement of the aerodynamic stability of the new vehicle. Kreiselgeräte must also make its control system simpler in most aspects, while more complicated in others. The company would remain with Boykow's basic concept, one fundamental to most inertial guidance systems: a stabilized platform that remains fixed in space regardless of the movements of the rocket. Such a platform provides a reference for measuring the position or acceleration of the vehicle. But for the A-5 it was imperative to add the third gyro for the roll axis, which had been omitted from the Sg 33 for reasons of simplification. To reduce complexity, Peenemünde and Kreiselgeräte decided to eliminate the nitrogen jets that stabilized the platform, as well as the electricity supply that powered the gyros after liftoff. The A-5 would be equipped with gyros large and heavy enough to hold the platform steady by themselves, even though they would be slowed down by friction during the forty-five seconds of engine firing. That solution was clearly a temporary stopgap and was inadequate to the much more demanding requirements of the A-4.

Eliminated as well were the small accelerometer wagons that measured horizontal movement away from the trajectory. The A-5 guidance system, which Kreiselgeräte soon called Sg 52, would concentrate only on maintaining the proper attitude of the rocket. To get more sensitive control and perhaps even stability, more inputs were needed to keep the rocket from yawing, pitching, and rolling and to return it to its desired position. In contrast to the A-3s, the gyros on the stabilized platform of the Sg 52 would feed in signals giving the attitude of the vehicle with respect to the platform. It would be necessary also to have a feedback of the positions of the jet vanes that deflected the engine's thrust. Those signals would then be mixed with the ones from the three rate gyros, which were not on the platform and which sensed the rate (velocity) of the rocket's turning. The mixed inputs would then be transmitted to the vanes via a mechanical system of rotating rods and gears.[40]

Most of the decisions about the new Kreiselgeräte system were made in short order, by the middle of January 1938. It took Dornberger and von Braun much longer to initiate a competing A-5 system with Siemens. The most relevant experience of Altvater's group was with autopilots that kept airplanes traveling on specific headings at set altitudes. In such devices, a "course gyro" maintained the heading. In some versions, one or two other gyros, in conjunction with rate gyros for stability, also kept the aircraft from rolling around its longitudinal axis or pitching its nose up and down. Those autopilots had no stabilized platform. It was thus natural for Siemens to try also to use position gyros fixed to the body of the rocket, although a system of that type is inherently less accurate than a platform, because a movement of the vehicle in one axis shifts the orientation of the other axes. Only the short period of active guidance during engine firing—less than 45 seconds for the A-5—and a control system that effectively limited missile motion would make the errors tolerable. A rocket-fixed system would, however, have the virtue of greater simplicity.

Despite the similarities between such a system and autopilots, Altvater's people had to start nearly from scratch in considering how to make it work on a rocket—and not just for the A-5, but above all for the A-4. Unlike an autopilot, a ballistic missile guidance system would have to work over a range of velocities from zero to about 5,400 kilometers per hour (3,350 miles per hour) and altitudes from sea level to 30 kilometers (100,000 feet), where the A-4 would burn out. Thus the external aerodynamic forces acting on the vehicle would be constantly changing, as would the rocket's center of gravity and moments of inertia because of emptying fuel tanks; yet in 1938 very little was known about any of the crucial quantities that would be needed to design an A-4 system. The high vibration and acceleration of launch also imposed a much more demanding environment on the equipment.[41]

In response to those difficult requirements Karl Fieber, one of Altvater's key subordinates, proposed a system using only two position gyros of a new specialized type. His idea would become the basic principle of the A-4 operational guidance system and thus was a crucial innovation for the missile. One gyro, the "Horizont," controlled the pitch axis; its name derived from aircraft artificial horizon gyros. The other, which Fieber called the "Vertikant," measured movement in both the roll and the yaw axes (like the Horizont, it was a two-degree-of-freedom gyro, that is, free to move in two axes). Siemens eventually came to apply the term "Vertikant guidance" to the whole system. According to Fieber, he filed a patent application for the idea at the end of 1938. When the patent was awarded, it was so secret that he was not even allowed to retain a copy.[42]

In Fieber's system, the Horizont had a further purpose besides measuring the attitude of the vehicle in pitch. It also sent the signals that tilted the rocket over from the vertical to an angle of about 45 degrees. The guided phase of a simple ballistic missile trajectory has the same function as an artillery barrel. The projectile reaches a certain velocity at a certain angle of firing at the end of the barrel or at the burnout of the engine, after which it coasts under the influence of gravity to the target (see Figure 3.3). Velocity and angle of fire are the primary determinants of range for a given projectile. As Becker and the Ordnance rocket enthusiasts had recognized from the outset, the price for getting rid of the gun barrel's limitations was the use of some complicated mechanism to stabilize and guide the rocket during burning. This mechanism must include some means to pitch the missile over as well, since a liquid-fuel rocket cannot easily be fired off an angled rail like a battlefield solid rocket. A missile using a liquid-fuel engine must take off vertically because its initial acceleration is low; if it were launched at an angle, the aerodynamic forces would be too weak to stabilize it against the disturbing forces of gravity and wind.

FIGURE 3.3
The Trajectory of the A-4

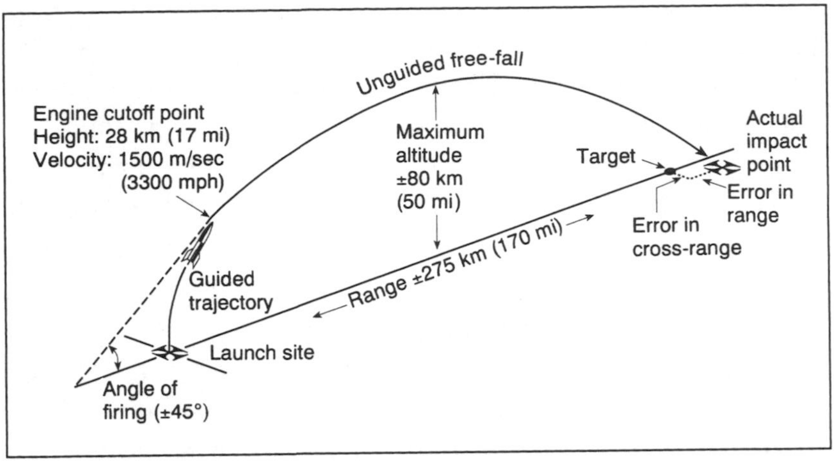

When Boykow was brought into the program, he outlined his scheme for carrying out the pitchover. After a period of vertical ascent, the missile would turn quickly to a 45-degree angle and stay rigidly on another straight trajectory. Immediately following the A-3 failures, the rocket group and Kreiselgeräte realized that the forces required from a control system by this scheme were highly unrealistic since the Sg 33 was too weak even hold the rocket on a vertical trajectory. During the ensuing discussions in early 1938 between Peenemünde and its two guidance contractors, von Braun and his assistants came to accept the need for a gradual pitching of the vehicle, at least for the A-5.[43]

The question was how to command such a maneuver. Fieber's patent is unavailable, so it is unclear how he proposed to do it, but a solution was soon found that could be employed on any rocket-fixed system. By using a clockwork mechanism to rotate slowly the pickoff (sensing mechanism) of the pitch gyro, it would essentially be fooled into thinking that the missile was pitching in the opposite direction to the desired trajectory. The correction sent by the gyro to the control system would push the vehicle's nose over. The stabilized platform of Kreiselgeräte's Sg 52 and successor systems required a more complicated mechanism that rotated the platform in pitch to get the same effect.[44]

Fieber's Vertikant guidance, particularly when it included this simple mechanism for carrying out the pitch program, was thus an imaginative and viable alternative to Kreiselgeräte's platforms. Actually developing the system at Siemens proved impossible, however. One company manager—perhaps Altvater himself—called the project "charlatanry." He was skeptical of the whole enterprise because of the rocket's rugged environment, the specialized character of Fieber's gyros, and Ordnance's desire for an A-5 system in less than a year from early 1938, which reflected the Army's belief in the urgency of the ballistic missile program. Additional opposition arose in Siemens because the aviation instruments division was strained to capacity producing for the Luftwaffe. Faced with resistance to the idea of producing even two dozen systems for the Army, Wernher von Braun seized on an "almost absurd" idea thrown out in desperation by Fieber. Von Braun asserted in a meeting that three ordinary aircraft "course gyros" could be used in the A-5 to indicate the position of the rocket, even though they lacked accuracy and were not normally exposed to such high accelerations. Afterward Fieber was furious, but von Braun slapped him on the shoulder and replied, "But, Dr. Fieber, boldly asserted is half proven!" It was one of his typically audacious moves to get around obstacles in his way.[45]

The date of this meeting is unclear, but by the fall of 1938 Siemens had laid down plans for an A-5 guidance system it called D 13. It used three course gyros for position and three rate gyros for stability, all attached to the body of the rocket. In August 1938 Peenemünde and the company also decided on the layout for transmitting commands to the jet vanes. Siemens had developed hydraulic servomotors to move aircraft control surfaces in response to its autopilots. Those systems translated electrical signals into variations in oil pressure in hydraulic pistons. The resulting mechanical energy could be used to rotate the jet vanes one way or the other. Hydraulic servos would in fact become one of the critical technologies for the A-4.[46]

By late 1938, then, Peenemünde had two A-5 guidance systems under way at two companies, and in Fieber's concept had a third possibility for the future, once better and more accurate gyroscopes were developed. But the situation was not satisfactory. Both Kreiselgeräte and Siemens were lagging farther and farther behind schedule. Siemens had indicated in early 1938 that its system should be available for launch no later than March 1939, which was already too late for Dornberger. In the event, no A-5 with D 13 guidance was actually launched until April 1940. When Kreiselgeräte promised in October 1938 to have a flight-ready Sg 52 by July 1939, von Braun wrote on the margins of the letter: "First promised: February, then: April 1, now July!" The feverish arms buildup of those years was undoubtedly part of the problem. The Army rocket project was only a sideline for both companies, since Kreiselgeräte worked almost exclusively for the Navy, and Siemens's aviation instruments division was completely tied to the Luftwaffe. In the circumstances, it is not surprising that Wernher von Braun, who was now devoting much of his time and energy to the guidance problem, should seek further ideas and corporate contractors.[47]

Sometime before the end of November 1938, von Braun came across the independent development by the Luftwaffe of another autopilot system. The project, based at the main air force test center of Rechlin in northern Germany, was headed by diploma engineer Waldemar Möller. Möller was to be transferred to the venerable precision instruments firm of Askania, where the autopilot could be taken from the laboratory to industrial production. After getting clearance from the Air Ministry, von Braun contacted Askania. The company advocated instead its simple Navy torpedo system, which used only position gyros and compressed air to convey the steering commands. Peenemünde immediately had to explain that the torpedo system would have to be extensively modified, because air damps out oscillations much less effectively than relatively dense water. Later investigations showed that a pneumatic control system was also far inferior to Siemens's oil-hydraulic one. In the end, the project was largely a waste of time and it was abandoned at the beginning of the war.[48]

In April 1939 the restless von Braun revived interest in the Rechlin autopilot. Möller's system had a special kind of rate gyro and a control system based on amplifiers and electric motors. Since Askania did not actually take over the project until the end of the year, Peenemünde received little help, other than consultation with Möller himself. The true significance of this, the third A-5 guidance system, was that it contributed further to the buildup of Peenemünde's own development capability. There was no one else to do the work on the "Rechlin guidance system." In any case, von Braun had to have the staff to supervise three or four parallel guidance developments at once, and there were many theoretical and technical problems about which the contractors knew nothing or were too busy to care.[49]

One such problem was establishing the extraordinarily difficult stability equations upon which the success of the A-4 depended. With the basic layout of the A-5 finished, the actual design of the A-4 began in January 1939, but this activity, like the aerodynamic work, only showed how much was still unknown. In 1937 the rocket group had hired a talented mathematician, Dr. Paul Schröder, to carry out the calculations, but von Braun soon became exasperated with him. Schröder asserted that certain guidance problems were impossible to solve—not an attitude that the brilliant, optimistic, hard-driving von Braun was likely to tolerate. In 1939, Schröder was shunted to the side and replaced by the highly capable Dr. Hermann Steuding from the Technical University of Darmstadt. Steuding and his staff, in conjunction with the aerodynamicists, were to make fundamental contributions to guidance theory.[50]

Steuding in turn recommended a friend from the university, Ernst Steinhoff, to be the head of a new guidance and control division that von Braun was building. Steinhoff, who started at Peenemünde on July 1, 1939, was a striking character. Only thirty-one years old, he held the world's record for distance in a glider as well as the honorary Luftwaffe rank of "Flight Captain" for his flying achievements. Steinhoff had been closely involved with the Air Ministry's German Research Establishment for Glider Flight, which was situated near Darmstadt, and was writing a doctoral dissertation on aviation instruments, which he completed in 1940. An enthusiastic Party member since May 1937, he acknowledged National Socialist doctrines as "ideals." According to von Braun, Steinhoff's task was to assemble a staff and to build a laboratory at a cost of at least a million marks. In addition to guidance and control, he took charge of on-board electrical equipment and Gerhard Reisig's measurement group. Steinhoff was less brilliant than Thiel and Hermann and contributed few basic innovations to the A-4, but he proved a good administrator and a "tremendous pusher," according to Fritz Müller, an engineer at Kreiselgeräte. Von Braun, meanwhile, continued to play a significant role in guidance and control. He not only exercised the management talent that was visible in all areas, such as his ability to cut through the discussion in a complicated meeting and seize on precisely what was critical, he also spurred intellectual productivity in the guidance group with his charismatic personality and his continual questioning.[51]

When Wernher von Braun created Steinhoff's new division, one effect was to reinforce the tendency toward an "everything-under-one-roof" facility in which corporate contracting was secondary. The new personnel in this area also contributed to the increasingly scientific tone at Peenemünde. Academically trained engineers from the technical universities took over more and more positions, in many cases overshadowing the mere handful of veterans from the early rocket groups, most of whom worked under Thiel. The center's technical leadership came to be dominated by a remarkably homogeneous group of young diploma and doctor engineers, almost all of whom were born between 1904 and 1914. (The aerodynamicists were the only significant group who were scientists by training; von Braun also had a doctorate in physics but was really an engineer.)

The war only increased the predominance of the academic engineers. The enactment of a civilian draft and the pressure to speed up A-4 development allowed many new specialists to be called to Peenemünde. Personal connections—often through the network of university institutes created after the easing of secrecy restrictions in the autumn of 1939—were frequently the determining factor in who was drafted. (Political connections played no role.) That was one reason why individuals from the Technical University of Darmstadt came to be so prominent in Steinhoff's division. Darmstadt was also dominant in university research for Peenemünde. Of 238 academic personnel working on contract for the program in January 1940, ninety-two were from Darmstadt. The next largest contractors were the Technical University of Dresden and the Institute for Oscillation Research (Berlin), with forty-five each.[52]

Research in guidance and control occupied more university personnel than the other two key technologies combined. This pattern reflected the number of problems Peenemünde had yet to solve in September 1939 if it was to make a guided A-4 feasible. Determining the basic gyroscope and control system configuration was only half the battle. Much additional radio equipment was needed beyond that developed to stop the A-5's engine manually and eject its parachute. The distances covered by the A-4 and the difficulty of recovering the vehicle meant that Peenemünde had to have a telemetry system that could send measurements to the ground instead of using a movie camera to record an oscilloscope screen, as was done with the A-5. A radio tracking system to measure velocity and position was also needed when the A-4 was out of sight. University institutes, particularly in Berlin and Dresden, made many contributions to the work. Of particular importance was the tracking system developed by Professor Wolman of Dresden. In its final form, it used a ground signal retransmitted by the A-4 to measure the velocity of the missile. Triangulation from multiple sites determined the trajectory. (Just as in the case of a siren, which is higher-pitched when coming toward one and lower going away, the direction and magnitude of the Doppler shift in frequency of the rocket's signal allowed its velocity to be determined.)[53]

University guidance research focused as well on a related problem: how to shut off the A-4's engine at the proper time to give it the correct range to hit the target. Boykow had wanted to use some kind of integrating accelerometer, that is, a device that integrated the acceleration of the missile over time to measure the velocity. One of the fundamental problems of that approach is that a small error in measurement of acceleration is increased exponentially as it is integrated. In July 1939 Kreiselgeräte revived the idea by proposing a gyro suspended in such a way that it could function as an accelerometer. Under Fritz Müller's leadership, the mechanism was developed into one of the primary means for determining the cutoff of the A-4. But in the fall of 1939 its potential for success was far from clear.[54]

With access to university researchers much increased after the war began, von Braun and Steinhoff launched a number of competing projects. At Darmstadt, two institutes began to develop accelerometers based on other physical principles. Wolman at Dresden, meanwhile, was able in 1940–41 to create an alternative radio method derived from his tracking work. A transmitter/receiver behind the launch site and in line with the direction of firing could determine the velocity of the missile and send the signal to cut the engine off at the proper moment. The signal was transmitted from the ground, and the missile doubled the frequency and then retransmitted it. Ground equipment compared the original with the received signal and automatically sent the command when the A-4 had reached the correct velocity. The Wolman radio and Müller accelerometer methods became the two A-4 engine cutoff systems and thus were essential innovations for the success of that missile.[55]

One final and critical piece was needed for the A-4, and it too emerged from Peenemünde's growing involvement with electronics and radio. Soon after Steinhoff's arrival, he decided that something had to be done about the effects of wind on accuracy. A wind blowing across the trajectory would push the missile sideways and thereby introduce a cross-range error at the target (see Figure 3.3). The planned guidance systems might successfully stabilize the vehicle and carry out the pitch program, but they could do nothing to detect horizontal movement away from the planned trajectory. Just before or after the September 1, 1939, attack on Poland, Steinhoff met with a member of the Air Ministry, who agreed to cooperate in the modification and improvement of Luftwaffe guide beams for navigation and bomber direction. In addition to the accuracy of the A-4, the difficult problem of guiding the descending glider missile was on the minds of the Peenemünders. Dornberger specifically mentioned in September and October the need to perfect a guide beam for that missile (later called the A-9) in justifying a request to the Luftwaffe for more test aircraft. The air service eventually lent Peenemünde-East three airplanes.[56]

Steinhoff agreement's with his Air Ministry colleague did not have much meaning, however. Neither the Luftwaffe nor corporate contractors had the manpower to shoulder the main burden of work during the war. Thus Steinhoff needed to build a guide beam section in his growing laboratory. How he did so is a good example of the role of personal connections in wartime recruiting. One night in October 1939 Helmut Hoelzer, an electrical engineer working at the electronics firm of Telefunken in Berlin, heard a noise under his window. He saw Steinhoff and Steuding, whom he knew from Darmstadt, plus a younger third man. They invited him out to a bar, and this third man, who was "whistling all the time," tapping his foot to the band music, and looking at the women, asked Hoelzer mysterious questions about how to guide a flying body—without, however, indicating what kind of body. Hoelzer said he could not answer. Within two weeks he received a civilian draft notice to go to Peenemünde-East, about which he knew nothing, and the first person who met him was the third man: von Braun, technical director of the place at age twenty-seven. Hoelzer was put to work researching guide beams.[57]

The one he finally developed, along with his staff, was derived from the blind-landing system of the electronics firm Lorenz. A transmitter 10 to 12 kilometers behind the launch site sent a signal alternately through two antennas a short distance apart. The missile could tell from the strength and character of the signals it received whether it was left or right of the line to the target and would steer its way back to the center. (Technically speaking, it was a guide-plane system, not a guide beam, since it did not control whether the missile was above or below the desired trajectory, only whether it was left or right.) Although this beam was simple in principle, the complications were considerable. A simple steering correction for the position relative to the guide plane would destabilize the missile. As the missile reached the center of the beam, the correction signal would go to zero, but it would still have the momentum to drift to the other side. The missile would then make a new correction in the other direction, with the result that each time the situation would get worse. Thus it was necessary that Hoelzer develop an electronic "mixing device" to calculate additional mathematical terms to modify the guide beam signal.[58]

By late 1940 Hoelzer and his assistant, Otto Hirschler, had managed to develop such a device. Beam testing was done with aircraft flying out across the Baltic to the occupied Danish island of Bornholm. Steinhoff piloted many of the flights himself. Even with that experience, perfecting a stable and workable system to be tested on the A-5 was difficult, and the first launch was not attempted until the spring of 1941. After working out innumerable problems in A-5 and A-4 launches, this guide beam was later used in some launches in the V-weapons campaign.[59]

The true significance of Hoelzer's work, however, lay elsewhere. By mid-1941 repeated launches with the A-5 had shown that stable flights could be achieved with all three control systems: Kreiselgeräte, Siemens, and Askania/Möller. But scaling up these systems for the A-4 presented new problems. The vehicle would be much larger, as would the aerodynamic forces; therefore the demands on the control system's vane servomotors would be much higher. Kreiselgeräte's elaborate systems of electrical motors and gears was inadequate to the task, and Möller's electric motors were inferior to Siemens's hydraulic systems. Even a hydraulic system might not be stable, however. Experiments with the engines running on the first static-test A-4s in the summer and fall of 1941 showed that the existing systems were indeed of questionable stability. Months more development and lots more money would be needed.[60]

In this critical situation, Hoelzer's "mixing device" provided the answer. By modifying it to take signals from the position gyros, the rate gyros could be eliminated altogether and the missile made stable. That crucial innovation was possible because, in his work for the guide beam mixing device, Hoelzer had to develop circuits for the mathematical operations of addition, integration, and differentiation. By differentiating the position signals from the gyros—that is, how they changed over time—it was possible to determine the turning rate of the missile in each axis. Thus rate gyros were unnecessary. With the calculation possibilities of this clever device, other stability terms could also be more easily introduced into the correction signals sent to the vane servomotors. Furthermore, the ratios of the various signals could be changed during engine firing to compensate for the changes in aerodynamic forces and missile characteristics. Finally, the elimination of the very expensive rate gyros meant large potential cost savings on a mass-produced A-4, which was becoming an urgent concern in 1941. The tradeoff was the need for more electronics, especially vacuum tubes, which were in short supply in the German war economy.[61]

Hoelzer's mixing device spawned yet another brilliant invention: one of the world's first fully electronic analog computers. He needed only to develop circuits for a few other mathematical operations to create such a computer. Beginning in 1942 his staff built one in the guidance and control laboratory for calculating and simulating A-4 trajectories. It could do in minutes what human "computers," typically women working with mechanical desk calculators, took weeks to accomplish. Hoelzer's computer added one more tool to the increasingly sophisticated simulation techniques pioneered by Peenemünde to overcome the vexing problems of ballistic missile guidance. But it was the application of the mixing device to the rocket's control system, not computer simulation or guide beam work, that proved to be the truly critical innovation developed by Hoelzer's group. Without it, an operational A-4 would have taken longer to develop and would have been more difficult to produce in large quantities.[62]

With the mixing device solution in late 1941, a final major piece had fallen in place for the creation of truly guided missile. By no means, however, did Steinhoff's division have a complete working guidance system. The limitations of A-5 flights and simulation techniques left many unknowns. Furthermore, the configuration of a feasible system was only beginning to jell. Launches with the Kreiselgeräte's Sg 52 beginning in October 1939—the first successful guided flights—and with Siemens and Askania/Möller systems starting in April 1940, had demonstrated that there were at least two feasible solutions to the gyro layout. Kreiselgeräte's stabilized platform was inherently superior in accuracy, but was complicated and expensive in comparison with the alternatives. The company also lacked capacity to manufacture a more advanced new version for the A-4 (the Sg 66) in large numbers. The second best solution was Fieber's "Vertikant guidance," with two rocket-fixed position gyros, because it used the lowest number of expensive components. It was tested in limited ways with both Siemens and Askania equipment on A-5s. Those flights demonstrated the superiority of the hydraulic servomotor over Möller's electric motors as well.[63]

The great difficulties in acquiring sufficient manpower at Siemens and the resulting delays in production had resulted in alienation between Altvater's division and Peenemünde. Because the Reich had the right to use Fieber's patent, the rocket group transferred development of the appropriate gyros and vane motors to Askania and the old gyroscope firm of Anschütz. Only the rapidly growing in-house development capability of Steinhoff's division—another consequence of the dramatic expansion of the Ordnance facility—allowed the Peenemünders themselves to take the lead and to mix components for the A-5 and A-4 from all three firms, as appropriate. Instead of being the developer of the Vertikant system, Siemens became just another contractor.[64]

Thus, on the eve of the first A-4 launches in 1942, it cannot be said that Steinhoff's guidance and control laboratory had a clear idea of what system would become the final one. Indeed, it would have been foolish to be sure of success. But the elements were there: one of two gyro arrangements, a mixing device, hydraulic vane servomotors, a radio or accelerometer engine-cutoff device, and possibly a guide beam. Considering how far from a long-range guided missile the Army rocket group had been only four years before, after the failure of the A-3s, this was a stunning achievement. Together with the eighteen-pot, 25-ton-thrust engine and the refined aerodynamic configuration of the A-4, the creation of missile guidance by Peenemünde had laid the foundations for a technological revolution in rocketry.

How could that revolution have been accomplished in five short years? A strong foundation had clearly been laid before 1936 by von Braun's group, and his charisma, intellect, and management talent continued to exercise a powerful influence thereafter. University research, especially after the war-induced loosening of security in September 1939, added another critical dimension. The key factors, however, were the investment of massive resources and Dornberger's government-dominated "everything-under-one-roof" concept for Peenemünde. Even when Ordnance tried to hand a problem over to corporate contractors, as in the case of guidance and control, industry's lack of capacity, interest, or technological capability had forced the rocket program to hire more specialists for its own staff. The resulting concentration of talented engineers in one place, under the inspired leadership of von Braun, Thiel, Hermann, and others, created a fruitful interaction of minds. Of course, brilliant ideas alone were not enough. Money, matériel, and manpower were needed to convert key innovations into working technologies. Massive expenditures on new facilities and personnel at Peenemünde were therefore required. The sheer size of those investments naturally contributed a great deal to speeding up development, but the "everything-under-one-roof" concept also helped to eliminate some of bureaucratic paperwork and delays created by contracting. Much testing and experimentation could be done in-house as soon as it was needed.[65]

Dornberger's organizational philosophy thus seems to have resulted in a fairly efficient use of research and development resources, judging by the technological breakthroughs achieved in only five years. But however those resources were managed, it is clear that laying the foundations for the ballistic missile in such a short time demanded a tremendous investment by the Reich, which would have been impossible without a high priority rating from the Army and support from within the Nazi leadership. To understand the political and financial foundations for Peenemünde's technological achievement, and to set the stage for later conflicts over the rocket program, we must therefore return to 1937–38 and the origins of the wartime battle for priority.

Chapter 4

Peenemünde's Time of Troubles

Until the outbreak of the war, the Army rocket program was largely sheltered from political interference. When it was only a small research effort, little support was needed beyond the level of the Chief of Ordnance. As the program grew into a multimillion-mark enterprise, it was necessary to secure the backing of the Army Commander-in-Chief for the creation of Peenemünde in 1936. But that task did not prove difficult. General von Fritsch, like many of his subordinates, retained a great deal of autonomy from Hitler in technical questions, had a blinkered vision of strategy, and came from the artillery, the branch out of which the rocket group had sprung. Ordnance's rocketry alliance with the politically powerful Luftwaffe further ensured that there would be few priority problems. Of course, Dornberger's engineers did not receive everything they wanted, but the construction of the wind tunnel, the creation of the guidance and control laboratory, and the tripling of Peenemünde's staff in less than two years indicate that program was generously funded and staffed even before the attack on Poland.

After Britain and France declared war in September 1939, however, the need to justify this expensive undertaking grew much stronger. Despite the lack of any serious challenge to the ballistic missile's military effectiveness, the voracious demands of the accelerated A-4 schedule collided with shortages of key resources like steel and skilled labor. Matters were made worse by a badly organized military command structure and war economy that provoked conflicts among the services over a range of issues. The resulting scramble for priority drew Hitler more and more into armaments decisionmaking as time passed. To sustain high-level support, Ordnance and Dornberger were therefore impelled by the pressure of circumstances, as well as by faith in the missile, to assert that the A-4 would be available in numbers sufficient to be militarily decisive—and soon. The problem of mass production thus loomed ever larger in the struggle to maintain the high priority of the rocket program. Central to the entire battle was a missile assembly facility to be built at the Baltic coast center.

THE PEENEMÜNDE PRODUCTION PLANT

Dornberger's idea for a missile factory was born—or first aired—on the Greifswalder Oie in December 1937. During the weather-plagued, ill-fated A-3 launch expedition, the group was temporarily trapped on the island by a violent storm. With no A-3s available for firing and with the water driven out of the harbor by high winds, there was little for the leaders of the Army rocket group to do but sit around debating the future. Discussions focused on A-4 construction. Arthur Rudolph described in a 1989 interview how the factory idea came up:

> And so one day von Braun and Dornberger and I were sitting together . . . out of the blue sky [Dornberger said], "I want to build a . . . production plant for the V-2 and the coming big rocket, and you will do that." I [replied], "Dornberger, for heaven's sake, I'm a development man, I'm not a production man, and you leave this up to industry, don't bother us fellows in development with your new ideas." And von Braun was of course saying the same thing, even harsher than I did. . . .
>
> There was an argument, there was a strong argument. I was so opposed to it.

In another interview Rudolph stated: "We tried desperately to talk him out of this idea. We told him we should concentrate on the development of the A-4 and leave any pilot production, or full production, to industry."[1]

Rudolph and von Braun were right to argue that the factory was premature, especially for a rocket group with no experience in manufacturing. Dornberger wanted to build a full missile assembly facility at Peenemünde, not merely a pilot production plant, yet the A-4 was only a rough concept on paper, and there was not even a vague sketch of the "coming big rocket," or A-10. Called simply the "100-ton device" before 1940, this larger missile was to have a range of 800 kilometers (500 miles) and would quadruple the A-4's 1-ton warhead and 25-ton engine thrust. As was the case in the planning of the development shops and test stands of Peenemünde-East, Dornberger wanted the rocket factory sized to accommodate the A-10, since the A-4 was viewed only as an interim weapon. All the objections of von Braun and Rudolph were to no avail. Dornberger warned Rudolph, then the head of the development shops, that he would be put under a project head and compelled to design the plant, if necessary. The chief of Wa Prüf 11 clearly felt that the factory was that urgent.[2]

For nearly a year nothing further came of the idea, presumably because of the failure of the A-3s and the need to demonstrate positive results. In the meantime, the program was aided by another stroke of luck. General (later Field Marshal) Walther von Brauchitsch, an artillerist, was promoted to Army Commander-in-Chief in February 1938 as a result of political intrigues by Göring and the SS against General von Fritsch and the War Minister, Field Marshal von Blomberg. Neither general had been enthusiastic about Hitler's turn toward a more aggressive foreign policy. To achieve greater control over the military, the Führer exploited those plots to force von Fritsch and von Blomberg out of office. He abolished the War Ministry and created an Armed Forces High Command (OKW), with himself as Supreme Commander. He persuaded von Brauchitsch to accept the Army position in part by giving him a large cash payment to facilitate a long-desired but costly divorce. Thereby compromised with the regime, von Brauchitsch further accelerated the Nazification of his service. He nonetheless remained an outsider in the highest echelons of the Third Reich.[3]

For the rocket group, however, his appointment was a boon. Von Brauchitsch promoted Becker to Ordnance Chief in March, undoubtedly with the approval of Hitler. In addition, the new Commander-in-Chief, who had been Dornberger's superior officer in the 1920s, looked upon the younger man's career with fatherly concern. Those advantages did not, however, result in an immediate go-ahead for Dornberger's rocket factory. Apparently only after the group successfully launched the first unguided A-5s in October 1938 did the Army Commander-in-Chief order the purchase of land for the "expansion" of

Peenemünde-East on November 21. It was, he stated, "particularly urgent for national defense."[4]

Planning began in January 1939, when Dornberger created a new Berlin subsection of Wa Prüf 11 headed by G. Schubert, a senior Army civil servant who had built camouflaged munitions factories in World War I. Rudolph became his chief engineer in Peenemünde, although not on a full-time basis until July. Their task was to construct by early 1943 something that had never existed before: a ballistic missile factory. The four-year time scale ordered by von Brauchitsch, no doubt on the advice of Becker and Dornberger, was the expected peacetime development period of either the A-4 or the A-10. In February 1939, however, the Peenemünde engineers gave a more realistic estimate of at least six years before the design of the bigger missile would be finished.[5]

Based on Dornberger's concept, the factory group quickly outlined a facility that dwarfed the existing development works. Judging by the new Heinkel aircraft plant at Oranienburg, north of Berlin, Schubert initially estimated the need for a workforce of almost five thousand. But there was no place to house that many people on the rural island of Usedom, so it was necessary to build a "Large Settlement," really an entire town, for them and their dependents. Moving them to work required a better system than the old, broken-down trains currently running to the Army and Luftwaffe facilities. Schubert's group planned a special electric train system, modeled on the Berlin S-Bahn (surface railroad). That system and the factory itself needed more electrical power than was available from the existing grid. A new power station had to be built. That requirement in turn necessitated a greatly improved harbor for coal delivery by ship.

As if these technical and financial challenges were not enough, Schubert perceived from the outset in January 1939 that Peenemünde's northern coastal location presented a number of strategic and practical problems. The island was more exposed to air attack than a site far inland, yet Usedom's sandy, water-logged soil was ill-suited for underground shelters. More expensive above-ground protection would have to be constructed. In addition, the soil problem, along with brine seepage into the water table, meant that water supply and sewage systems for the workforce would be difficult to construct. The land would have to be pumped out and a dike built to protect the area from high water. Finally, the relatively remote location of Peenemünde complicated construction and labor problems generally.[6]

The massive scale and the difficulties of the production plant project would soon make it an onerous burden on the rocket program. Why, then, did Dornberger choose to build there, rather than find an industrial contractor for missile production? Clearly, the factory fitted his "everything-under-one-roof" concept, which had worked well in research and development. But Dornberger's specific rationale was that the Army's missile technology was so new and exotic that it was important for the plant to be situated next to the development works for easy consultation. Moreover, extensive testing and checkout would be required for production rockets, necessitating stands to test-fire them; combustion chambers would have to be tested and calibrated as well. Only at Peenemünde, he argued, were there sufficient safety and security to carry out that continual operation, since explosions were likely. Schubert's group therefore planned three test stands surrounded by circular earthen blast walls west of the factory complex.[7]

Dornberger's extensive testing regime shows how premature it was to rush the ballistic missile into production. Only the conviction that this weapon could be militarily decisive, and that Germany must exploit its lead in a race with other powers, explains why Dornberger and the Army leadership accepted the necessity of initiating manufacture while development was far from complete. The approval of the missile factory by von Brauchitsch and Becker also reflected the Army's desire to build an empire in armaments production. Since 1934 Ordnance had been constructing its own munitions plants on the grounds that private companies would never build sufficient capacity for wartime needs. Those "Army-owned enterprises" were then grouped under a large holding company to disguise Ordnance's control from the outside world. Many were in fact leased to explosives firms for management, but the Army's tendency toward state ownership was deeply rooted in the Prussian tradition of the officer corps. The sociopolitical environment created by National Socialism may have had some influence too. While Hitler took little interest in economics and discouraged the "socialist" elements in his own party, there is little doubt that the role of nationalized corporations grew during the Third Reich.[8]

The Army-owned enterprises certainly influenced Schubert's plans for the Peenemünde Production Plant. In 1941, while lamenting the disruptive effects of the war on that factory, he stated that his original plan had been to create, along the lines of the Army munitions factories, a "model facility" in which "expense should not be the sole determining factor." The buildings, the machines, the fire and air raid protection systems, the facilities for worker eating and relaxation, would all be first-rate. He also incorporated into his plan the German Labor Front's requirements for good social provisions and the "beauty of labor." Thus he included sports fields, cinemas, and attractive apprenticeship training shops. Together with the facility's scale and construction problems, this extravagance resulted in a hefty bill for the factory: by one 1939 estimate, 180 million marks, or roughly quintuple the expenditure on the development works up to the end of 1938.[9]

In return for that investment, the Reich was to receive five hundred A-10s or 1,500 A-4s a year, numbers that look, in comparison to later wartime production, unimpressive. The low output resulted from at least two factors. First, the German armed services in the 1930s emphasized quality over quantity in armaments production. That meant an excessive dependence on skilled labor and a poor level of economic efficiency in the early years of war. This high-skill philosophy, which was taken for granted by Army planners like Schubert, could only have been reinforced for the Peenemünde facility by the exotic and highly specialized character of the technology to be manufactured.[10]

A second factor lowering output was an emphasis on making as much of the rocket at Peenemünde as possible. The Production Plant was supposed to build almost all the major missile components itself, except for the guidance system and the turbopump–steam generator combination. The rationale for that approach probably arose from a combination of the exacting standards to be applied during production, Dornberger's general preference for "everything under one roof," and the limited manufacturing capacity available elsewhere in an economy apparently strained to its limits. In defense of the original factory concept, Rudolph later claimed that the facility had been designed as a "pilot production plant" to pioneer the technology before handing it over to private firms for mass production. Although that was the name the factory later carried, there is no evidence that Dornberger's originally intended it to function that way. The pilot production label was not applied until the fall of 1941 and became meaningless within a couple of months, because the factory was once again designated as a regular production site.[11]

The planned output of 1,500 A-4s a year (125 a month) was thus the maximum missile production that Army Ordnance foresaw in 1939, at least in single-shift operation. As it turned out, about 6,000 A-4s would be built in much more difficult conditions over fifteen months at the end of the war. Although it is hardly fair to blame Schubert, Dornberger, and their superiors for failing to see all the shortcomings of the Nazi economy or their own planning, it is an indictment of the leadership of the rocket program and the Army that they committed such huge resources to a project to launch fewer than three dozen high-explosive warheads a week at enemy targets—and in the expectation of spectacular military results. The decision to build the Production Plant is comprehensible only in the light of Ordnance's romantic infatuation with rocket technology, the predominance of artillerists in the Army leadership, and German officers' tendency to offer tactical solutions to strategic problems.

Von Brauchitsch's November 1938 order reflected the Army's remaining autonomy in the armaments sector as well. In the last years before the war, a free-for-all existed among the services. Hitler consciously attempted to play off individuals and groups against each other, while demanding as rapid a reamarmament as possible without regard to economic feasibility or effective interservice coordination. By 1938–39, skilled labor and other crucial resources were stretched to their limits, and even a "Führer order" did not necessarily have much effect. For example, in January 1939 Hitler ordered that the Navy's grandiose "Z-Plan" for a massive battle fleet be given priority over the plans of the Army and Luftwaffe. But that had little practical impact, because the resources were not there given the prevailing inefficiency, and the other services were not about to cooperate. Thus it was entirely possible for the Army to order a huge increase in the scale of the Peenemünde project without consulting anyone, but it was quite another thing actually to carry it out. At the outset of the planning, Dornberger told Schubert that he could not hire a large staff. Even labor to build the construction workers' barracks on Usedom was difficult to find.[12]

The solution arrived at by Becker and the Army leadership was to convince Hitler of the value of the rocket program by giving him a demonstration. Blinded by the Führer cult that reached new heights after the Munich agreement and by Hitler's increasing control over the military, the Army generals sought a "Führer order" emphasizing the priority of missile development. On a rainy day in March 1939 von Brauchitsch brought Hitler to Kummersdorf, which was only a short distance from Berlin. (The Supreme Commander never set foot in Peenemünde.) The two were accompanied by Becker and were met by Dornberger, Thiel, and von Braun. The Führer saw firings of the 300- and the 1,000-kg-thrust motors, with their ear-splitting roar and impressive exhaust jets, but his expression scarcely changed. When von Braun explained the workings of the liquid-fuel rocket using a cutaway A-3, Hitler listened closely but then walked away, shaking his head in incomprehension. A test stand demonstration of the A-5's guidance system in action left him similarly unmoved. Finally Dornberger described the A-4 to the Führer in a smaller, closed circle. During lunch afterward, Hitler gave the group only a backhanded compliment: "Well, it was grand!" He went on to describe his acquaintance with Max Valier in Munich a decade before but wrote him off as a "dreamer."[13]

To Dornberger, the "whole visit seemed . . . strange, if not downright unbelievable." Previous witnesses to rocket firings, he claims, had all been "enraptured, thrilled, and carried away by the spectacle," just as Göring would be a few weeks later. Yet Hitler had seemed singularly unimpressed. That did not prevent Dornberger from referring on March 31, in a request for more construction labor, to the Führer's alleged command to finish the Peenemünde Production Plant in four years. By seizing on "a brief nod" Hitler had given to his description of the A-4's development period, the chief of Wa Prüf 11 showed the instincts of good Third Reich politician. A "Führer order" was often just a passing comment by Hitler exploited for the purposes of whoever was present. But the net effect of the Kummersdorf visit was precisely zero: It neither hurt the rocket program nor helped it. The buildup of the development organization at Peenemünde continued apace; the planning of the production facility proceeded unchanged, but there were few construction workers available to do the work. Not until after the outbreak of war would there be significant changes, both positive and negative.[14]

THE BATTLE FOR PRIORITY BEGINS

On September 5, 1939, two days after the British and French declarations of war and one day after the announcement of full economic mobilization, Dornberger and Becker saw von Brauchitsch at Army headquarters in Zossen, near Kummersdorf. Military callups threatened to drain Peenemünde of manpower. At the meeting von Brauchitsch signed an order, apparently drafted by Dornberger in advance, that the "Peenemünde Project (Army Experimental Center, Production Plant and Construction Office) is to be pushed forward with all possible means as particularly urgent for national defense." It was hoped that this decree would forestall the loss of further important people, especially skilled workers, to military units.[15]

General von Brauchitsch demanded something in return for his protection. The construction of the factory and the development of the A-4 were to be accelerated so that missile production could begin in September 1941 instead of the projected date of 1943. Presumably that was the earliest date Dornberger thought possible, but it was still a risky gamble. From now on, the program would be carried out under the pressure of unrealistic deadlines imposed for political reasons. The Army Commander-in-Chief probably wanted to make the missile project defensible as one that would produce results soon enough to affect the course of the war.[16]

Von Brauchitsch issued the order guaranteeing Peenemünde's priority without consulting either Hitler or the Armed Forces High Command (OKW), just as he had done with his November 1938 order for the Production Plant. His accelerated schedule meant that missile production would have to begin at least two years earlier than planned, which in turn implied rounding up thousands of new construction workers in a labor market even tighter than before the war. Yet in 1939, as in 1938, the Army Commander-in-Chief assumed that he had the power to proceed without consulting anyone. Only afterward, on September 15, was a priority order obtained from the OKW Economics Office.[17]

There was, in fact, no coherent priority system in Germany before the summer of 1940. Hermann Göring, as head of the Four-Year Plan set up in 1936 to prepare the nation for war, was allegedly the Third Reich's "economic dictator," but he lacked the time and competence to carry out his huge task while simultaneously acting as Commander-in-Chief of the Luftwaffe. The Economics Minister, Walter Funk, was a Nazi Party lightweight and creature of Göring who controlled the civilian economy. General Georg Thomas, who headed the OKW Economics Office, officially controlled military production and priority. His concept was to pick up where the Army leadership had left off in 1918 and use his national network of armaments inspectors to impose a stifling military control on industry. That divisive and confused situation resulted in a botched mobilization of the war economy and what General Thomas called, as early as October 1939, "a war of all against all." To top it off, Hitler's personal tri-service staff, the OKW, was headed by the subservient General Wilhelm Keitel, who received little respect from the service commanders.[18]

For the moment, however, the Ordnance rocket group remained unaware of the storm that was brewing as a result of the prevailing disorganization and inefficiency. On October 11 Thomas presented for Göring's signature a priority order for construction in an attempt to sort out the services' incompatible demands for labor and resources. In order to keep all three armed forces happy, Thomas put the Army's Peenemünde project, the Navy's U-boats, and the Luftwaffe's aircraft production program together in the first rank. Göring issued the order, but it had little effect. Thomas later commented on the services' "construction mania" in the fall of 1939, a phenomenon in which Army Ordnance was a full participant. On October 9 Becker had asked for a further acceleration of the A-4 schedule, perhaps because of the quick victory over Poland, with a goal of completing the Production Plant by May 31, 1941. Nine thousand construction workers would be required for Peenemünde. By mid-November there were already five thousand on the island, including those working for the Luftwaffe. So many workers had arrived so suddenly that housing, cooking, and sewage facilities were strained to the breaking point.[19]

Meanwhile, a number of steps were taken to meet the urgent new deadlines for A-4 development. One was to open the research process to university institutes. Another was to eliminate or postpone less important projects, particularly the interservice rocket-aircraft program. The Luftwaffe had itself canceled the pure-rocket Heinkel He 176 on September 12, because it was too small and its performance was disappointing. The airplane had first flown with a Walter hydrogen-peroxide motor on June 20, but its speed, stability, and endurance were poor. A second He 176 using the Army's liquid-oxygen/alcohol system was under construction but was brought to an abrupt halt by the Luftwaffe order. What remained of the joint aircraft program was only the He 112 with the redesigned Army rocket engine in its tail. Since it had been finished and delivered to Peenemünde-West in June, it required little attention from the eastern side of the complex. The aircraft flew until June 1940, when it crashed, killing the pilot. Thus all that was really left of the Army–Luftwaffe rocketry alliance was the collaboration in guidance technology fostered by Steinhoff, plus the project to build a liquid-oxygen/alcohol takeoff-assist system for heavily loaded bombers. Cooperation had reached a new low, but that was entirely in keeping with the desire of both services to focus on more critical projects after the start of the war.[20]

For the development engineers at Peenemünde, almost all efforts now centered on the A-4 and related activities, such as A-5 launches and the design of the long-range winged A-4 (soon dubbed the A-9). The "100-ton device" (A-10) remained on the books too, but it was a mere paper concept as long as the smaller missile was unfinished. On the production side, Dornberger applied pressure to finish the final plans for the factory in a hurry. Its completion was still about four years away, in all probability because of manpower shortages in Schubert's group and at the job site. Now an enormous acceleration of the schedule was required, although as Schubert himself noted on November 16, "in the current stage of the [A-4's] development . . . too little is fixed even now."[21] As events would show, two more years of fundamental breakthroughs and basic testing would be required to complete even an untested configuration of the missile. Schubert's comment demonstrates again the fundamentally political character of the deadlines that had been imposed on the program by von Brauchitsch, Becker, and Dornberger.

Only days later, Hitler's intervention began to undermine Ordnance's overly optimistic planning. The speedup in the construction of the Production Plant greatly expanded its demand for steel. In order to satisfy that demand, von Brauchitsch had added 4,000 metric tons a month from his own quota to the 2,000 allotted by the Ordnance Development and Testing Division. At a time when this commodity was rationed and in ever shorter supply, General Keitel, the OKW Chief of Staff, felt it necessary to obtain the Führer's acquiescence to this decision. According to a later Dornberger memorandum, Keitel made an "error" regarding the character of the weapons being developed at Peenemünde, causing Hitler to see the project as "not so urgent." (Keitel's "error," if there was one, was in not conveying with sufficient enthusiasm the Army's belief in the rocket.) Subsequently, on November 20, Becker asked for and received permission from the Führer to give a presentation on the missile program. It did not do much good. Hitler decided "that development and expansion must proceed as originally planned, but he could not give his permission for an accelerated expansion."[22]

The content of this decision needs to be analyzed carefully, especially in view of the later assertions of Dornberger and others that Hitler delayed the A-4's military deployment by up to two years. The Führer clearly indicated that the prewar schedule should apply, at least as far as the steel quotas were concerned. He in no way "dropped Peenemünde from the priority list," nor did he give the program "just enough money to continue on a very small scale," to cite only two particularly inaccurate postwar statements by Dornberger.[23] There was no meaningful priority list at that time; all that existed was Göring's ineffective construction order, which was not reversed by Hitler, plus rationing of key resources such as steel and coal. The Führer did support development of the missile, although not on the new urgent schedule. As events would prove, however, that did not prevent the Army from continuing to push the program as fast as it could. Immersed in the Third Reich's Hitler cult, Dornberger and his contemporaries consistently exaggerated the Führer's ability to control events, especially in the early years of the war, when he was often indecisive or uninterested as far as the details of the war economy were concerned.

Hitler's decision on Peenemünde's steel quotas was probably influenced by concerns beyond his lukewarm opinion of the rocket program. The Supreme Commander was on very bad terms with the Army leadership after the Polish campaign, because the generals considered his demands for an immediate offensive against France reckless and passively resisted them. Without the Führer's knowledge, a military opposition movement had sprung up after the Blomberg–Fritsch affair. During the tense months of October–December 1939, indecisive coup plots flourished again in the highest circles of the Army. Chief of the Army General Staff Franz Halder went to meetings with Hitler with a pistol in his pocket, but he did nothing.[24]

Hitler's Polish gamble had led to a major war far sooner than the military planning date of 1942–44. That contributed to a severe ammunition shortage after the Polish campaign. The shortage, plus the Army High Command's desire for further troop training to eliminate operational shortcomings seen in September, were the main reasons for the generals' resistance to a fall western offensive. By November a full-fledged "munitions crisis" broke out in the leadership of the Third Reich. Hitler fixed upon Army Ordnance, which produced most of the munitions for all three services, as the main scapegoat. Cutting back Peenemünde's quota was Hitler's way of telling Ordnance to concentrate its scarce steel resources on more pressing construction projects—a perfectly valid decision.[25]

Hitler's intervention nonetheless shocked Dornberger, because it undercut his assumption that the Production Plant would be ready by the time A-4 development was scheduled to be finished, that is, by May 1941. The chief of the rocket program unleashed a campaign in December to reverse the decision, especially after he realized that Peenemünde would receive only 2,000 tons of steel a month beginning in January. The net effect, he indicated, would be to cut back and delay the factory greatly. Based on studies by Schubert's group, two of the three huge assembly buildings, each with dimensions of 120 by 240 meters (400 by 800 feet), would have to be canceled. By eliminating or postponing many other buildings and by cutting down construction costs, it would be theoretically possible to salvage a manufacturing date of September 1941, but only at the cost of very low production figures and a slow buildup even in a two-shift operation. Dornberger gave a schedule of eighteen A-4s a month at the outset, growing to ninety in July 1942. Full capacity would therefore be only 1,080 missiles annually.[26]

Dornberger's figures and dates were completely hypothetical and assumed that development would be finished on time. Structuring the plan that way did, however, protect the early deadline for missile deployment at the cost of very low initial output. But Dornberger naturally held out higher figures at earlier dates if the full steel quota were to be restored. It was all to no avail. On December 19 Becker met with von Brauchitsch. The Army Commander-in-Chief confirmed that he could not alter Hitler's steel cutback but ordered that development continue on the accelerated schedule.[27]

About the same time, a new, more serious threat emerged. Severe steel shortages and the "munitions crisis" led to an ineffective campaign to cancel construction projects not classed as "important for the war." By early January 1940, however, Becker was able to reassure the rocket group that Peenemünde would not lose its status; von Brauchitsch's decision remained in effect. The turn of the year 1939–40 was thus a nerve-racking time for the rocket enthusiasts, even though the actual impact of the priority crisis on the Army's missile research—as opposed to the production facility—was minimal.[28]

Although Dornberger was appalled at the cutbacks to his Production Plant, his defense of the program had not been without effect. One of his key arguments was the claim that Germany was in an international missile race that it could ill afford to lose. In a December 14, 1939, memorandum, he stated:

> It also must be remembered that intensive development in the area of the long-range rocket has been carried out with the support of the relevant armed services in all larger states, e.g., France, England, the United States of America, and Russia. Germany has an unquestioned lead of a few years in the long-range rocket field, which will be lost if the Peenemünde Production Plant is stopped. Because enemy states will accelerate the development of this perhaps decisive weapon even more in the current war, we must expect its deployment by the enemy side starting in early 1942 at the latest.[29]

The last estimate, which was based on no solid intelligence, shows Dornberger to be a true believer, and one willing to exaggerate for the sake of the cause. But his magnification of the foreign threat was not unusual for military-industrial project managers, then or since. A similarly flawed, although arguably more justified, estimate of Germany's potential for building an atomic bomb spurred the Anglo-American Manhattan Project in 1941–42.

The most ironic feature of Dornberger's December 14 memorandum is the claim that the A-4 could be decisive. In his memoirs, this comes up only in the context of a famous July 1943 meeting with Hitler, where he allegedly became worried when the Führer asserted that the weapon would be decisive. One must again wonder how Dornberger thought that relatively small numbers of conventionally armed missiles would have such an effect, since they would be launched, a January 1940 document says, against "valuable area targets (warehouses, airports, military-industrial facilities, railyards, and so forth)." To understand his thinking, one must accept his (unrealistic) assumption, based on his engineers' theoretical calculations, that the accuracy of the missile would be less than a kilometer. But this targeting also appears to reflect the limited perspective of an artillery officer, with no influence from discussions of the aerial terror bombing of cities.[30]

With the onset of the harsh winter of 1939–40, which halted construction for three months, and the beginning of the steel cutbacks, the Production Plant became bogged down in a morass of difficulties that were to last most of the next three years. The shifting priority and quota situation of the facility made planning chaotic. Manufacturing equipment arrived, and there was no place to put it; contracts had to be broken because of the construction cutbacks; the inherent problems of building on a low-lying barrier island caused further delays. To those problems were added constant difficulties in maintaining the construction workforce, especially with the housing shortages and the primitiveness of the barracks. In addition, the construction and manpower resources of the factory were constantly being raided to finish the more pressing projects in the development works, especially the A-4 launch facility (Test Stand VII). But the higher priority of that project reminds us that the Plant's problems had little impact on the rate at which missile research and development went forward under von Braun.[31]

For Dornberger and his subordinates, the situation nonetheless looked threatening in early 1940. At the beginning of March Wa Prüf 11 was informed that it would receive only 80 percent of its steel quota for the second quarter of the year. The uncertain political climate also unnerved at least one major contractor, Kreiselgeräte, which had to be reassured in mid-February that high-priority status for rocket development was continuing. Not coincidentally, Göring initiated a new campaign to eliminate unnecessary building projects about this time. Fritz Todt, the autobahn czar and head of construction in the Four-Year Plan, had urged that action. As a part of the campaign, Hitler wanted to reduce the Production Plant from first to second priority on Göring's list, but the intervention of Albert Speer, the Führer's chief architect, prevented it.[32]

Speer made that effort because, over the preceding year, the Ordnance rocket program had forged a growing alliance with him—an alliance which was especially fortunate for Ordnance in view of his unexpected rise to Armaments Minister in early 1942. Speer had been consulted on the plan for the factory's "Large Settlement" since February 1939 and had accepted a supervisory role over Peenemünde construction at the outbreak of the war, when he and Todt had taken over most Army and Luftwaffe domestic building projects. On first visiting the rocket center in January 1940, Speer had established a personal bond with its young engineers, if his memoirs from the late 1960s are to be believed. Their work "exerted a strange fascination upon me. It was like the planning of a miracle. I was impressed . . . by these technicians with their fantastic visions, these mathematical romantics. Whenever I visited Peenemünde I also felt, quite spontaneously, quite akin to them." Speer was only seven years older than von Braun and like him had made a meteoric rise in the Third Reich to direct massive projects: the Nuremberg party rally buildings, the reconstruction of Berlin. His intervention would serve the rocket program well over the next few years, but his later assertion that he continued to build Peenemünde behind Hitler's back, even though its priority was cut off, is clearly false.[33]

In the troubled months of early 1940 a second personality began to loom large for the program: Fritz Todt. During the endless rounds of meetings produced by the "munitions crisis," the name of the masterful builder of the autobahns and the West Wall fortifications had figured more and more in the search for solutions to the paralysis in the war economy. On March 17 Hitler made him the first Minister for Armaments and Munitions, with the specific task of eliminating the bottlenecks in Army weapons production. It was typical of the divided character of the Third Reich that Todt received no power over the Navy or the Luftwaffe, and even his ability to influence Army Ordnance and the OKW Economics Office was doubtful. He immediately issued orders stopping government experimental projects that could not produce results by October, but those injunctions were ignored in Peenemünde and elsewhere.[34]

To the Army leadership, Todt's appointment was an open rebuke. As a countermove, Becker momentarily persuaded Hitler to create a single more powerful *Wehrmacht* (Armed Forces) Ordnance office, which the Army would inevitably dominate. But an important director of the Krupp concern, appalled at the attempted reassertion of military control over industry, induced the Führer to reverse the decision on the same day, April 8. In the process the industrialist did not hesitate to hint at unspecified scandals in Becker's family. Demoralized by months of attacks on Army Ordnance during the "munitions crisis," Becker shot himself to death that evening after hearing of the latest assault on his person. It was a shocking blow to the rocket group, but the very next day Dornberger and von Braun visited Todt in his office and were reassured of his support for Peenemünde, whatever his orders had said. Moreover, General Emil Leeb, Becker's successor and another artillerist, turned out to support the rocket program as enthusiastically as the former chief of Ordnance.[35]

Leeb's enthusiasm for the ballistic missile was amply demonstrated two months later. In the national euphoria over the lightning defeat of France, he thought he saw his moment. So convinced was Leeb of the A-4's decisiveness, he wanted to give it the highest possible priority even if it could no longer affect the war against Britain, which appeared to be virtually over. Larger numbers of A-4s should still be produced, he said on June 20, "in order to have the possibility of keeping England under pressure even after the conclusion of peace." To accelerate development and production, Leeb therefore approved measures outlined by Dornberger that would give the rocket program a kind of national superpriority: Peenemünde would be rated higher than the U-boats, armored vehicles, military aircraft, and all other armaments programs of the Third Reich. His immediate superior, the Chief of Army Armaments and Commander of the Replacement Army, General Friedrich Fromm, supported this misguided plan. Von Brauchitsch, although sympathetic too, quashed it because it would require him to go to Hitler to reverse the steel cutback.[36]

The Army Commander-in-Chief doubtless saw that it was not the right moment to approach the Führer, who was confident that the Luftwaffe would defeat Britain. He was also aware of Hitler's and Göring's demand for the "redirection of armaments production" away from the Army and toward the Navy and Luftwaffe. The "redirection" reflected the two leaders' naïve belief that a huge economy could be switched from one direction to another in a moment. In the end their demand accomplished little but to spread confusion in a war economy already losing momentum because of the widely held belief that victory was near. To make matters worse, by late July Hitler was reemphasizing the buildup of the Army and its Panzer divisions for an attack on the Soviet Union.[37]

As a part of these twists and turns, a new—and largely futile—attempt was made to straighten out the mess in the war economy. Fritz Todt's modest advances toward a more rational system were vitiated by the victory over France, which strengthened the hand of the military. His two main competitors, Göring and General Thomas of OKW, now attempted to adapt the priority system to the "redirection of armaments production." As the Peenemünde case has shown, there were two uncoordinated systems: the construction priority levels of Göring's October 1939 order and steel rationing. The latter had the greater impact but stifled production instead of encouraging efficiency. Göring (for whom Hitler had just invented the rank Reich Marshal, equivalent to a six-star general, to keep him above the many new Field Marshals like von Brauchitsch) issued a revised system incorporating steel rationing on July 18. His order created only two priority levels (I and II), reflecting Hitler's penchant for avoiding difficult choices. As a result, the system was ineffective, but it shocked Army Ordnance because Peenemünde was not mentioned at all. After two weeks of panic, the rocket program was reinstated in priority level "I" "through the back door" under the category of "munitions in short supply."[38]

Only one measure discussed during the June euphoria was actually carried out. Leeb had decided that the Army's own inefficient bureaucracy was a hindrance to the construction of the Production Plant. He approved negotiations to have Albert Speer's Peenemünde organization take total responsibility for building sites at the center. Construction Group Schlempp, named after its local chief, formally displaced the Army construction office on September 15.[39]

To speed the project further, Ordnance went to Todt looking for construction workers to supplement the basic force of 3,000 that had been working for the Army at Peenemünde since late 1939. In mid-July, Dornberger asked the Minister to double that number. Todt promised a thousand from his own projects and another thousand from Speer's, but Speer soon said that he could not fulfill his end of the bargain. Todt did supply a thousand, but they may only have replaced some 900 Polish workers who were apparently withdrawn after OKW forbade the use of non-Germans on secret projects. Unlike the situation after 1942, the security threat posed by forced and foreign labor still outweighed the need for manpower in the early years. A few hundred Czech workers had been used at the Peenemünde factory site in mid-1939 but were sent away at the start of the war, presumably to protect secrecy. The Polish workers appear to have arrived in the spring of 1940. Whether or not they were actually withdrawn that summer, the exploitation of East European laborers and prisoners of war would soon become normal in Peenemünde and elsewhere. It was a harbinger of the eventual enslavement of concentration camp inmates for A-4 production.[40]

All the uncertainty and bureaucratic maneuvering of the early summer was not, however, the end of the program's troubles. At the end of August the priority system entered a new state of flux, throwing Peenemünde's status again into question. So ineffective was the July priority order that Thomas and the service ordnance chiefs agreed to split level "I" into "Ia" and "Ib." On August 20 Hitler ordered the establishment of a "special level S" for "Operation Sealion," the invasion of Britain, above both. The problem was that the services were all jockeying to have their programs in the top level, while stealing resources from each other or blaming the OKW for their problems. So competitive and disorganized was the situation that Todt was forced to complain to Hitler. The Army had issued armaments orders directly affecting his Ministry that he heard about later only through "unofficial channels."[41]

Dornberger responded to the new period of uncertainty by again arguing for the disastrous impact of a priority reduction on Germany's position in the alleged international missile race. He mentioned the discovery by occupation authorities of secret military experiments by the French rocketry pioneer Robert Esnault-Pelterie, suggested that Vichy France might still be pursuing a missile(!), and noted that there had been no news of Goddard since 1938, when the U.S. War Department had supposedly intervened. (In fact, the American physicist and inventor began working on small projects for the military only in 1941.) Dornberger believed that Goddard was "about two years behind German development" in 1938. That estimate was not too far from the truth, but once again Dornberger had no intelligence to support the assumption that the Americans had made huge investments in a ballistic missile program. His claims were "fully" supported by the Army Ordnance leadership nonetheless. In the midst of the Battle of Britain, however, the argument was ineffective, and the rocket program was reduced at the end of September to level "Ib," that is, third priority.[42]

Although there had been some temporary loss of workers in the manufacturing of key electronic components during the uncertainties of late July, that priority assignment produced the first real threat to the *development* of the A-4. Firms began responding with letters saying that their contracts for the program could not be fulfilled because of the demands of projects at higher priority levels. Dornberger's protests went up the line to the Army Commander-in-Chief, and after only two weeks Peenemünde was unofficially bumped up to second priority ("Ia") by the OKW. Dornberger naturally still wanted to get all the way to the top level ("S"). On November 19 Keitel informed von Brauchitsch of Hitler's decision. The Führer confirmed the rocket program's assignment to second priority, including the steel quotas imposed one year before. The Führer may have been restating his interest in the project after the Luftwaffe's failure in the Battle of Britain, but his endorsement remained lukewarm at best.[43]

In early 1941 Peenemünde's priority level took on new urgency with the invention of yet another level above all others, "special level SS," which had nothing to do with *the* SS. A process of priority inflation was at work: The services would all try to crowd their contracts into the top level until there were not enough manpower and resources to satisfy even that level's demands, given the German war economy's poor efficiency. Since the OKW was often too weak to refuse the demands of the services, and Hitler preferred to avoid hard decisions, it was easier to invent a new level than to impose priority reductions. As a result, lower levels suffered "a certain degeneration," according to an OKW staff member. By February 1941 levels below "SS" and "S" were phased out as meaningless.[44]

In this light the rocket program was reevaluated, and on February 5 the head of the Development and Testing Division, General Koch, told Dornberger that its priority would be split: Development would go to the top level ("SS"), and the Production Plant would be put in "S." That measure, which became official toward the end of March, presumably corresponded to Hitler's view at the time. In a phone call to Leeb, General Fromm said on May 7 that "in line with the Führer order, only development is allowed in Peenemünde, therefore at most a test series" can be produced. If, by this order, Hitler meant that the missile had to be proved to work before manufacturing it, his position was entirely reasonable. It is also possible that this had been his thinking since the autumn of 1939.[45]

Thus after six months of lower priority, missile development was once again supported as most urgent. For the production facility, however, machine tools and laborers remained difficult to acquire. Dornberger still hoped that the A-4 would be finished and production could begin on a small scale in early 1942 if the Army found more resources. Plans called for the factory eventually to turn out five hundred A-4s a year, the official goal since the cutbacks of early 1940, unless the grandiose original plans for three assembly buildings could be revived. But all those ideas remained unrealistic. Even if top priority for the production facility could be obtained from Hitler—as it would be in the late summer of 1941—developing and producing the A-4 in less than a year was a fantasy. Dornberger and Ordnance had become trapped by their own political salesmanship and by their unavoidable ignorance of the difficult technological challenges that lay ahead.[46]

The first and most difficult phase of the priority battle had thus come to an end. The year and a half since Hitler's steel cutbacks had truly been Peenemünde's time of troubles, in large part because it was also the era, as Speer later said, of "incompetence, arrogance, and egotism" in the war economy. Weak leadership, competing bureaucracies, interservice rivalries, and shortages of key resources had produced paralysis and inefficiency. In the process, Peenemünde's priority was changed numerous times, or at least had to be upheld in numerous battles. Those battles no doubt wasted Dornberger's time and certainly delayed the ill-conceived and extravagantly planned Production Plant, thereby shaping the rocket chief's perception of the pace of the whole A-4 program. Yet there were ways in which the existing political division and irresponsibility helped the program too. The remnants of the Army's autonomy and power, in conjunction with Ordnance's alliance-building with Nazi leaders like Speer, allowed the senior service to sustain the Peenemünde project while avoiding any systematic examination of the missile's military effectiveness.[47]

While the Production Plant was hard hit by the priority crises of 1939–40, ballistic missile development was much less affected. The uncertain situations in July and October 1940 undoubtedly did cause some lost time on contracts for important equipment, especially critical components like gyroscopes and transmitters. It is also possible that the reduction to second priority for six months in the winter of 1940–41 had an impact on contract deliveries. Under the rules, however, contractors were not supposed to lose any workers to first-priority projects.[48]

Further delays in the delivery of guidance, radio, and turbopump components were caused by a lack of available industrial capacity even when rocket development was in first priority, which was most of the time. Siemens's aircraft instruments division, for example, was none too cooperative, because it had so many pressing commitments to the Luftwaffe for mass-produced items. The national superpriority that Dornberger and Leeb had requested in June 1940, however illusory and selfish, might indeed have helped to speed matters up a bit in such cases. But, contrary to to Dornberger's view, delays caused by a lack of capacity had nothing to do with Hitler's downgrading of Peenemünde's priority. The Führer's most important decision—to cut back steel quotas—affected factory construction only. If he had been more supportive of the project, of course, it would have helped, but the Army managed to circumvent his occasional attempts to curb accelerated development.[49]

On the other side of the ledger are the stunning breakthroughs in key technologies achieved between 1939 and 1941, especially in guidance and control. Those impressive accomplishments are grounds for skepticism about critical delays in missile research and development in that period. Further evidence on that point comes from Ernst Steinhoff. In his annual report for 1940, the single worst year for priority difficulties, the guidance and control chief wrote that "the desired goals could almost everywhere be reached." Difficulties in finding specialists had moderated in the course of the year. Only in the area of skilled craftsmen did Steinhoff see a significant continuing problem. Thus, when Dornberger claimed in October 1940 that development had already slipped three-quarters of a year to the end of 1941, it really represented the evaporation of the utopian deadline of May 1941.[50] As a result of the first phase of the priority battle, missile development lagged by only a few months, and capacity problems may have added another couple of months.

In the minds of Dornberger and his associates, the impact was much worse. But, inspired by an enthusiastic belief in the technology and backed by the Army leadership, they naturally fought on with all available means to advance their project. As the summer of 1941 approached, they were confronted with three principal challenges: winning Hitler's favor, launching new cooperative projects with the Luftwaffe, and bringing the first A-4s to the test stand.

Chapter 5

Hitler Embraces the Rocket

The sixteen months from June 1941 to October 1942 were a transitional period for the Army rocket group and for the Third Reich. Hitler's gigantic, long-desired assault on the Soviet Union, "Operation Barbarossa," was launched on June 22. Months of spectacular victories (and unimaginable brutality) followed, leading the German populace to expect a triumphal peace within a year or two. At the end of 1941, in a shocking turnabout, the offensive ground to halt in the Russian winter, and the United States entered the war after the Japanese attack on Pearl Harbor. The altered situation forced the Nazi leadership into a major reorganization of a war economy already mired in serious difficulties. Not only would the Reich have to match the industrial output of three great powers, instead of only one (Britain), it also had to satisfy the Eastern Front's insatiable appetite for more and more soldiers. The resulting manpower crisis, in combination with Hitler's vicious racial ideology, led directly to the exploitation of forced and slave labor on a huge scale.

The shifting winds of the war inevitably had an impact on the rocket program. The priority battle continued, but in an altered form; the changes in the direction of the war economy distracted Dornberger and Ordnance, but Hitler embraced the project more enthusiastically, with a corresponding increase in the materials and manpower devoted to it. Because of the changing strategic situation, Army–Luftwaffe relations in rocketry changed as well, to a pattern reminiscent of the late 1930s: more intense cooperation combined with greater interservice competition. This time, however, the cooperation focused on anti-aircraft defense, while the rivalry was less hidden. In the spring of 1942 the air service initiated a competing project to build a cruise missile later called the V-1. Von Braun's development group also had to struggle through a difficult transition between June 1941 and October 1942: from building and launching the A-5 to constructing and firing its much larger, more complicated successor. Bringing the A-4 to the launch pad, getting it to fly, and making it suitable for large-scale production proved to be much more difficult than expected. A successful launch became essential to the high priority of the project. Peenemünde could not go on indefinitely consuming huge quantities of resources; the promises made for the ballistic missile were beginning to wear thin.

THE FÜHRER APPROVES—BUT THE BATTLE CONTINUES

After a relatively quiet interlude, the second phase of the A-4 priority battle began with yet another "Führer order" for the "redirection of armaments production." This time Hitler ordered the reorientation toward an air force and fleet aimed at Britain in the days *before* Barbarossa was launched—so arrogant was his belief in quick victory over the Soviet Union. As in the preceding summer, the result was confusion and infighting, but Armaments Minister Fritz Todt did make some marginal progress toward a rationalization of the war economy. Although Göring and OKW Economics chief Thomas had managed to stymie Todt's efforts to carry out a reform since May 1940, the Armaments Minister did obtain from the Führer the right to control all armed forces construction under the new "redirection" order. Todt set out to cut back Army construction in order to shift resources to the Luftwaffe and Navy.[1]

The Minister's decrees immediately threw into doubt Peenemünde's 50-million-mark construction budget for the next fiscal year. Since 90 percent of that money was earmarked for the Production Plant, and the facility was still bogged down in manpower shortages and contract delays due to its second priority ("S") status, Todt's campaign threatened to set the factory even farther back. Greatly disturbed by this prospect, Army Ordnance chief Leeb quickly secured a meeting with Field Marshal von Brauchitsch on June 28. He carried with him a memorandum originally drafted by Dornberger.[2]

That memorandum included a novel argument for the ballistic missile. It is the first extant Ordnance document that clearly advances a terror bombing rationale for the A-4, which corresponds to Dornberger's own recollection that he began to think in those terms only after the Luftwaffe's costly losses over Britain in 1940–41. He now presented the missile as "significant relief for the employment of aircraft against England, and especially London and the port cities." The lack of any "means of defense," the "accuracy" of the missile, and the ability to launch "day and night at irregular times and regardless of weather" would make it a particularly effective contribution to "the defeat of England." The obvious implication was that it would be employed as a psychological weapon against civilian populations.[3]

Although von Brauchitsch must have been preoccupied with the huge battles then unfolding in the East, he accepted the gravity of the threat to Peenemünde and appreciated the new rationale for the missile as well. Interservice rivalry further contributed to the Commander-in-Chief's sense of urgency on this issue. Already disturbed at the favor shown to the Luftwaffe in the "redirection" orders, von Brauchitsch no doubt learned of Ordnance's paranoid reaction to an expression of interest in the rocket program by the air force, after Air Ministry officials visited Peenemünde in mid-June to investigate new joint projects in anti-aircraft defense. The Commander-in-Chief immediately began looking for an opportunity to make a presentation to Hitler. Reaching the Führer could fend off threats from both Todt and Göring, while simultaneously giving the Army a bigger role in the defeat of Britain.[4]

For most of July to September, the priority struggle proceeded on two unconnected levels. In one, Todt became ever more threatening in his tone, while demanding cutbacks in what he now saw as an extravagantly built production facility at Peenemünde. In a tart July 30 letter to General Fromm, who stood between Leeb and von Brauchitsch in the chain of command, he complained:

> In Peenemünde they are building a paradise. The housing, the social provisions, the clubs and apartments, the warehouses and factory halls,

> all present the highest amount of expenditure one could possibly imagine. I am convinced that the 5,500 construction workers [there] are quite sufficient to carry out a far larger program in a relatively shorter time, if one were only to build in a manner appropriate to the achievement of a wartime objective.

As a means to that end, he demanded that rough wooden barracks-style construction be applied as much as possible, in line with the guidelines he had issued earlier in the month. That prospect particularly distressed Schubert, the head of the Production Plant, who had laid out the factory in peacetime as a model manufacturing facility.[5]

Even before the Armaments Minister had made those demands, von Brauchitsch prepared to make an end run around Todt. He decided that the Führer would be more likely to take an interest in the rocket program if his protégé, Dornberger, were to visit headquarters. On July 31, one day before the chief of Wa Prüf 11 had a chance to see Todt's letter, Dornberger drafted a memorandum for presentation to Hitler. On von Brauchitsch's order, he avoided complaining about the previous troubles of Peenemünde. Instead, he emphasized that the A-4 could, "in addition to the material damage, have a particularly large impact on morale, even when air superiority is no longer present." Dornberger also discussed joint projects with the Luftwaffe, a winged A-4 (the A-9) for longer ranges, and a two-stage missile to hit the United States.[6]

A possible design for an "America rocket" (in modern terminology, an intercontinental ballistic missile or ICBM) had emerged during the preceding year in the studies of the center's Projects Office. That division, headed by Ludwig Roth, had investigated placing the A-9 on top of the proposed 100-metric-ton-thrust A-10. To reach the United States, however, it would be necessary to increase the A-10's power almost twofold to 180 tons (about 400,000 lbs of thrust), and even then the A-9 would be only marginally able to hit East Coast cities after a hypersonic glide. The concept was actually far beyond Peenemünde's technological grasp: The guidance requirements were too extreme, the aerodynamics were unknown, and the materials did not yet exist to prevent the upper stage from burning up during reentry into the atmosphere. But the idea appealed to the vivid imaginations of von

Braun and his engineers. It also provided Dornberger and the Army with another weapon against Luftwaffe competition, since the idea of hitting America undoubtedly appealed to the Nazi leadership at that time. Confidence in the imminent defeat of the Soviet Union was high, while worry about the possible entry of the United States into the war was rising; the Luftwaffe had its own plans for a bomber that could fly to America and back.[7]

After a wait of nearly three weeks, the Army Commander-in-Chief's campaign finally paid off. On August 20, 1941, Dornberger, von Braun, and Steinhoff met the Führer at his *Wolfsschanze* ("Wolf's Lair") headquarters in East Prussia. Also present were Fromm and the head of the OKW, Field Marshal Keitel, but von Brauchitsch was absent. Dornberger began by showing a propaganda movie that included footage of the A-2 and A-3, the construction of Peenemünde, launches of the A-5, and rocket-plane and rocket-assisted takeoff experiments. The film, which had been made in October 1940, ended with an ominous warning about foreign competition. The rocket program's chief followed with a lecture in line with his script. According to his August 21 memorandum—the only existing account of the meeting—the Führer said that "this development is of revolutionary importance for the conduct of warfare in the whole world. The deployment of a few thousand devices per year is therefore unwise. If it is deployed, hundreds of thousands of devices per year must be manufactured and launched." The "if" was significant; in spite of Hitler's newfound enthusiasm for the A-4, he continued to withhold a mass production order until the outcome of its development was known.[8]

The Führer's comments were an interesting mixture of perceptiveness and absurdity. In the long run the ballistic missile was a revolutionary weapon, just as its advocates had been saying since the early 1930s, but not in the limited, conventionally armed form of the A-4. The Germans seem not to have made the connection between atomic weapons and the missile, because their nuclear project never proceeded much beyond preliminary reactor experiments and theoretical studies of a bomb. In any case, by early 1942 the German leadership decided that the gigantic industrial effort required for an atomic weapon was not feasible during the war. (The bombs later produced by the Manhattan Project weighed more than 4 metric tons—too heavy for an A-4.) Although Hitler never understood the concept of nuclear weapons, he appears to have correctly perceived, on August 20 at least, the minimal strategic impact of small numbers of A-4s, so he asked for hundreds of thousands. But the raw materials and manufacturing capacity needed to fulfill that demand made it absurd. Despite Hitler's often masterful command of the details of weaponry, he had once again demonstrated his inability to comprehend the liquid-fuel missile. He saw it as merely a big artillery shell, bearing out Speer's observation that the Führer was comfortable only with World War I technology.[9]

Notwithstanding Hitler's faulty understanding of the missile and his refusal to issue a mass production order, von Brauchitsch's carefully planned effort to bypass Todt and outflank Göring was clearly a big success for the Army. Keitel approved the promotion of the production facility to first priority ("SS"), the same as development. The whole center was provided with the further protection of being included among the "special enterprises," a category invented in 1941 to introduce more gradation into a poorly differentiated priority system. To cement the factory's new status, Ordnance merged it into one center with what was now officially called the "Development Works." The acting Commander, Major Gerhard Stegmaier, assumed the same title over the whole center, while remaining the military head of the development side. (Peenemünde's Commander, Colonel Leo Zanssen, was absent from mid-1941 until April 1942, leading a solid-fuel rocket battery on the Eastern Front.) Schubert's factory was renamed the "Pilot Production Plant," with the task of pioneering large-scale manufacturing for later transfer to industry.[10]

Hitler's backing allowed von Brauchitsch to alleviate many of Peenemünde's manpower problems. The Commander-in-Chief ordered the founding of a unit with active duty status that could pull engineers and craftsmen from the regular Army. When the formation of the "Northern Experimental Command" was completed in November, it had 641 officers and men, plus fifteen others for administrative supervision. Because rank was ignored and the men were placed in jobs and paid civilian salaries according to their training, if they wished, some rather unusual situations arose. Privates could be supervising captains in the laboratory. Helmut Hoelzer, head of the guide beam division, solved this problem by ordering everyone to wear white lab coats buttoned up to the neck. The placement of the soldiers in line with their civilian accomplishments worked very well. The only problems arose from the narrow-mindedness of the unit's commander and the NCOs, who found the situation discomforting. In later years the Command expanded considerably and rescued not a few talented engineers and skilled workers from bullets and frostbite on the Eastern Front. The fact that they had been there at all indicates both the shortsightedness of the Reich's original draft policy and the increasingly desperate manpower shortages faced by the Germans.[11]

The formation of the Northern Experimental Command and the top priority rating for the whole facility clearly were important gains from the Peenemünders' audience with the Führer. But even while that was going on, Todt was still fighting with Ordnance for a cut in the center's construction budget. In a September 13 letter to Dornberger, he threatened to recommend a complete construction stoppage to Hitler if Ordnance did not accept his budget figure of 20 million marks. Leeb was amazed; Todt did not even appear to know about the Führer's order! In fact, the Minister did refer in passing to that order, but his lack of knowledge about it speaks volumes about the disorganized and competitive character of the war economy before 1942 and about the ability of the Army, however much its power was in decline, to exploit the situation. Dornberger, backed by the OKW priority order issued on September 15, wrote to Todt discussing the visit to Wolfsschanze. In his letter, the chief of Wa Prüf 11 indicated his belief in the "decisive importance of this weapon for the war" and his fear of "the progressive development of this same area by the USA. We must maintain our lead if we want to beat the Americans." (He had made a similar comment to Keitel on August 20.) There is no record of the Minister's reaction to those assertions, but the two sides soon worked out a compromise budget of 25 million marks. The Army also managed to rebuff Todt's attempt to take over Peenemünde construction by emphasizing that this was Speer's responsibility. Speer had indicated his opposition too but had otherwise kept a low profile throughout the conflict.[12]

The net effect of all maneuvering was to give the rocket program new impetus while at the same time forcing construction shortcuts on the production facility. The problem now confronting Peenemünde was to prepare for the mass production of the A-4, contingent upon Hitler's order. In the first rush of enthusiasm after August 20, numbers in the range of 50,000 to 150,000 a year were bandied about, and the Peenemünde factory received its Pilot Production Plant label. The idea of building an A-4 "test series" or "zero series" to ease the transition to mass production was not new. Dornberger had mentioned it as early as August 1940 but had mostly invoked it as an excuse to keep the factory going. For a few weeks in the fall of 1941 the facility's new name actually corresponded with its intended purpose. But in short order feverish studies at Peenemünde and in the OKW Economics Office showed the absurdity of manufacturing 150,000 missiles annually. For one thing, the entire aircraft manufacturing capacity of Germany would have to be taken over! The Führer's huge numbers had to be given as inconspicuous a burial as possible, and the Production Plant returned to its original purpose as the main assembly facility, notwithstanding a test series of 585 missiles it was to construct first (the development shops were to build fifteen). Based on the limited supply of liquid oxygen, Dornberger set an annual production goal of 5,000—without anyone telling Hitler.[13]

Only on the margins did the idea of producing tens of thousands of missiles live on. For nearly a year in 1941–42, Dornberger's preferred solution was the A-8, a simplified, longer-range or heavier-payload A-4 with a 30-ton-thrust engine powered by nitric acid and diesel oil. The propulsion chief, Thiel, had begun investigating this propellant combination in the spring of 1941 and favored it as a way of getting rid of the problems of handling and manufacturing liquid oxygen. In 1942, however, the A-8 fell out of favor because of questions about its aerodynamic stability at higher cutoff velocities, the pressing need to concentrate on the A-4, and Hitler's lack of interest in the concept. The Führer's reasons are unknown, but it is possible that Germany's oil supply problems were a factor.[14]

The figure of 5,000 missiles a year therefore remained the operative one. Although this was at least within the bounds of feasibility, it was about triple the 1939 target for the Peenemünde factory and ten times the goal set after the cutbacks of 1940. The new urgency of the program also seemed to demand a faster transition to production. Beginning in late October, Dornberger launched an Army Ordnance "Working Staff" to plan the process; it was most noteworthy for ignoring the Armaments Ministry altogether. After the confrontation with Todt, that is not surprising, but the decision also reflected the continuing divisions in the war economy. The Minister had power over Army munitions and armored vehicles, but Ordnance fought to exclude him from further gains. Throughout the budget conflict it had been Todt's power as construction czar in the Four-Year Plan that had counted, not his title of Minister. Dornberger's Working Committee therefore included only Development and Testing personnel (including Peenemünders), plus representatives of other Ordnance divisions.[15]

In order to speed up production of both the planned test series and regular manufacturing, further factory capacity was needed. The Production Plant could assemble more missiles only by manufacturing less in-house. The rocket group therefore moved quickly in the fall of 1941 to find new subcontractors in private industry for the steam generator, the fuel tanks, and large sections of the fuselage.[16]

The most important new subcontractor was Zeppelin Airship Construction Ltd. At Stegmaier's instigation, he, von Braun, design bureau chief "Papa" Riedel, and Eberhard Rees, von Braun's deputy for the development shops, had traveled down to Friedrichshafen in early September to see the firm's management, headed by Dr. Hugo Eckener. Eckener was the spiritual heir of Count Zeppelin and the world-famous captain of the airship voyages of the 1920s and early 1930s. Because Eckener's company had appropriate experience with lightweight aluminum manufacturing and its capacity was underutilized by the Air Ministry, it had been only too happy to accept contracts for propellant tanks and various fuselage sections. During the visit, the Peenemünders had also raised the idea of using empty Zeppelin hangars to assemble A-4s. In early December the rocket group revived that idea, because the Peenemünde plant did not have the capacity to assemble all 5,000 missiles a year. The Friedrichshafen company was designated as the second missile factory at the end of 1941.[17]

As the winter set in, however, it became clear that the troubles of Peenemünde were far from over. There was a crisis situation in the war economy. In many categories production was falling below the already unsatisfactory levels of the summer because of to shortages, military callups of workers, and inefficiency. The failure of Operation Barbarossa to knock the Soviet Union out of the war became apparent in

November as well. A new sense of desperation gripped the German leadership; as if to confirm the urgency of the situation, the Soviet counteroffensive before Moscow and Hitler's declaration of war on the United States occurred within days of each other in the first half of December. Confronted with a terrible situation on the Russian front, Hitler dismissed von Brauchitsch on the nineteenth and appointed himself Army Commander-in-Chief. The public excuse was heart trouble, but in reality the field marshal was the scapegoat for the disastrous effects of the Führer's own megalomania. Von Brauchitsch nonetheless deserves no sympathy, because he and the Army were deeply implicated in Hitler's race war of mass extermination in the East.[18]

The departure of Peenemünde's most powerful advocate was obviously a worrisome portent for the leaders of the rocket program, but there were more immediate threats. Notwithstanding a top priority rating and the August Führer order, in early December the Pilot Production Plant was forced to accept large cuts in its steel and gasoline quotas because of severe shortages of those commodities. In early January the center also lost its status as a "special enterprise" for a few weeks, until notices began to pile up from contractors that they could not hold to their schedules. Moreover, unlike the more open-minded Dornberger, Schubert was appalled by Todt's new plans for the rationalization of the war economy, since they seemed to undermine further his original vision of the factory.[19]

The Armaments Minister was in fact the big winner in the latest crisis. Despite his unwelcome warnings to Hitler that the eastern war was a national calamity, the Führer saw no way ahead other than Todt's plans for rationalization. The only alternative would have been to give in to Göring's campaign for a dictatorial role over all armaments production, but the Reich Marshal's star was already on the wane for the failures of both the Luftwaffe and the Four-Year Plan. In a series of decrees from December to early February, Hitler extended and strengthened the Armament Ministry's system of committees, which were organized on the principle of the "self-responsibility of industry"—that is, industrial enterprises were to coordinate and reorganize production among themselves with reduced intervention from the bureaucracy.[20]

Suddenly, at the height of his influence, Todt was killed. On the morning of February 8, 1942, following a tumultuous meeting with

Hitler over new decrees for the economy, the Minister's plane blew up in the air after taking off from the airfield near the Wolfsschanze. Rumors immediately circulated that Hitler had eliminated him, but there appears to be no reason why the Führer would reorganize the economy in Todt's favor and then get rid of him. It is possible that Göring or the SS was behind it, but in all likelihood the plane's self-destruct mechanism was accidentally triggered after takeoff, because the pilot was making a desperate attempt to return to the runway at the time of the explosion.[21]

By midday the Führer had already settled on a surprising choice as Todt's replacement: Albert Speer, age thirty-six. The architect had coincidentally been passing through headquarters and had planned to be on the same plane as Todt but had delayed his departure because Hitler had kept him up until three in the morning. As a result, Speer's appointment has often been described as fortuitous—a case of being in the right place at the right time—but he really was the logical candidate to carry forward Todt's work. To have given Göring the post would have been to reopen many of the battles so recently fought; Speer had in any case accumulated much relevant experience in armaments construction. As it turned out, he also had a more ruthless drive for power than Todt: Within a year he had absorbed most of Thomas's OKW Economics Office, driven the general into retirement, and secured the acquiescence of the Army and Navy to his system. Only the aircraft industry was kept beyond his reach, but Speer coordinated his activities with Field Marshal Erhard Milch, Göring's number-two man in the Air Ministry. In short order, the productivity of German industry began to increase dramatically.[22]

For the rocket program, Speer's appointment was another stroke of luck; in fact, it more than compensated for the forced retirement of von Brauchitsch. The ambitious Minister was, by his own later admission, an uncritical enthusiast of the program, at least until the end of the war, when he all too suddenly realized its lack of military sense. Speer also possessed immediate access to Hitler and was determined to make the war economy more responsive to the Führer's wishes.[23]

Regarding Peenemünde, the Führer's wishes in the spring of 1942 still seemed to be favorable, despite his lack of interest in the priority problems that beset the production facility. On March 5 or 6, Hitler asked Speer to investigate the raw materials requirements of producing 3,000 A-4s per *month,* although Speer's record of this meeting may reflect his own advocacy of the project. Two weeks later the Führer did, however, reject the Luftwaffe's transparently political request that it be allowed to do a "theoretical investigation" of the value of the A-4. Although Hitler's enthusiasm for the rocket seems to have waned in the spring and summer of 1942, claims that he was hostile to the project (Army adjutant Gerhard Engel), or at least "exceedingly skeptical" (Speer), must themselves be regarded with skepticism. Those stories reflect Speer's disappointment at the Führer's unwillingness to commit to mass production before the missile had proved itself.[24]

Thus, while the August Führer order may not have been the final breakthrough that the Peenemünders had thought it was, the rocket program had seen many of its manpower and material difficulties lessen after mid-1941. Even more than in the first phase of the priority battle, it is therefore difficult to find much evidence that Hitler significantly delayed the A-4's development. It was the Production Plant that absorbed the bulk of the priority setbacks in the winter of 1941–42. But those delays would ultimately prove irrelevant, because technological problems had began to mount that would push the operational readiness of the missile back by as much as two years. That winter also saw increased competition from the Luftwaffe, which complicated the lives of the program's leaders and made them feel that they were struggling against endless political obstacles, even when the competition had little effect on the A-4 schedule. Ironically, the new interservice struggles came on the heels of renewed attempts at cooperation that only demonstrated how contradictory was the Army–Luftwaffe relationship in rocketry.

FLYING BOMBS, ROCKET PLANES, AND ANTI-AIRCRAFT MISSILES

The air force's March 1942 attempt to secure permission to do a "theoretical investigation" of the A-4 was only one sign of its growing insecurity regarding the Army rocket program. The Luftwaffe's loss of prestige and its failure as a strategic bombing force over Britain were bad enough without the senior service presenting its ballistic missile as "relief" for the bomber, a claim Dornberger repeated in late March to General Thomas. In the same letter, the chief of Wa Prüf 11 made an argument for effectiveness of the A-4 as a terror weapon that was even more blunt than his assertions of mid-1941. A months-long, day-and-night missile campaign against British cities, he contended, would, "by creating panic and disorganization, make an important contribution to ending the war." Furthermore, the unit cost of the missile would be cheaper than the cost of a bomber and its well-trained crew, which would be lost in just a few missions. That sort of argument did not sit well with Luftwaffe officers, who began to grumble that "the Army is beginning to fly."[25]

The first manifestation of the air service's new opposition to the program had come somewhat earlier, in December 1941, when the Air Ministry obstructed Ordnance's request to dismantle and move Zeppelin's largest airship hangar in Friedrichshafen. The Ministry had wanted to blow it up, because it blocked a planned runway extension at Dornier Aircraft, another company in the Zeppelin group. Suddenly, the Ministry found that it needed the hangar when the Army requested it for use as an A-4 assembly plant. After three months of wrangling, Erhard Milch finally decided in favor of Ordnance, if it would bear all costs and provide the labor for the move. Notwithstanding this decision, Milch then went on to do more than anyone else to sustain competition with the Army by backing the creation of the "flying bomb" project, later dubbed "Vengeance Weapon 1" or V-1 by the Propaganda Ministry. That cheap alternative to the A-4 (V-2) was, according to Speer, "more than anything else a prestige object" for his friend Milch. In other words, it was a way to get back at the Army.[26]

The origins of the V-1 go back to the Paul Schmidt pulsejet experiments, which the Air Ministry and Army Ordnance had co-sponsored since 1935. By 1938, Dornberger was convinced that Schmidt was making too little progress toward a usable air-breathing propulsion system based on rapid, intermittent explosions. The work seemed of little interest to the Army anyway. In spite of those doubts, Ordnance did not actually withdraw its financial contribution until 1940. Meanwhile, the Luftwaffe had become convinced by 1939 that Schmidt was moving too slowly and asked an aircraft engine firm, Argus, to develop a pulsejet based on principles it outlined. Only later did Argus hear of Schmidt's experiments, and its final engine incorporated only one element from the original inventor's system. The Argus engine was tried out on aircraft in 1941–42, including a small fighter prototype, but its incredible noise and vibration made it virtually worthless for manned airplanes. (The characteristic sound of the engine later inspired the well-known Allied nickname, "the buzz bomb.")[27]

How that engine came to be linked to the flying bomb idea is not entirely clear. The concept of an unmanned, explosives-laden airplane that could automatically dive on its target was by no means new; experimental propeller-driven flying bombs had been tried but never deployed by the United States and Britain in World War I and later. A jet-propelled "aerial torpedo" had also been proposed independently by Hellmuth Walter and Paul Schmidt in 1934. Argus outlined a similar concept in 1939. At any rate, in a few months between March and June 1942, undoubtedly in response to the Army's A-4, the Air Ministry pulled together a team to design what today would be called a cruise missile, based on the already available Argus pulse jet. Another factor was Hitler's demand for vengeance following the Royal Air Force's first effective night raids against civilian populations, beginning with the historic city of Lübeck at the end of March. Milch picked the small aircraft firm of Fieseler to design the missile, which received the official designation Fi 103 and the code-name "Cherry Stone." On June 19, 1942, immediately after Milch had witnessed the first A-4 launch attempt, a conference in the Air Ministry formally gave the go-ahead to build "Cherry Stone" on a crash basis.[28]

Given the ensuing interservice battle, most writers have pictured the Army–Luftwaffe relationship after 1942 as one of rivalry. But the story is actually much more complex and mirrors the larger relationship between the two services. From the day of its creation, the Luftwaffe had striven for independence from the Army, but as a primarily tactical air force it was closely tied to its parent service nonetheless. Thus we need to look at the other side of the story: the ongoing collaboration in rocketry. That is particularly important because 1941–42 was a transitional period that set the pattern for the rest of the war.

Despite the end of the formal interservice rocket-aircraft program in 1939–40, informal collaboration had continued and perhaps even increased, especially in guidance and control. Ernst Steinhoff, with his close relationship with the Luftwaffe, had cultivated his contacts in the

Air Ministry out of both necessity and inclination. In the areas of guide beams, radio equipment, servomotors, and gyroscopes, the benefits of cooperative development and exchange of data were considerable. For the A-4's guidance system, it was also a matter of life and death that the Luftwaffe collaborate if large numbers of gyros and vane motors were to be manufactured in firms like Siemens's aircraft instruments division, now spun off as a subsidiary. Because those components were to be taken from regular aircraft manufacturing lines, and because Peenemünde needed Siemens's mass production capability, Steinhoff had to and did secure the Air Ministry's acquiescence in the fall of 1940 to ordering thousands of gyros and hydraulic servomotors.[29]

In addition to guidance, another area of informal but intense Luftwaffe–Army collaboration was the daily operation of Peenemünde. Luftwaffe aircraft used for drop tests, guide beam experiments, and transportation flew from the airfield at Peenemünde-West. Among the pilots using the runway was Wernher von Braun, who was provided with a fast single-engine plane for business trips. In planning and building the housing settlement, the electrical power plant, and the new commuter train system, the Production Plant also had to coordinate its activities with the smaller Luftwaffe test center. According to Gerhard Reisig, head of the measurement group until 1943, interpersonal relations were cordial as well and were largely untouched by battles at higher levels. In the later war years, when the two competing long-range missiles were being launched, East and West cooperated in tracking them.[30]

During the interlude from 1940 to 1942, however, the only formal interservice rocket project was the liquid-oxygen/alcohol takeoff-assist system. The Air Ministry had contracted with Army Ordnance in January 1939 to develop two egg-shaped pods, each containing a 1,000-kg-thrust motor, to be strapped under the wings of heavily loaded bombers. After burning for thirty seconds, the pods were to be dropped off and parachuted back to earth sufficiently undamaged for reuse. The Air Ministry imposed stringent requirements for safety and simplicity of function, since explosions or failures of one unit to ignite were clearly to be avoided at all costs. The first aircraft drop test was carried out at the end of August 1940, followed by dozens of experimental takeoffs from Peenemünde-West over the next year and a half.[31]

But the project ran aground in the latter half of 1941, when the designated manufacturer, Schmidding, failed to produce the preliminary models of the mass production version. Instead, the company's engineers redesigned the units according to their own ideas, encouraged by the responsible person in the Air Ministry Technical Office. That was precisely the sort of slipshod administration that had flourished in the office under the tenure of General Ernst Udet. The famous ace had been installed by Göring in 1936 in part to undercut Milch, whose administrative competence the "second man in the Reich" found threatening. In the end, Udet's bungling of aircraft development and production led to his dramatic suicide. On November 17, 1941, Udet shot himself, "scrawling on the wall before he died that Göring had betrayed him to the Jews in the Air Ministry."[32] One week later, Wa Prüf 11 warned the Luftwaffe that Schmidding's new version of the takeoff-assist system was dangerous and that much time had been lost. The following February the Army rocket group washed its hands of the whole affair and left it to Schmidding and the Air Ministry to decide which version would be produced. In fact, no liquid-oxygen systems ever saw active service with the Luftwaffe, which had more sensible but less powerful solid-propellant and hydrogen peroxide units anyway.[33]

The failure of the takeoff-assist project was the low point of formal Army–Luftwaffe collaboration in rocketry. In 1942, however, an important new field of cooperation would open up: anti-aircraft missiles. The story of *Wasserfall* ("Waterfall"), the missile that became the second most important Army rocket project in the latter half of the war, had its origin in the spring of 1941. On May 7 of that year, Dornberger phoned Stegmaier and asked him to study the possibility of a liquid-propellant anti-aircraft missile with a maximum altitude of 15 to 18 kilometers (about 50,000–60,000 feet).[34]

Using rockets to shoot down airplanes was an old idea and had been considered by the Ordnance group from the outset. But unguided solid-fuel rockets could never match the accuracy of anti-aircraft guns, the responsibility for defense against airplanes became the preserve of the Luftwaffe, and the Army's liquid-fuel rocket program focused on the creation of an offensive weapon that might justify its exorbitant costs with "war-winning" results. It is therefore certain that Dornberger's request originated in the Air Ministry. A few advocates inside the

Luftwaffe had pushed the anti-aircraft missile repeatedly over the years, and Rheinmetall-Borsig, an artillery and munitions manufacturer with its own line of solid rockets, had made a proposal. In February 1941 the Inspector-General of Flak (anti-aircraft artillery) had called for missile development because of the failure of guns at night against British bombers. Probably in response to that call, Henschel Aircraft, a builder of rocket-assisted glide bombs, presented the Ministry with a missile proposal in June 1941.[35]

Although Dornberger had asked Peenemünde to study the anti-aircraft missile, Wernher von Braun quickly moved toward the manned rocket interceptor instead. On May 13 he drafted a letter to Ordnance chief Leeb requesting permission to bring Professor Willi Messerschmitt, head of the aircraft firm of the same name, to Peenemünde to discuss such an aircraft. Von Braun's fascination with rocket planes was intense and longstanding. At the beginning of July 1939, in the wake of the first He 176 flights, he had written a proposal for a rocket fighter would take off vertically. The Luftwaffe ignored the document, being more interested in its own program at Messerschmitt, the hydrogen-peroxide-fueled Me 163 Komet. In late May 1941 the young technical director produced a new version of his proposal. It discussed both missile and rocket-fighter solutions to the specific problem of nighttime air raids, which were becoming an RAF specialty. In the new proposal von Braun came down firmly on the side of the manned aircraft, because the missile's burden on war production would be justified, he felt, only if the probability of hitting an enemy bomber was 100 percent. He noted the demand on the overloaded radar and aircraft-instruments sector that would be imposed by a throwaway device with very expensive components. As a way around that difficulty, he proposed what was in effect a manned missile: an interceptor launched vertically and guided automatically to the target. The pilot's responsibility would be only to carry out the actual attack and to land safely. The propulsion system for either the rocket plane or the missile would be based on nitric acid and diesel oil, a combination that BMW's Berlin aircraft engine plant had first developed at the instigation of the Luftwaffe.[36]

Those ideas excited immediate interest in the Air Ministry. By the time officers from anti-aircraft artillery development arrived for a tour in mid-June, the Ministry had already decided that Peenemünde-East would be paired with Messerschmitt to build an "Interceptor" (the English term was used) and that anti-aircraft missiles would be studied through tests carried out on modified A-5s. Three weeks later an even higher-level delegation came through, led by the chief engineer in the Technical Office, Roluf Lucht. Although Lucht saw a successful A-5 launch from the Oie, he did not share the enthusiasm of the Flak people for the anti-aircraft missile, nor did he accept their argument that its production demands could be satisfied from the existing Flak radar and instruments industry. He agreed with von Braun and the Peenemünders that the Interceptor was a more feasible concept but announced that Messerschmitt was too burdened to take on another rocket fighter. Fieseler, the same company that would later receive the V-1 airframe contract, was picked in its place. Guide beam tests with A-5s would serve to test the early versions of the Interceptor's automatic approach system.[37]

Through the fall of 1941, Fieseler concentrated on a design study of the Interceptor, while the Army rocket center looked into the guidance problem. In spite of Lucht's pronouncements, the Luftwaffe considered the idea of an anti-aircraft missile as well, perhaps because Lucht in the meantime had been fired for incompetence. At the end of November Fieseler produced its study of the Interceptor or Fi 166, as it was dubbed. The company outlined six possible configurations, some using rocket propulsion only, others with an enlarged A-5-type rocket booster and a turbojet engine for cruising at high altitudes. Notwithstanding the excessive complexity and doubtful safety of the proposed designs, an early December meeting between Army Ordnance and the Air Ministry again reached the conclusion that efforts should concentrate on the Interceptor, since the resources needed to develop and produce an anti-aircraft missile were too great. Dornberger had become reluctant to make any commitment to Luftwaffe projects, but he and von Braun agreed with that decision. By January 1942, however, the air force came to its senses and shelved the Interceptor proposal.[38]

There the matter rested for some months. But inside the Luftwaffe, the advocates of the anti-aircraft missile were plotting another comeback. The two individuals who appear to be most involved were the new Inspector-General of Flak as of January 1942, General Walther von Axthelm, and a junior officer in the Luftwaffe General Staff, Major Friedrich Halder. Probably at von Axthelm's request, Halder wrote a blistering memorandum in May 1942 attacking the Flak development division, which had been transferred from Army Ordnance to Air Ministry control in 1940. Halder called the division a collection of out-of-touch Army traditionalists who failed to see the potential of radical new technologies like the rocket—hardly a fair charge in view of the events of the preceding summer. Looking ahead to the future, Halder foresaw the day when aircraft speeds and altitudes would be such that anti-aircraft artillery would no longer be able to keep up. As it was, in firing against Allied bombers, gunners already had to allow for lead times of twenty-five seconds or more for the shell to reach altitude. Only missiles or unusual gun designs promised a solution to the problem in the long run.[39]

After Halder's superiors vetted his document to make it more politic, it became the basis for General von Axthelm's new program for anti-aircraft artillery in June 1942. Along with accelerated rocket development, this program proposed various improvements to conventional guns and radar, plus cooperation with the Navy on a new superheavy 24-centimeter Flak gun. After an unexplained delay, von Axthelm was able to get Göring to agree reluctantly to this document, and it was issued as an order from the Reich Marshal on September 1. Hitler dismissed the document as "utopian" but did nothing to stop its implementation. Neither did the inheritor of most of Udet's powers, Milch, although the Field Marshal was equally unconvinced that defensive missiles were a sound idea. Only later would his opinion change.[40]

In view of all that skepticism and the Luftwaffe's own 1941 decision, why did the air force reverse its position in 1942? Göring, like Hitler, had always overestimated the importance of Flak over fighters for defense. But what was new was the rise of an effective RAF bomber threat, symbolized by the first thousand-plane raid on Cologne in May 1942. Attacks of this type provoked harangues from Hitler against Göring and damaged the Luftwaffe's prestige even further. Adding to such worries, the Germans knew of the American B-29 Superfortress then under development, although it was ultimately used only against

Japan, and they saw their own jet programs producing aircraft of potentially revolutionary performance. In those circumstances, the anti-aircraft experts were right to say that the effectiveness of conventional artillery against high-altitude, high-speed aircraft would be nil in the not too distant future.[41]

Yet there is little doubt that the anti-aircraft missile decision was another major blunder in German weapons development. In his memoirs, Speer asserts that if Hitler had not delayed the Me 262 jet fighter, and if the Army had concentrated on Wasserfall instead of the A-4, the Allied bomber fleets could have been defeated in 1944. Leaving aside the persistent mythology about the jets, even if the Germans had not lost a year changing their minds about the missile, it would not have made any difference. The 1941 warnings about the burden on the war economy were correct—and those estimates were made without any clear idea of the overwhelming technical problems that would be faced in the guidance sector. The fact that the air force had slowed down radar development in the first two years of the war only made matters worse. Germany now found itself in a dilemma that only an earlier and more energetic program of conventional fighter defense could have prevented. Instead, Göring and the Luftwaffe leadership waited too long and then gave in to the missile enthusiasts, who promised a magic answer to their problems.[42]

In the aftermath of this decision, the Air Ministry followed its usual philosophy of competition by initiating programs at Henschel, Rheinmetall, *and* Army Ordnance, which fragmented its overstretched research and development capability even further. As for the Army, Peenemünde-East was promised Luftwaffe personnel to develop both a smaller solid-fuel rocket and a large nitric acid-fueled one (the later Wasserfall). It was a significant new commitment to Army–Luftwaffe collaboration and one that would overshadow the rocket-aircraft program of the 1930s in scale. Thus, while the two services would compete in long-range missiles between 1942 and 1945, they would simultaneously cooperate in anti-aircraft missiles. For Dornberger's rocket engineers, however, the amount of time spent on formal inter-service projects was still minor in 1941–42. During that period, and to a large extent thereafter, one problem dominated all others: getting the A-4 to work and getting it into production.

THE A-4 REACHES THE LAUNCH PAD

Until the fall of 1941, the schedule for launching the world's first ballistic missile remained hypothetical, with the result that it was highly politicized and highly optimistic. During the struggle to win Hitler's favor that summer, Dornberger was still promising to complete the development of the A-4 by the end of 1941 and to begin firing on Britain in the second half of 1942. By October he was telling the Army General Staff that the first launch would come "at the latest in January," the completion of development no sooner than the autumn of 1942, and military deployment "is not to be expected before the end of 1942"—in other words, 1943. At the beginning of November his technical people in Peenemünde were estimating the first launch no sooner than mid-February; by January it had slipped to March.[43]

At least three things were going on here. First, Dornberger was confident of greater political support and may have felt less need to exaggerate, although his behavior from beginning to end fits a classical pattern of the military-industrial complex: schedule and performance predictions that were overly optimistic, if not actually dishonest. The second factor in the lengthening schedules was the increased emphasis on mass production, with the accompanying A-4 "test series" of six hundred. The purpose of that series was above all to determine statistically, in good artilleryman's fashion, a "firing table" that would allow launch crews to set the guidance system for a specified target. Firing hundreds of rockets would naturally take some time, even if the project coincided with some of the preparations for deployment. Most important was the third factor: coming to terms with the enormous difficulty of making a radically new technology work reliably. Even the relatively routine A-5 launch operations on the Greifswalder Oie did not adequately prepare the rocket group for the quantum leap in performance the A-4 promised and demanded.

The extent of the delays became apparent only after the first missiles reached the test stands. The initial static-test model, "Injection Aggregate 1," was hand-built in Peenemünde's workshops and then moved to Test Stand V in October 1940 for completion. It remained there for the entire first half of 1941. Welding problems and inexperience were part of the problem, but the basic difficulty was creating a reliable and workable system of valves, controls, and switches for the engine assembly. If the eighteen-pot engine was a "plumber's nightmare," the brand-new steam generator–turbopump system that moved the propellants from the tanks to the injectors was no less so. (The A-5 still used nitrogen gas pressurization to empty the tanks.) Unless care was taken in the design and installation process, fires and explosions could be triggered in any number of ways, for example, by leaky lines or by propellants reaching the combustion chamber before ignition in the wrong quantities and with the wrong timing. Riedel's design bureau, working with Thiel's propulsion group, created at least two valve systems and tried them on Injection Aggregate 1 and its successors by running the fuels through without igniting them. Not until the summer of 1941 were the second and third static-test A-4s finished, and only in late September was the engine actually fired on the first one.[44]

Those difficulties were harbingers of worse to come. On October 21 at the A-4 launch site (Test Stand VII) and on November 5 at the engine test stand (I), two missiles exploded. The first caused significant damage to the facilities; two days after the second, Dornberger sent a blistering and revealing memorandum to Peenemünde. He called the leading managers of the center "irresponsible" for leaving "young, inexperienced test engineers with the leadership of tests" on new rockets, especially "in the present life-and-death situation for HVP [Peenemünde-East]." He demanded that von Braun, Thiel, or "Papa" Riedel be present for the first thirty experiments on each vehicle. He reproached them as well for traveling the country trying to arrange mass-production contracts singlehandedly, for indulging in "futuristic dreams," and for engaging in endless negotiations with the Air Ministry over the Interceptor and anti-aircraft missiles. Until the air force had made up its mind as to what it wanted and supplied the manpower needed, "these totally useless meetings over future hopes will cease and you gentlemen will alone concentrate on the *development* of the A-4."[45]

Dornberger was annoyed about more than the Luftwaffe discussions. The summer and fall of 1941 was the apogee of planning at Peenemünde for the A-9/A-10 ICBM and, on the margins, for even more exotic possibilities like a manned A-9 and the employment of atomic reactors for rocket propulsion. A small contract to study the latter possibility was given in 1942 to the Research Institute of the Reich Post Ministry, notwithstanding Dornberger's admonition. Those dreams did not divert many resources from the main task, but they were in line with the character of Wernher von Braun, who, in the words of Dornberger, "reveled in any project that promised to be on a gigantic scale, and, usually, in the distant future. I had to brake him back to hard facts and the everyday."[46]

From late 1941 until late 1942, Dornberger had to do a lot of "braking" in order to concentrate all of the energies of von Braun and his group on the A-4, its production, and the most important follow-on projects. Only six days after his fiery memorandum, Dornberger sent another to Peenemünde, indicating that he had had an argument with von Braun about whether the A-8 or the A-9 should be the next missile. The chief of Wa Prüf 11 ordered a concentration on the A-8, because he felt that developing its nitric acid–oil propulsion system was more feasible than building the glider missile. Much was uncertain about the A-9's aerodynamics and guidance, yet the leading Peenemünde engineers seemed to prefer the A-9 because it was a more interesting problem. But Dornberger could not make his order stick; some A-9 research continued, as did the design of the A-7, an A-5 with wings that was to serve as a test vehicle for the glider missile concept.[47]

Just before Christmas Dornberger sent another memorandum to the leadership of Peenemünde, this time appealing for the concentration of all efforts on launching the first A-4 by the end of February 1942. He also lectured them on the slowness with which scarce materials were being eliminated from various components of the production version. Ever since the audience with Hitler and the mass production planning of the fall, that problem had become urgent because of the severe shortages of many metals, especially aluminum. Dornberger's demand for a quick changeover, however, was rather unreasonable. The task facing von Braun's engineers—making the A-4 fly, even though its exotic technology was suffering from innumerable and inevitable "childhood diseases"—was daunting enough without the redesign in steel of critical components like the engine and the turbopump. For anything that came into contact with super-cold liquid oxygen, the change proved extremely difficult because of the brittleness of steel at cryogenic temperatures. Imposing that challenge on top of the one of creating a mass produced ballistic missile threatened to make the design process chaotic.[48]

The seriousness of Peenemünde's technological and organizational difficulties that winter is revealed by yet another furious memo, penned by Dornberger on February 5. This time his outburst was caused by an accident with the first flight vehicle, "Launch Aggregate 1" (later called "Test Model 1" or V1), while it was suspended for an engine firing on Test Stand I. After being fully tanked, the missile slipped out of its "corset," fell 2 meters, smashed three fins, and came to rest on the rim of the engine nozzle. At fault was a design oversight that had all the hallmarks of inexperience: The corset was not able to bear the full weight of the fueled vehicle after the liquid oxygen's temperature caused shrinkage of the fuselage. Dornberger was enraged that such an apparently stupid mistake had been made, delaying the first launch by a few more weeks. He reminded his subordinates about the absolute political necessity of showing some practical results. He also ranted against what he saw as their Raketenflugplatz mentality, "i.e., the propagandistic exploitation of an idea only to receive money and priorities, while not even building a usable test device practicable for eventual mass production." It was hardly a fair charge; of the thousands of engineers and craftsmen at Peenemünde, only about a dozen came from the old rocket groups.[49]

Most interesting, however, is his discussion of the problems with the A-4 design and testing process, made even more fascinating by the responses of Thiel, who neatly scribbled pungent marginal comments on his copy. First, Dornberger was annoyed by the interminable stay of missiles on the test stands, which he felt revealed a deeper problem:

> The whole design has apparently been done as if hundreds of people had the time to spend weeks going all over the rocket on the test stand, installing valves, doing assembly work, moving cables, and generally fumbling around. Cooperation between the Test group and the TB ["Papa" Riedel's design bureau] is lacking . . .

He accused the engineers of forgetting the demands of simplicity and ease of access that would be necessary for mass production and field use; instead the missile was a "flying laboratory."[50]

Thiel's marginalia confirm that he indeed had serious conflicts with Riedel. His comment on the above quote was: "The TB fights tooth and nail against any influence from Tr [Propulsion] or Vers [Test]." On an earlier page, Thiel asked why the test stands should take the blame for the "garbage" produced by the design bureau and the shops. But the propulsion chief also noted that there was a contradiction between Dornberger's demands for results in the shortest possible time and his demands for a well-planned, fully developed vehicle.[51]

A Thiel comment on the last page of Dornberger's memorandum reveals another serious difficulty: No complete "assembly drawings" for the first flight vehicle existed. The design bureau was responsible for producing blueprints of all parts of the A-4, except for the guidance and electrical equipment (only Steinhoff's division had the requisite knowledge for those). Yet a complete set of drawings had not yet appeared, and indeed two years later still had not, in spite of innumerable promises by the Peenemünde leadership. The question is: What blame does Riedel deserve for this mess? Decades after the fact, this is difficult to determine, but the old hand from Heylandt and Kummersdorf was stubborn and, as an engineer with only a two-year technical-school education, was in over his head with the A-4 project. He resented criticism from the increasingly dominant diploma and doctor engineers at Peenemünde. Riedel's difficult personality, combined with the drawings fiasco, would lead to his being shunted into another job by the end of the summer. Yet there is little doubt that the endless changes on the test stands, which were often done without any formal documentation, made "Papa" Riedel's job extremely difficult, and the confusion imposed by the elimination of scarce materials only exacerbated matters further. In addition, the entire organization was dominated by development engineers with little or no manufacturing experience. For all his real shortcomings, Riedel ultimately was the scapegoat for the difficulties experienced in trying to transfer an exotic technology into production on a very short time scale.[52]

A final revealing insight is provided by Dornberger's angry memorandum about the Launch Aggregate 1 accident. Apparently von Braun had asked him the day before (February 4) to close down the Pilot Production Plant and transfer its personnel to the Development Works, if priority problems imposed any new personnel cutbacks. Dornberger's response was only to go into a rather silly harangue about how it was not the number of bodies that counted for a successful design, but rather the "deliberations of a single superior mind." Thiel's marginal comment was: "The VW [the Pilot Production Plant] will once again be the death of us. It eats up people and produces nothing!" Ironically, it was the factory that had suffered most, since skilled workers were constantly being taken away to other projects, such as finishing Test Stand VII. But, in the sense that the plant was an ill-conceived, unproductive project, Thiel and von Braun were right. By June 1943 relations deteriorated to the point that Schubert complained to Dornberger about "the hitherto existing rivalry" between the two sides of the Army center and about the factory's alleged role as the "milch cow for the EW [the Development Works]."[53]

Clearly, the situation at Peenemünde in the winter of 1942 was tense, nor was it eased by the fate of the first flight vehicle. After repairs, the now renamed A-4/V1 was returned to Test Stand VII. On March 18 Thiel drew up the organization of the launch crew for an attempt in the next five to ten days. The vehicle would have a unique trajectory: it would have a pitch program of only 10 degrees plus half-full propellant tanks, so that it could be recovered by parachute, presumably to salvage recording instruments. But fifteen minutes before midnight that evening, while Thiel was observing the first burn test with the guidance system running, a reddish flame exploded from the side of the missile just over the engine. The steam generator and many lines were wrecked, and the engine shut itself off automatically. The propulsion chief soon concluded that leaks in the fuel and oxidizer lines, caused by vibration and structural stress, allowed an explosive mixture to build up over the head of the motor. Fortunately the test stand was undamaged, because the tanks did not rupture, but this time the vehicle was junked for spare parts. The first launch of an A-4 would have to be put off a couple of months longer.[54]

No new outburst from Dornberger is to be found in the records; perhaps he accepted that this accident was unavoidable. In any case, planning continued for a very optimistic launch schedule. One document from April 20 predicted A-4/V2 on May 12, A-4/V3 on June 9, A-4/V7 on July 4, and four more by September 29. The out-of-sequence launch of the seventh flight vehicle was a specific request of Dornberger, who wanted the earliest possible attempt with a missile light enough to reach the promised range of 270 kilometers. (The V2 to V6 vehicles could go only 180–190 km because of excessive weight.) The manipulation of the schedule, von Braun admitted in July, was done "for propagandistic reasons."[55]

Notwithstanding the best hopes of the rocket group, endless problems plagued the second flight vehicle too. In late April A-4/V2's fuselage was damaged while it was being erected on Test Stand VII because of a mismatch in the dimensions of the rocket and its *Meillerwagen* transporter-trailer. Further weeks were lost, and the vehicle was not finally ready for launch until June 13.[56]

Attending that day was a powerful delegation of observers: Speer, Milch, General Fromm, General Leeb, and his Navy counterpart, Admiral Witzell, along with many others. Tension and excitement ran high, as the Nazi armaments elite, Dornberger, and the center leadership waited for the first launch from Peenemünde itself. The visitors, who were perched on the roofs of the main buildings, must also have been stunned by the mere sight of the world's first large rocket, just as Dieter Huzel would be in the summer of 1943. Finally it lifted off, but it disappeared into low-hanging clouds only 80 meters above the ground. (It is possible that the launch was rushed because the presence of the VIPs.) The vehicle made it through Mach 1, but soon thereafter its battery failed and the engine quit because of the missile's rapid rolling, which began immediately after launch. The A-4/V2 was next sighted falling through the clouds tumbling, with no visible fins. It crashed into the sea only 600 meters (less than half a mile) from the shore.[57]

That disappointing result seems not to have discouraged those in attendance. In the high-level meeting that took place afterward, Albert Speer announced his intention to give the A-4 the newly created superpriority rating of "DE" (the German initials for "urgent development"). The Armaments Minister also promised Dornberger help in acquiring more manpower, as did Erhard Milch, who said he would provide some designers and production planners from the aircraft industry. But Milch was being somewhat two-faced, for he announced in the same breath the Luftwaffe's "Cherry Stone" missile based on the Argus pulsejet.[58]

At the meeting Dornberger also requested Speer's help in finding a director for a Production Planning Directorate, which von Braun had suggested as a way to get the problem-plagued transition to manufacturing moving again. Almost a month later the Armaments Ministry nominated Detmar Stahlknecht, an engineering manager with experience in the quantity production of aircraft. The task of Stahlknecht's Directorate was to produce a new set of production drawings and to organize the manufacture of parts and subassemblies. At the end of the summer Walter "Papa" Riedel was placed under Stahlknecht's command to supervise the creation of the drawings. The design bureau chief was replaced in his former job by an energetic diploma engineer who, by an odd coincidence, had a virtually identical name. Walther Riedel had worked on alternate 25-ton engine designs at the Technical Universities of Dresden and Berlin.[59]

Speer's support made him an indispensable ally in the Führer's personal entourage, but Dornberger continued to view the Minister's role as one of giving advice and setting overall priorities, while the details of A-4 production remained in the hands of Army Ordnance. Following the Führer's request to Speer to study the manufacture of 3,000 missiles a month, Dornberger had concluded, in April discussions with Ministry officials, that 1,000 per month might be feasible, if more liquid oxygen plants were built and if the ethanol fuel, which was produced by fermenting potatoes, was diluted with methanol, at the cost of a slight loss of performance. Under pressure to produce a launch success, however, he and the Peenemünders pushed that rather optimistic manufacturing goal into the background as the responsibility of the Production Planning Directorate and the Ordnance A-4 Working Committee. Speer himself did not intervene during the spring and summer of 1942, presumably because he saw that the time was not yet ripe. What was needed to secure Hitler's go-ahead was a demonstration that the rocket actually worked. As a way of pushing the rocket group, in August Speer formally approved a "DE" rating, but for the completion of twenty test missiles only.[60]

It nonetheless took time to analyze the failure of the A-4/V2 and to modify the V3 for firing. The problem was soon traced to a roll-rate gyro malfunction, which occurred immediately after the June launch. Wind may have contributed to the problem by rolling the vehicle faster than the guidance system's ability to compensate, just as in the case of the A-3s. Von Braun's group decided that the best solution was to incorporate a version of Helmut Hoelzer's electronic "mixing device" and to eliminate the rate gyros altogether, as had been planned for the later "test series" missiles. That change added two more weeks to the schedule. Also eating up time were other alterations to the A-4/V3 to increase the vehicle's control over the roll axis.[61]

Finally, on August 16, 1942, the second launch attempt was made—this time for a more restricted audience. Shortly after noon, the missile lifted off in conditions of much better visibility. Once again, awesome rumbling filled the sky and hope skyrocketed with the V3's flight. The black-and-white markings on the missile showed that it did not roll. For the first forty-five seconds, things appeared to go very well, but at a speed of 2,345 kilometers per hour (about Mach 2) the engine suddenly stopped. (Full burn time for an A-4 was about a minute.) As a result, the control system was no longer effective, since the jet vanes could not exert any force when the engine was off. At that point the atmosphere was still sufficiently dense to overpower the vehicle's marginal stability. The rocket began to veer away from the direction of flight, its nose was ripped off by aerodynamic forces, and flames erupted from both ends. The fins came off, and what was left of the vehicle tumbled into the Baltic 8.7 kilometers from the launch site. Postflight analysis suggested that acceleration had slowed abnormally after thirty-seven seconds and then the engine had stopped altogether because of a failure of the steam generator or turbopump. Unfortunately, the new telemetry system for radioing data to the ground had quit only four seconds after launch, leaving ground-based movie film as the only evidence.[62]

That flight was an improvement over the first, but a new round of modifications to the next flight vehicle, the A-4/V4, was necessary, including strengthening the nose. (Dornberger's plan to launch the V7 first had been dropped in the meantime as holding up the schedule yet further.) It was not until the end of September that the missile was ready to be launched. On the eve of the next attempt, the chief of Wa Prüf 11 sent a memorandum to Peenemünde summarizing in dramatic terms what was at stake:

> After presentations to the Reich Minister for Armaments and Munitions [Speer], the Chief of Army Armaments and the Commander of the Re-

> placement Army [Fromm], and the Chief of Ordnance [Leeb], the situation for HAP [Peenemünde-East] is, at the moment, the following:
>
> 1) the Führer does not believe in the success of the guidance system and therefore in its ability to achieve the desired accuracy,
>
> 2) the Reich Minister . . . doubts our success. He is supported in his opinion by *Hauptamtsleiter* Sauer [Karl Otto Saur, Speer's deputy] and Field Marshal Milch,
>
> 3) [General Fromm] has lost trust in our ability to meet our deadlines because, at the end of 1942, we still have not achieved a long-range shot, [and]
>
> 4) the Chief of Army Ordnance is beginning to doubt our word.

Dornberger went on to outline a number of specific measures for speeding up operations, with the overall goal of "launch, launch, launch." He wanted all twenty first-priority missiles fired by December 31(!). As a means to that end, he asked the management and labor force of Peenemünde to work long hours, seven days a week, and to push all other projects into the background.[63]

Weather and other problems delayed the next launch a few more days. Finally, on October 3, at two minutes before four in the afternoon, the A-4/V4 lifted off and arced out over the Baltic on a perfect fall day. The rocket, which carried on its side a *Woman in the Moon* logo, continued straight on its course until all that was visible was a glowing dot at the end of a white exhaust contrail. When the shifting winds at high altitude turned the contrail into a zig-zag of "frozen lightning," many thought that the missile had gone awry. Nonetheless, it continued unperturbed, and at fifty-eight seconds the engine made a normal, if slightly early, cutoff. On the roof of the guidance division's Measurement House, Dornberger and Zanssen wept and hugged each other with joy. As the measurement tone whined over the loudspeakers in the background, conveying in audible form the Doppler tracking of the rocket's velocity, the two rushed down to meet the launch crew celebrating at the Test Stand. Just before the five-minute mark, the tone suddenly stopped. A-4/V4 had smashed into the sea about 190 kilometers (120 miles) away, after brushing the edge of space at an altitude of nearly 80 kilometers (fifty miles). All world records for altitude and velocity had been obliterated. It was a marvelous achievement and—as subsequent failures would show—also a lucky one.[64]

That night, at the gala celebration in the Officers' Club, Dornberger delivered a moving speech in which he stated that "the space ship is born." In his A-4 rhetoric he was never modest in equating the rocket to the wheel, the steam engine, and the airplane as a fundamental new invention in transportation, nor did he shrink from saying that it opened the road to spaceflight. In assigning to the missile this elevated historical role, he was not far from the truth. But Dornberger probably also spoke of the A-4 as the weapon that might change the course of the war for the Third Reich. There is little doubt of his enthusiasm for Hitler and the system, nor can his exaggerated view of the missile's military impact be denied. In any case, he did say one thing that was incontestable: Peenemünde's troubles were not over—they were just beginning. It was not enough to have the key technologies in hand, or even to get them to work together once. The rocket group must "launch, launch, launch." It also must secure Hitler's approval and find a way to turn its "flying laboratory" into a mass production weapon. If it accomplished those latter objectives, however, the program would become an even more valuable political property. In the wings, the power brokers were gathering.[65]

Chapter 6

Speer, Himmler, and Slave Labor

Dornberger was not slow to exploit the October 3, 1942, success. Within a day or two, he and General Leeb were in Speer's office reporting on the flight and requesting approval for mass production. On the eighth, the chief of Wa Prüf 11 put into circulation a propaganda document laying out the details of the launch and explaining the manufacturing plans for the A-4. Apparently he and Army Ordnance did not have the slightest hesitation about charging ahead before the V4's feat could be duplicated. But experience would prove to be a bitter teacher. V5 would at least be a partial success, going 147 kilometers on October 21. The next four launches, from November to early January, would be complete and utter failures.[1]

The joy that Dornberger felt after October 3 was not, however, completely unsullied. On the ninth he asked Wernher von Braun to report within a week on the Luftwaffe's "Cherry Stone" (V-1) project. According to Dornberger's memorandum:

> Lately there have been many remarks from government offices, firms, etc., that the A-4 program no longer possesses the importance ascribed to it by the people who work on it. In that connection, confidential hints have been made that "Cherry Stone" . . . is far more valuable and has every chance of catching up and passing the A-4 program, if not making it altogether illusory.

The A-4 obviously had its enemies and skeptics in the air force and elsewhere. Von Braun was soon able to reassure Dornberger, however, that the poor accuracy and shorter range of the cruise missile made it less than competitive with the ballistic missile, although it had every chance of being a success. Dornberger reported those conclusions to Leeb but expressed his amazement at the Luftwaffe's ability to initiate a competing program in apparent secrecy from a sister service.[2] In view of the many priority battles the Army rocket program had already endured, this evidence of growing interservice rivalry was certainly a disquieting sign. But just as von Braun surmised, the V-1 would never prove a decisive threat to the survival of the A-4 project, although it did at times worry Dornberger.

Meanwhile, V4's result and Dornberger's demands for an acceleration of the launch schedule forced Peenemünde to concentrate even more decisively on getting the A-4 into production. "Special Program S," issued on October 10 by Development Works chief Stegmaier, suspended all activity on most other research projects, most notably the A-9 glider missile and its subscale version, the A-7. Only two engineless, drop-test versions of the A-7 were to be finished, and only because they were subject to corrosion. (They were soon glide-tested, with mediocre results.) Ten people from Ludwig Roth's small Projects Office, that embodiment of the rocket group's "most cherished desires and hopes for the future" (Dornberger), were to be sent to other divisions. In line with the new interservice program approved by Göring in September, Roth's roughly thirty remaining staff members were to draw up plans for anti-aircraft missiles. It would take time for the Air Ministry to mobilize people and resources for those projects, however, so the application of rocket technology to air defense remained only a minor distraction for von Braun's group in the winter of 1942–43.[3]

Although the Wasserfall anti-aircraft missile eventually became Peenemünde's second major development project, the crash status of the A-4 program was clearly transforming the center into an organization predominantly concerned with missile production. Until December 1942 the transformation occurred with little outside intervention, because of Dornberger's clout and because of the remnants of the Army's autonomy within the Nazi system. However, once Albert Speer gained Hitler's approval for the full-scale manufacture of the A-4, he was no longer satisfied with a limited role. More ominously, Reichsführer-SS Heinrich Himmler began to take an interest in the A-4 as well. As in the war economy as a whole, Army Ordnance's power over

Peenemünde began to decline as first the Armaments Ministry and then the SS began to intrude into its domain. By the summer of 1943 Himmler's blackshirts had made their presence felt through the supply of concentration camp labor and through the infiltration of agents, but it was Speer and his managers who had come to dominate the organization of A-4 production. As a result, the Army center at Peenemünde reached the apogee of its institutional growth in August 1943, but it had already lost much of its autonomy. The Third Reich's "muscle men," to use von Braun's later term, now set the tone. The first and foremost of these was Speer's designated commissioner for the missile program, Gerhard Degenkolb.[4]

THE SPEER MINISTRY INTRUDES

For Albert Speer to extend his power over the A-4, he first had to secure the Führer's long-sought approval for mass production. Hitler's attitude toward the ballistic missile in 1942 had been frustrating to its advocates. After having embraced the A-4 so enthusiastically in August 1941, he had again become moody about it. He continued to refuse a production order, although he did nothing to slow down the rate of its development. In June, after hearing Speer's report on the first launch attempt, he expressed his skepticism about the A-4's guidance. The Führer's reaction to October 3 was more positive but still not entirely satisfactory. In Speer's official minutes, Hitler is described as agreeing that the parallel development of the A-4 and "Cherry Stone" was a "valuable" suggestion, but the ballistic missile "only makes sense if 5,000 projectiles are available simultaneously for an initial mass attack." That absurd comment again demonstrated Hitler's failure to understand that the A-4 was not a simple projectile that could be manufactured and fired in huge quantities. When von Braun examined the mass attack in December, he could foresee only a one-time maximum effort of 108 operational launches in twenty-four hours—and a normal rate of twenty-seven a day.[5]

The next time Speer presented the issue to Hitler, something had changed. According to the meeting minutes of November 22—three days after the Soviets launched their Stalingrad offensive and two weeks after the western Allies landed in north Africa—"the Führer takes a great interest in A-4 production planning and believes that, if the necessary numbers can be produced promptly, one can make a very strong impression on England with this weapon." Apparently, in his search for ways to reverse the rapidly deteriorating strategic situation and to exact "vengeance" on Britain for the RAF's "terror raids" against Germany, he had overcome many of his well-founded reservations about the impact of a small number of missiles on British morale.[6]

The imaginary threat of American missile attacks from the periphery of Europe may also have influenced Hitler's decision to order the A-4 into production. In early January 1943 he told Speer that it was "absolutely essential, in view of the rocket development going forward in America, that the most urgent experiments be made to find out if the jamming of the guide beams can provide a defense against such rockets." The repeated warnings from Ordnance about foreign competition, reinforced by faulty German intelligence reports, had had their effect. The Army rocket program had become an ironic mirror image of the Manhattan Project: While the Germans were racing a virtually nonexistent American missile program, the Americans (with British and Canadian help) were racing a virtually nonexistent German atomic bomb effort.[7]

How large a factor that was in the Führer's change of heart is difficult to determine, although it was clearly quite secondary to "vengeance." But, whatever the reason, Speer had finally received Hitler's approval for A-4 production. That good news was immediately conveyed to Dornberger, along with another order from the Führer: massive air-raid-proof bunkers must be built along the Channel coast for firing the missiles at London.[8]

That demand sparked a fascinating controversy between the chief of Wa Prüf 11 and his engineers. From his war experience in the heavy artillery and from the unwieldy character of the Paris Gun's railroad-borne equipment, Dornberger knew that missile batteries needed to be lightly equipped and mobile. In December 1939 he had proposed a train with minimal equipment that could carry and launch eight missiles. By 1942 Dornberger's group still planned a railroad battery, but the primary deployment mode was to be a motorized regiment, consisting of three batteries of trucks and other road vehicles. Klaus

Riedel, the veteran of the Raketenflugplatz, had been given the task of planning the vehicles after he was removed from the leadership of the test stands in 1940. When Hitler demanded bunkers, however, von Braun immediately jumped at the idea. He asserted in a November 27 memo to Dornberger that this launch mode was more suited to a missile still far from being a mass production item. The road-mobile system, he felt, required a more refined rocket and better-trained crews. The heart of the matter was that he and his engineers felt more comfortable with something resembling their test-stand operations. Dornberger disagreed heatedly, but had little choice but to go along with Hitler's wishes.[9]

While the Peenemünders were considering the design of a bunker to be used in conjunction with the mobile regiment (the railroad concept was shelved), Speer was considering how best to push through A-4 production with the maximum possible speed. Hitler still did not want to give missile manufacturing the superpriority rating of "DE," a grade that was supposed to be reserved for urgent new weapons development, not for production. No later than December 5, the Armaments Minister decided that the best solution was the creation of a "A-4 Special Committee" headed by one of his most energetic subordinates, the "dictator" of locomotive production, Gerhard Degenkolb.[10]

Degenkolb was representative of the new breed of corporate managers who had come to the fore under Todt and Speer. On leave from his position as a director in the heavy industry firm of DEMAG, the heavy-set, totally bald Degenkolb was a fanatical Nazi and a ruthless and authoritarian personality. He had made his reputation by shaking up the production of locomotives in 1942, when Germany's vast continental empire was desperately short of transport.[11]

Dornberger probably first met Degenkolb three days before Christmas during a crucial meeting in Speer's Berlin office. Among others in attendance were Ordnance chief Leeb and Karl Otto Saur, Speer's forceful, even rude deputy, who was apparently still unconvinced of the A-4's value. The chief of Wa Prüf 11 lectured on the plans for military deployment and the possible sites for a bunker inland from Calais and Boulogne, France, using a cutaway model and a diorama of a mobile battery, complete with toy vehicles. Speer approved the plans and assigned bunker construction to the Organization Todt, the huge construction organization he had inherited from the former Armaments Minister. A site selection group composed of Stegmaier, guidance chief Steinhoff, and Lieutenant Colonel Georg Thom, Dornberger's chief of staff for liquid-fuel rocketry, was to leave immediately after Christmas. Most important, Speer formally approved A-4 production "on the order of the Führer." He may even have brought with him a decree signed by Hitler. At the same time he dismissed the suggestion, probably from Dornberger, that "Cherry Stone" was a threat to the A-4. The Minister felt that it was not far enough along in its development. Indeed, when the Luftwaffe attempted the first launch of the missile from Peenemünde-West two days later, it was a complete failure.[12]

In January 1943 Degenkolb began to organize the A-4 Special Committee, which was modeled on the committee system of economic administration that Todt had founded and Speer had extended. Under the supervision of Degenkolb's Berlin office at "Locomotive House," the new body would eventually comprise approximately twenty "working committees" of industry, Ministry, and Ordnance representatives (including Peenemünders). The task of the subcommittees was to coordinate the production of the missile and its various components, ancillary equipment, and propellants. Von Braun, for example, became chairman of "Final Acceptance," responsible for the engine and missile test stands located at each production site for calibration and quality control. (Since the A-4 had no feedback control regulating engine thrust, it was necessary to calibrate and match combustion chambers and turbopumps through systematic testing.) Detmar Stahlknecht, on assignment from Speer since mid-1942 as head of the Production Planning Directorate in Peenemünde, was made chairman of a similarly named subcommittee. That move was an attempt to coordinate the latest organizational innovation with the preceding one, but the boundary lines between the activities of Stahlknecht and Degenkolb initially remained fuzzy. An even earlier attempt to sort out the confusion in the missile production program, Dornberger's Ordnance A-4 Working Staff, had in the meantime faded into irrelevance because of his attempt to proceed without the Armaments Ministry.[13]

Notwithstanding that experience, the chief of Wa Prüf 11 naturally resented the A-4 Special Committee's intrusion into his bailiwick, particularly when it came in the form of an individual as tactless and overbearing as Degenkolb. In his very first meeting with the locomotives czar, Dornberger had realized with a shock that Degenkolb had been one of Ordnance's most vituperative critics during the "munitions crisis" of 1940, leading to the suicide of General Becker. Degenkolb embodied the Armament Ministry's reorganization of the war economy in favor of industry and at the expense of the military. In short order, the relationship between the two became tense. Matters only worsened at the end of February 1943, when Degenkolb tried to arrange for the sale of the Peenemünde Production Plant to the giant electrical firm AEG. He was justifiably unimpressed by the way Dornberger and Schubert had laid out the factory, but he also brought with him a general contempt for military management. In order to stop the AEG initiative, Dornberger proposed the creation of an "Adolf Hitler Ltd." instead, presumably under Army ownership. Ultimately the power of the senior service was still sufficient to stymie Degenkolb's blatant power grab; AEG pulled out of the deal in early March.[14]

Within days of that decision, a senior AEG director and electronics expert, Professor Waldemar Petersen, presented an even more radical proposal: The whole of Peenemünde-East should be converted into a company. Surprisingly, General Leeb had made the same comment at the end of 1941, while turning down Dornberger's proposal to militarize the center's personnel completely, perhaps by expanding the Northern Experimental Command. Leeb had apparently accepted many of the criticisms of the bureaucratic manner in which Army Ordnance operated. Nothing came of those discussions, and Peenemünde East remained what it was: a military research and development facility with a large civilian component. In November 1942 Ordnance began a new round of discussions on lessening red tape in the A-4 production program. Normally, all contracts for rocket parts and subassemblies, tools, and jigs for the factories, and the like, had to be let through Ordnance's procurement and price control bureaucracy in Berlin. The outcome of those discussions, reinforced by pressure from the Ministry and Degenkolb, was the creation in the first half of 1943 of a Peenemünde purchasing and contracting operation working on more commercial lines. That measure streamlined administration and allowed the Army to defeat—temporarily—the attempt of Speer's industrial managers, with their close ties to the Party, to take the Peenemünde center away.[15]

It was not long before the engineers at Peenemünde shared Dornberger's anger and irritation with Degenkolb. As soon as the chairman of the A-4 Special Committee had finished filling out his organization, he moved to force a dramatic speedup in missile production. The last and most ambitious program outlined by Stahlknecht, in February 1943, called for Peenemünde and Friedrichshafen to begin assembling five A-4s in April and July, respectively, and to increase their monthly output in steps to a maximum of three hundred rockets each from September 1944 on (an annual production rate of 7,200). In the wake of the October 3 success, Dornberger had proposed adding a third A-4 assembly plant in the General Government (that is, rump Poland). This colonial appendage of the Third Reich lay under a vicious, genocidal occupation, but it was beyond the range of Allied bombers and promised a supply of cheap Polish labor. Speer gave oral approval to that plan in late December, but Degenkolb soon cast it aside, probably because the factory would take too long to construct. At the end of March Stahlknecht announced that the Rax-Werke, a locomotive and railcar factory in Wiener Neustadt, Austria, would be the third site. A couple of days later Degenkolb issued his first program: each of the three sites was to begin producing three hundred rockets a month by December 1943![16]

Degenkolb's schedule bordered on the absurd; he had clearly failed to understand that a missile was much more complex than a locomotive. Although his energetic action, combined with the backing of the Armaments Ministry, had done much to accelerate the A-4 production program, the practical difficulties of the new schedule were insurmountable. Degenkolb apparently thought he could create a third factory by fiat, even though Rax would need test stands and a liquid-oxygen plant to supply the engine firings, as was the case in Peenemünde and Friedrichshafen. (In mid-1942 the Army had started building a new test-stand area for Zeppelin near the village of Oberraderach.) It was also unclear how the original two sites were to be ready in time, or where the program was going to find the parts manufacturers, alcohol supply, and liquid-oxygen capacity needed to assemble, test, and launch ten thousand missiles a year. Von Braun had to exert all his authority just to keep his people cooperative.[17]

The A-4 was also still far from ready for manufacturing in the spring of 1943. Only two or three of the eleven launch attempts from late October to the end of March were even partially successful. Crucial guidance equipment like the radio cutoff system was not available for launch testing until the spring, and the first rocket even to approach the designated range of 270 kilometers did not fly until April 3. Moreover, to the shock of the non-Peenemünde members of the A-4 Special Committee, many of the drawings of electrical components were still not ready in March, nor were subcontractors lined up for many components. In virtually all areas, the drawings and parts lists remained in a state of confusion, and excessively complicated development models had been forced into production for lack of time.[18]

The situation drove Walter Thiel, among others, to the brink of despair. Before going on a short rest cure in March, he wrote a letter to Wernher von Braun from Friedrichshafen, indicating that he was mentally and physically exhausted. He had not yet been able to make the "mixing nozzle" injector work as a replacement for the eighteen-pot motor, the turbopump–steam generator system was nightmarishly complicated and unreliable, and liquid oxygen was a poor propellant choice for a weapon. A combination of nitric acid and a hydrocarbon fuel, on the other hand, could be made to ignite on contact and would lack the handling problems of a supercold fluid. Stegmaier's marginal comment said it all: "The war is not going to wait for Dr. Thiel." The order of the day was send weapons into production, whatever the drawbacks.[19]

In the end Thiel and his colleagues had no choice but to tolerate the myriad interventions of Degenkolb and his friends on the A-4 Special Committee. It was the price Peenemünde had to pay for the earlier excessive salesmanship of Dornberger and the Army. The leadership of the Third Reich had been repeatedly told that the A-4 was a weapon that could change the course of the war. Now that Hitler and many of his subordinates were coming to embrace that proposition, the rocket group had to produce results—or face the end of their program.

HIMMLER TAKES AN INTEREST

Besides Speer, one other top Nazi leader began to pay close attention to the Army rocket program: the Reichsführer-SS, Heinrich Himmler. He had been informed of the October 3 success and had shown some previous interest in rocketry. In addition to promoting the development of solid rockets for the Waffen-SS (the military wing of the SS, which was growing into a rival of the Army), he had promoted the career of Helmut von Zborowski, an enthusiastic SS officer who was pioneering nitric acid rocket technology at BMW Aircraft Engines in Berlin. But it was probably Hitler's endorsement of the A-4 in late November 1942 that induced Himmler to take a look at the Army's new weapon. Less than three weeks later, on December 11, he traveled to Peenemünde for his first visit to the complex. While there he witnessed the embarrassing failure of the A-4/V9, which blew up and crashed after only four seconds.[20]

Five days after the visit, one of Himmler's chief deputies, Gottlob Berger, wrote the Reichsführer-SS at the request of an old friend from southwest Germany, Gerhard Stegmaier. Berger headed the SS Main Office, which was responsible, among other things, for recruiting. The letter reads:

> The Peenemünde Army Establishment is deeply impressed with the visit of the Reichsführer-SS even today. Lt. Col. *Stegmaier,* who was happy as a small child about his special greeting from the Reichsführer-SS, asks to convey the following message:
>
> The section chief, Col. *Dornberger,* wishes to make an official presentation to the Führer, together with the developer [of the missile], Dr. *v. Braun,* in order to hold discussions with the Führer and to hear his views and wishes, especially regarding the deployment possibilities of the device.
>
> The decisions that the Führer would lay down would then give a clear direction to the already ongoing partial preparations for deployment.
>
> I await further orders.[21]

What Dornberger hoped to accomplish through such a visit is not entirely clear, but perhaps he sought to express as diplomatically as possible his objections to the proposed bunker.

Himmler did not convey the request to Hitler until his visit to East Prussia on January 23. The Führer turned down the proposed audience. A few days later Stegmaier visited Berger, accompanied by Dornberger's chief of staff, Thom. The head of the Development Works carried with him a letter he wrote conveying a new Dornberger request: that Himmler intervene with Hitler because of problems in the electronics industry.

> It is urgently necessary that A-4 program be given a higher priority than the radar program through an order of the Führer. This can be justified with the argument that the A-4 has the character of an offensive weapon while radar is a part of the defense. [The preference for offense over defense was basic to Hitler's strategic thought.] All previous attempts to achieve this status for the A-4 have either failed or been ineffective.

Dornberger was dissatisfied because Speer refused to follow through on a promise to favor the A-4 over radar; perhaps the Minister realized that the anti-aircraft defense of the Reich was in enough trouble already. In any case, Himmler told Berger to do what he could to support Dornberger's request. On February 10, a week after the devastating news of the surrender of the encircled Sixth Army at Stalingrad, he asked the Führer again if the Peenemünders could have an audience. As before, he had no luck.[22]

Hitler's attitude toward the rocket program in this period is opaque, to say the least. His approval of A-4 production and deployment hardly suggests that he was skeptical of the missile. Yet Dornberger's memoirs make much of the Führer's alleged dream in March 1943 that "no A-4 would ever reach England." There is no evidence that the dream had any impact on the missile program, and the only evidence that it even happened is Dornberger's recollection. The actions of Speer and Degenkolb clearly demonstrate that the A-4 had become one of the highest priorities of the Third Reich. Himmler also wrote to Milch on February 3 that the Führer was keenly interested in rocket development because it would be "a very decisive weapon for the future."[23]

It is nonetheless true that Hitler still refused to accede to Himmler's requests or to give A-4 production a "DE" rating—hardly a questionable decision in view of Germany's strategic dilemmas and the rocket program's lack of military sense. Of course, Hitler never really grasped the latter point, nor did he gain from his many meetings with Speer any further understanding of the technology. He apparently still thought of the A-4 as a large artillery shell, because in early January he asked that his Minister investigate the possibility of blasting it from the barrel of the giant 80-cm siege mortar "Dora." Dornberger was left with the delicate task of explaining in writing why the idea was fatuous.[24]

Thus Dornberger's attempts to circumvent Speer and reach Hitler through the Stegmaier–Berger–Himmler connection were ineffective. That outcome, however, was not the true significance of his action. The chief of the Army rocket program had brought the power of the SS into his own domain, even though he would complain loudly about it in his memoirs. Given Himmler's relentless empire-building and his conviction that his organization was the most zealous and efficient tool for carrying out Hitler's wishes, he would probably have taken an interest anyway. But Dornberger's actions, bolstered by Stegmaier's enthusiasm for the SS, gave the Reichsführer-SS every reason to hope that he might eventually bring the program into his orbit.[25]

Himmler might have taken further encouragement from another fact: Wernher von Braun was an SS officer. In 1947 von Braun offered this explanation to the American authorities:

> In spring 1940, one SS-Standartenführer (SS Colonel) Mueller from Greifswald, a bigger town in the vicinity of Peenemünde, looked me up in my office . . . and told me that Reichsfuehrer SS Himmler had sent him with the order to urge me to join the SS. I told him that I was so busy with my rocket work that I had no time to spare for any political activity. He then told me that my being in the SS would cost me no time at all. I would be awarded the rank of a "Untersturmfuehrer" (lieutenant) and it was a very definite desire of Himmler that I attend [sic] his invitation to join. . . .
>
> Realizing that matter was of highly political significance for the relation between the SS and the Army, I called immediately on my military superior . . . Dr. Dornberger. He informed me that the SS had for a long time been trying to get their "fingers in the pie" of the rocket work. I asked him what to do. He replied on the spot that if I wanted to continue our mutual work, I had no alternative but to join. He added that he

> hoped our old cordial relation of confidence would avoid any future difficulties that could arise.
>
> After having received two letters of exhortation from Mueller, I finally wrote him my consent. Two weeks later, I received a letter reading that Reichsfuehrer SS Himmler had approved my request . . . and had appointed me . . . [to] the staff of Obergruppenfuehrer [General] Mazow [Mazuw], [in] Stettin (Whom I did not even know).

Dornberger must have been concerned about offending Himmler at a time when the priority of the program was threatened. Whatever his reasons were for pushing his subordinate to join, von Braun's membership officially commenced on May 1, 1940. After receiving promotions in late 1941 and late 1942, the Development Works' Technical Director held the rank of Hauptsturmführer (Captain).[26]

Von Braun's statement is self-serving, but other evidence confirms the impression that he was motivated primarily by a desire to continue working on his lifelong obsession, rocket development, whatever the moral cost. Gerhard Reisig, head of the measurement group, remembers that von Braun asked a small group of engineers: "Should I do it or not do it? Would it help me or would it be bad for me?" His previous actions also show him to be an opportunist. The brilliant young aristocrat had a conservative and nationalist upbringing, but he had showed no particular ideological enthusiasm for National Socialism in his youth. In November 1933, while he was studying at the university and working at Kummersdorf, he had joined a Berlin SS unit that offered horseback riding. He may have done so only as recreation and as a demonstration of political loyalty at a time when the Nazis were consolidating their control. He was given the lowest rank, "SS candidate," but dropped out after half a year.[27]

At the end of 1937 an unnamed authority "officially demanded" that he join the Nazi Party. During that year membership had been reopened for individuals of social significance. At least two or three other leading Peenemünders were also asked to join at that time. In general, the Army was able to keep the Party from playing an important role in the complex, but a sample of twenty-eight prominent Peenemünde engineers and scientists shows that thirteen or fourteen became Party members and four, including von Braun, were in the SS. A few, like

Arthur Rudolph (a member since mid-1931), were ideologically committed Nazis, but the survey confirms the impression that von Braun was fairly typical. He and his colleagues were motivated first and foremost by a desire to advance themselves and their work, compounded perhaps by enthusiasm for the foreign and domestic "accomplishments" of National Socialism. Anti-Semitism may also have been important for a few individuals, but most, like von Braun, were apathetic about the issue. During the war it was also not wise to ask too many questions about the fate of the Jews.[28]

Whether von Braun's SS rank influenced Himmler's interest in Peenemünde in early 1943 is unclear. The promotions notwithstanding, von Braun is known to have worn his uniform only once, and that was during Himmler's second visit in June 1943, when Dornberger allegedly ordered him to do it. In March and April of that year, the thirty-one-year-old Technical Director also used his rank, as he needed to, when applying to the SS Race and Settlement Office for permission to marry. (For unknown reasons, the marriage to a Berlin physical education teacher never took place.) On the second of two letters to the Office he added a handwritten salutation to Himmler: "Führer!" But he never used his rank on any other correspondence and appears not to have been proud of his SS membership. Himmler may well have perceived von Braun's lack of ideological commitment, since he was energetically pushing the BMW rocket engineer Helmut von Zborowski at the time. Only later would von Braun's rank and position become of greater significance in Himmler's plans.[29]

At the end of March 1943 a new opportunity to meddle in Peenemünde's affairs presented itself to the SS. The Gestapo had arrested an anti-Nazi group led by Catholic priests in the nearby city of Stettin (now Szczecin, Poland). Because the Commander of Peenemünde, Colonel Leo Zanssen, was a Catholic and was politically suspect, the local SS and police commander, Emil Mazuw, thought he might be part of the conspiracy. Himmler telegraphed Gottlob Berger to use his personal connection (Stegmaier) to obtain more information on Zanssen. Berger asked the chief of the Development Works to visit him clandestinely over the weekend. Stegmaier could not confirm the charges, but he gave the SS some of what it wanted. According to Berger's report of April 5, Zanssen was an old-fashioned officer out of touch with the common man and National Socialism. He dampened his subordinates' enthusiasm for work and was a member of a leading Catholic society. More than that: "*Zansen* [sic] is an alcoholic. When he is overcome with a howling depression, which happens often, particularly under the influence of alcohol, he writes religious and sentimental letters and gives religious and sentimental speeches."[30]

Stegmaier had stabbed his superior in the back. It is true that Zanssen was not universally popular; he was excitable and could get very angry. Some leading Peenemünders were friendly with him, others less so, including, apparently, Stegmaier. Zanssen was also known to be a social drinker, but the charge that he was an alcoholic was false, as was the claim that he was a member of the "Catholic Action." In the tradition of the German officer corps, Zanssen was not outspokenly religious; his wife and children were in fact Protestant. The real problem was his increasingly apparent alienation with the Third Reich. In 1933 Zanssen, like many junior officers, had been enthusiastic about the "national revolution." When the Nazis came to power, he even made his children stand up at the dinner table and sing the Party anthem, the "Horst Wessel Song," although he was never a member. But at some point he started to grow more and more disillusioned. His wife later attributed that to his service on the Eastern front in 1941–42. He did not go out of his way to pay obeisance to local Party officials and, under the influence of a drink or two, he would become careless about expressing his views. In Mazuw's original telegram to Himmler, the SS commander had noted that he and the Gauleiter (regional Party leader) for Pomerania had known Zanssen for years and "had never placed great trust in him."[31]

Himmler told Berger, who was in Berlin, to bring the matter up with Speer and also to inform Ernst Kaltenbrunner's Reich Security Main Office, of which the Gestapo was a part. Apparently the ensuing investigations produced no evidence that Zanssen had anything to with the Catholic priests or any substantial proof of the Peenemünde Commander's disloyalty. But before those investigations were even completed, the Reichsführer-SS decided to try to undermine Zanssen through another route, perhaps in the hope that Stegmaier would be made Commander. On April 24 Himmler wrote to General Rudolf Schmundt, Hitler's OKW adjutant and head of the Army Personnel Office since

October 1942, when it had been moved directly into Führer headquarters. He disingenuously reported Stegmaier's accusations to Schmundt without claiming to have verified them. After a two-week delay, on May 6 Army Ordnance suddenly received a call from the Personnel Office that Zanssen was to be removed from his position immediately as a security risk.[32]

Dornberger and Zanssen were thunderstruck. The Commander of Peenemünde-East officially withdrew from the Catholic Church the same day and spent a sleepless night pacing the floor. Dornberger frantically attempted to find out what was behind the demeaning treatment of his close friend. Until that time, neither had been informed that Zanssen was under suspicion, although the Gestapo had indicated to the Commander that it had planted an agent in Peenemünde to investigate possible connections with the anti-Nazi group. As Zanssen told a police official a few days later, he suspected that his dismissal was due to the machinations of private industry (meaning Degenkolb's cronies) to get rid of Dornberger and himself and take over Peenemünde.[33]

Pending a final resolution of Zanssen's fate, Dornberger issued an order on May 11 appointing himself Commander and assigning his friend to Berlin to act as his representative at Wa Prüf 11. Backed by the Chief of Army Armaments, General Fromm, who was outraged by the way the order had come down from Führer headquarters, Dornberger then set out to clear Zanssen's name. In short order Dornberger discovered who was behind the accusations: Stegmaier. He probably surmised this from the copy he had received of Himmler's letter to Schmundt. Dornberger's reaction to the information news is surprising: He merely gave Stegmaier a reprimand. As he explained in his letter to Ordnance Chief Leeb in early June, Stegmaier had been afflicted by a "real contradiction" between his loyalty to his superior officers and his loyalty to Nazism. Because of his friendship with Berger and his "lack of trust" in Dornberger, Stegmaier had therefore repeated his hostile views about Zanssen to Berger.[34]

Dornberger was angry, but he gave Stegmaier some leeway because his own loyalty to Hitler and the Third Reich was unquestioning, which could not be said of Zanssen. In his notebooks one can find a draft of the pep talk he gave to his senior subordinates on assuming the Commander's post on May 12. He states: "My National Socialist beliefs should be widely known." He goes on to say that his sole aim was "put in the hands of the Führer sufficient numbers of this weapon [the A-4], which—it is my conviction and unshakable belief—will decide the war." In the pseudo-apolitical tradition of the officer corps, neither he nor Stegmaier was a member of the Nazi party or its principal organizations, but because he was also afflicted with conflicting loyalties to Nazism and the Army, he gave Stegmaier rather mild treatment, even though the Development Works chief had betrayed one of his closest friends.[35]

In the end, Zanssen's name was cleared and Himmler had to back away from the affair. Fromm sent a letter to the Reichsführer-SS, based on a Dornberger draft, which strongly contested the charges. Because there was no solid evidence of Zanssen's disloyalty, Himmler asked Berger what he should reply. The chief of the SS Main Office responded that Thom and Dornberger had visited him a number of times and he had offered this explanation: "In view of the special importance of secrecy, it was a purely comradely gesture" to make the Army aware of Zanssen's questionable statements in letters and elsewhere! Himmler should take nothing back, Berger suggested, but rather should leave it to Berger to give the same explanation to Fromm. It was an incredibly cynical letter that revealed in its very first line Himmler's attempt to exploit the affair: "The Col. *Zansen* case has not turned out quite the way we expected."[36]

On the evening of June 28, Himmler arrived at Peenemünde for a second, more formal tour of the center, to be held the following day. He was accompanied by Mazuw and Leeb. According to Dornberger, he was finally able to confront the Reichsführer-SS about the Zanssen accusations the next afternoon, while on a boat to the Oie to witness the second, and this time successful, A-4 launch attempt from Peenemünde. The new Commander got nothing but a cold brushoff. Himmler indeed had every reason to be disgruntled about the way the whole thing had turned out. But there were other satisfactions to be found in the visit. The sight of Wernher von Braun in a black uniform must have inspired Himmler, because the Technical Director received an early promotion to Sturmbannführer (Major), backdated to June 28. The Reichsführer-SS must also have been excited about the progress of the missile, which the Führer's whole entourage now believed could knock Britain out of the war. In the meantime, he would have to bide his time and await another opportunity to exert his influence over the program.[37]

THE DECISION TO USE SLAVE LABOR

While the Zanssen affair was still being cooked up behind the closed doors of the SS, Peenemünde became more deeply involved with Himmler's organization by another route: through the decision to use concentration camp labor for A-4 production. After the war the Peenemünders made the SS a scapegoat for all the crimes associated with the rocket program, yet the initiative for this action came not from the SS but from the A-4 Special Committee and Arthur Rudolph, who were searching for a way around the prevailing labor shortage.

The problem of finding workers for the missile assembly plants had long plagued Dornberger, Schubert, and the production planners. The original Peenemünde site had been designed as an "everything-under-one-roof" factory with a highly skilled workforce, but wartime compromises had turned the misnamed Pilot Production Plant into an assembly facility for parts mostly produced elsewhere. That change at least allowed for the possibility of using fewer skilled workers. Immediately after the October 3 success, Dornberger still proposed transferring the labor needed for the Peenemünde plant from the Development Works, but the second site at Zeppelin could use "mostly foreigners and prisoners of war," as was becoming common in German industry. His plan to build a third site in Poland also implied that the labor force there would be dominated by non-German labor under some form of compulsion. In any place where Poles, Soviet prisoners of war (POWs), or other Eastern Europeans were employed, they stood under extremely discriminatory and exploitative Nazi regulations. Most of their wages were confiscated, and they had to wear a distinctive badge, as did the Jews. Vicious racial laws dictated transfer to a concentration camp or even execution for such "crimes" as sexual relations with "Aryan" women. Food, housing, and health care were often wretched.[38]

Poles had worked under those conditions at Peenemünde since 1940, but their numbers had remained rather small, and responsibility for them lay in the hands of Speer's Construction Group Schlempp. No more than a thousand Polish workers had been employed at any one time in 1940; in December of that year there were 633 Poles out of 4,780 construction workers on both sides of the Army facility. (Foreign workers were not allowed into the Development Works until about 1943.) Beginning in the spring of 1941, 700–1,000 Italian contract workers were employed on site during the construction season. The number of Polish and, later, French workers was apparently only a few hundred. As elsewhere in Germany, voluntary foreign laborers with a right to vacation were less favored, because they were not as easy to control as the forced labor of POWs and civilian workers drafted from the occupied countries. But the ultrasecret character of Peenemünde, combined with the priority problems of the Production Plant, greatly limited the use of both forced and voluntary foreign labor. Only after Hitler's go-ahead for A-4 production did that barrier begin to come down. At the end of April 1943 there were more than three thousand foreign laborers in both Peenemünde-East and Peenemünde-West.[39]

When Degenkolb formed the A-4 Special Committee earlier in the year, he also created a "Labor Supply" subcommittee headed by a Mr. Jaeger from his office. Its task—to find production workers—was organizationally distinct from that of the construction labor bureaucracy, although both drew on much the same pool of workers. Arthur Rudolph, who was the Peenemünde factory's chief engineer, met Jaeger at the beginning of February and was told that at least the problem of finding unskilled workers for A-4 production could be solved through the use of "Russians," mostly Soviet POWs.[40]

That plan was entirely in keeping with the general movement of the German war economy. Until early 1942 few soldiers captured by the Germans on the Eastern Front had been exploited for work, because the armed forces (not the SS) had intentionally left these "subhumans" to die in huge numbers. In one of the forgotten Holocausts of the Third Reich, more than 2 million of some 3.35 million Soviet prisoners were dead by February 1942 as a result of mass starvation and disease. (The death toll by 1945 was more than 3 million out of 5 million captured.) Only after Germany failed to knock the Soviet Union out of the war in 1941 did their labor become desirable, particularly because the Eastern Front's insatiable need for German manpower made the labor shortage in the Reich ever more severe. With the further reorganization of the economy under Speer, an attempt was also made to rationalize the employment of foreign workers. On March 21, 1942, Hitler appointed Fritz Sauckel, Nazi Gauleiter of the central German state of Thuringia, as General Plenipotentiary for Labor Supply, with power over the bureaucracy of the labor exchanges. At the same time greater efforts were made to salvage surviving Soviet POWs for work. They were reinforced by the large number of Soviets captured in the summer campaign of 1942 and by the forced recruitment of "Eastern workers," mostly young women.[41]

Thus, for a short time in early 1943 the use of "Russians" became the basic plan for the Peenemünde Production Plant. On April 8 and 9, however, Jaeger visited Peenemünde and recommended another solution to the factory's labor problem: the employment of SS concentration camp prisoners. That proposal was in consonance with a related shift in the war economy. Until the spring of 1942 Himmler's organization had attempted, with limited success, to exploit its prisoners in camp-based enterprises. The camps' primary purpose remained suppression of opposition to the Nazis plus, beginning in late 1941, the mass extermination in Polish camps of Jews and gypsies. The SS had turned down requests by the almost equally cold-blooded Army Ordnance leadership to use camp prisoners as slave laborers in armaments production. But even before the appointment of Sauckel, Himmler decided to secure his own place in the war economy by reorganizing his camp system to emphasize greater economic efficiency. Through mergers, he created the SS Economic and Administrative Main Office and gave it the camp inspectorate. The SS began, in effect, a rent-a-slave service to firms and government enterprises at a typical rate of four marks a day for unskilled workers and six marks for skilled ones. In return, the SS supplied guards, food, clothing, and shelter, usually in a manner that led to a heavy death toll from starvation, disease, and overwork. The lives of camp inmates were, by definition, expendable.[42]

After Jaeger suggested that concentration camp prisoners be used in the Peenemünde Production Plant, Rudolph went on a tour of the Heinkel aircraft works in Oranienburg, north of Berlin, on April 12. He returned enthused about the advantages of concentration camp labor. In only eight months Heinkel had increased the number of prisoners in

Oranienburg from zero to four thousand, because it had only the "best experiences" with them. The mixture of multiple nationalities—mostly Russians, Poles, and French—helped to undermine resistance activity, and the prisoners could be accommodated in the locker rooms of the plant, where their beds were crammed "very close together." The area in which the prisoners worked and lived was surrounded by an electrically charged barbed wire fence with watchtowers. Rudolph continued: "This system has worked well, and the employment of detainees *[Häftlinge]* in general has had considerable advantages over the earlier employment of foreigners, especially because all non-work-related tasks are taken over by the SS and the detainees offer greater protection for secrecy." The memo concludes: "Production in the F1 [the main assembly building at Peenemünde] can be carried out by detainees." The Production Plant would have to make an application through Stahlknecht to the A-4 Special Committee, which would approach the SS.[43]

This document is of extraordinary importance for two reasons. First, in conjunction with Jaeger's original suggestion, Rudolph's memo initiated the systematic exploitation of slave labor in the rocket program months before the creation of the infamous Mittelwerk underground facility near Nordhausen in Thuringia. In the traditional accounts of Wernher von Braun and others who followed him, it was Himmler who forced the program to use concentration camp prisoners following the first large-scale air raid on Peenemünde. (Dornberger, on the other hand, *never once* mentions the prisoners in his memoirs, despite his central role in decisions regarding their use.) Second, the Oranienburg memorandum is also relevant to the recent, much-publicized Rudolph case. Because of his role as a manager in the underground facility, Rudolph, who was Project Manager of NASA's Saturn V moon rocket in the 1960s, was forced to leave the United States in 1984 and give up his citizenship. Yet the U.S. Justice Department made its case without knowing about the April 1943 document, which shows that Rudolph was not just the manager of slave labor but also an advocate of it.[44]

Jaeger's and Rudolph's suggestion was enthusiastically taken up by Dornberger and the A-4 Special Committee. In the third week of April, when the chief of Wa Prüf 11 went on a trip to Friedrichshafen and Vienna with unnamed Committee members, they discussed the acquisition of 2,200 skilled SS prisoners for the Rax-Werke assembly plant.

Meanwhile, the Zeppelin company had apparently begun using concentration camp labor on its own initiative. The Friedrichshafen camp, which was a branch camp of Dachau (near Munich), is first mentioned on February 23, 1943, and its prisoners were soon put to work on Peenemünde contracts. Zeppelin had been manufacturing A-4 propellant tanks, midsection fuselages, and other components since 1942, as well as preparing for assembly operations.[45]

In the spring of 1943 it was still not clear, however, that SS prisoners were the complete solution to the production labor problem. More German skilled workers would be needed in any case. The Army and the Committee therefore invited the labor supply czar, Sauckel, to Peenemünde for the royal tour, which took place on May 13–14, one day after Dornberger had become Commander. Sauckel went away thrilled by the successful A-4 launch he had seen and promised full support. Some Russian POWs and other laborers did later arrive, but it is unclear if Sauckel's promises ultimately amounted to much. On June 2 Degenkolb's deputy, Heinz Kunze, complained that the labor bureaucracy had done nothing to exchange French workers with vacation rights for forced laborers with no such rights. German skilled workers, moreover, were simply unavailable. Rudolph was told that there was no recourse in the latter case but to transfer personnel from the Development Works. Although the number of workers and engineers under Stegmaier and von Braun had expanded considerably, particularly through the diversion of soldiers into the Northern Experimental Command, that suggestion was hardly realistic in view of the heavy demands imposed by A-4 production and the Wasserfall project.[46]

In the same June 2 meeting the Peenemünde group, headed by Rudolph, presented a formal request for 1,400 concentration camp prisoners, subdivided by skill. Jaeger was to complete negotiations with the SS, while Rudolph would be the liaison person with the camp commandant when the first installment of prisoners arrived. The chairman of the labor supply subcommittee set the maximum number of prisoners for Peenemünde at 2,500. They were to be temporarily housed on the lower floor of F1, where locker rooms, storage areas, air-raid shelters, and workshops for the assembly of subsections and components were located. On the main floor of the building, an "assembly line" for whole missiles was being finished. The rockets would be assembled in a horizontal position on movable railcarts, in contrast to the development shops, which put them together vertically more or less by hand. Only two weeks later, on June 17, Schubert's chronicle of the Peenemünde factory reports "arrival of the first 200 detainees, half German, half Russian, who will be housed in the F1 and who will be employed first on [the construction of] the barbed wire fence around the F1." Three days after that, the infamous Mauthausen camp delivered about five hundred prisoners to the Rax-Werke in Austria. Those men immediately began finishing the main assembly hall and performing other construction work.[47]

The new arrivals in Peenemünde had been shipped from the Buchenwald camp near Weimar in Thuringia, accompanied by about sixty SS guards and staff. Virtually the only information now available about their conditions comes from the postwar testimony of a German criminal convict, Willy Steimel, who was involved in the prisoner-run section of the camp administration at Buchenwald, Peenemünde, and Mittelwerk. His testimony is not entirely trustworthy, because he was in a privileged situation and was probably an SS informant. By Steimel's account, conditions in the F1 were very good by camp standards. The facilities had been designed with German workers in mind and were thus new and clean. The admixture of a small number of prisoners with many civilian and Army people also inhibited some of the usual "chicaneries" and beatings by guards, leading to "a partly bearable situation." Steimel's health, he said, began to improve. Still, the F1 was no rest camp; an eleven-hour day and a six-day week were normal for prisoners at the time. According to Steimel, during the four months spent at Peenemünde, three prisoners died of tuberculosis, two of injuries, and one was shot trying to escape, while four others died from drinking alcohol rocket fuel that was part methanol.[48]

On July 11 four hundred more, mostly French prisoners, arrived in Peenemünde. Schubert was disgruntled because few of them had any specialized skills; he asked the SS to exchange them for German craftsmen. Five days later prisoners began to work on the assembly of A-4 midsections and on the construction of jigs for the main assembly line. SS detainees also flowed into Wiener Neustadt and Friedrichshafen during that same period. Clearly, A-4 production was about to begin in a big way at long last through the exploitation of slave labor.[49]

MISSILE MANIA

As the hectic preparations of summer 1943 indicate, the A-4 had finally become the highest priority of the Third Reich. Speer had secured the Führer's backing for such a declaration by carefully preparing the ground politically. In February he formed a high-level "Long-Range Bombardment Commission," probably at Hitler's request. Chaired by AEG director Petersen, its task was to consider the merits of the competing Army and Luftwaffe missile projects. Yet it scarcely functioned, aside from a handful of meetings, most notably at the "comparison shoot" of May 26, 1943. On that day, the greatest collection of Third Reich dignitaries ever assembled at Peenemünde saw two launches each of the A-4 and "Cherry Stone" (now more often known as the Fi 103). Among those present were Speer, Saur, Degenkolb, General Fritz Fromm, Grand Admiral Karl Dönitz (Commander-in-Chief of the Navy), Field Marshal Erhard Milch, and a host of other high-ranking officers. The nominal purpose was to compare the performance of the two weapons and decide whether to proceed with both. After witnessing the successful launch of A-4/V26 around noon, those present saw the two Fi 103s crash ignominiously into the Baltic after lunch. Late in the afternoon, A-4/V25 took off successfully in better visibility than had the earlier vehicle but fell very short when the engine cut off too early because of a "calculation error." Milch nonetheless slapped Dornberger on the back and exclaimed: "Congratulations! Two–nothing in your favor!"[50]

The Commission's decision—which may have been taken even before the launches—was to proceed with both weapons at all possible speed. Any skeptics left among the assembled group were drowned out by the collective euphoria of the civilian and military leadership of the Third Reich, which now hoped that secret weapons would reverse the increasingly worrisome course of the war. Approval of both the A-4 and the Fi 103 was justified on the plausible grounds that their advantages and disadvantages were complementary: The cruise missile was cheaper and simpler but had to be launched from fixed catapults vulnerable to air attack, while the ballistic missile would be more accurate but would require components and propellants in short supply. Employed together, they would also have a greater effect on Britain.[51]

As the perfunctory character of the "comparison shoot" indicates, the competition was basically a sham. Speer had been pushing Hitler since October 1942 to accept the two programs in parallel, probably with the aim of protecting both the A-4 and his friendly relations with Milch. Because the Third Reich was so riven by competing empires, it was unlikely that the Commission would have had the power to shut down one program in favor of the other anyway, barring a decisive ruling from Hitler. The Long-Range Bombardment Commission appears to have been nothing but a Speer maneuver to patch up interservice relations, neutralize opposition, and secure top priority for A-4 production. As if to confirm that everything had been settled in advance, Speer announced during the May 26 meeting that the A-4 would receive not merely a "DE" rating (at least for the first 1,500 missiles), but the highest possible "DE" rating. To implement that decision, a new numbering system would be introduced similar to that already subdividing the second priority level, "SS." The A-4 would also get a very high number in the "SS" series.[52]

On June 2 the Armaments Ministry issued the necessary order, and a follow-up document from June 11 stated: "The Führer has ordered that the A-4 program rates in priority before all other armaments production." The next day Dornberger received a promotion to Brigadier General. More dignitaries streamed into Peenemünde, including retired Army Commander-in-Chief von Brauchitsch, Reichsführer-SS Heinrich Himmler, chemical-industry leader Dr. Carl Krauch, and OKW Chief Field Marshal Keitel. The biggest reward, however, came on July 7, when Dornberger and von Braun were suddenly called to East Prussia to meet the Führer. They were accompanied by Ernst Steinhoff, who piloted their He 111 bomber-transport. The fact that the three flew together alarmed Himmler, when he found out. He asked Hitler to forbid a repetition of this incident, lest the leadership of the rocket program be wiped out in a single crash.[53]

After keeping the visitors waiting for hours at the Wolfsschanze, the Führer finally appeared, accompanied by Speer, Keitel, and other aides. Dornberger was shocked at the unhealthy, hunched-over appearance of Hitler, whom "I had not seen . . . since March 1939"—a patently false statement in light of the 1941 visit to the Wolfsschanze. As on that occasion, the Peenemünders began with a propaganda film, in this case one made immediately after the October 3 success. For the first time Hitler saw the impressive images of an A-4 launch. Von Braun narrated the film; Dornberger lectured and showed models of the deployment systems to the visibly fascinated Führer. According to Dornberger, he attempted to persuade Hitler not to construct a launch bunker, but that cannot possibly be true. Organization Todt had been building one for months at Watten (Eperlecques) in the Pas de Calais. It is more likely that he indicated that the mobile system was less vulnerable to Allied air power, but Hitler was so infatuated by the success of the U-boat pens in France and by the thought of gigantic quantities of concrete that he rejected the argument out of hand. Two weeks later the Führer ordered that another bunker be constructed, this time in Normandy, for the bombardment of cities in southwest England and Wales.[54]

In Dornberger's account of the July 7 audience, Hitler went on to demand a 10-ton warhead and a production rate of two thousand missiles a month. When the new general told him that those ideas were impossible, a "strange, fanatical light flared up in Hitler's eyes. I feared he was going to break out into one of his mad rages. 'But what I want is annihilation—annihilating effect!'" Dornberger relates that he had to keep his composure and find a way to appease the Führer. He allegedly went on to plead with the dictator not to let the Propaganda Ministry spread more rumors of war-winning "wonder weapons"; the A-4 was no such weapon. His account is difficult to believe in view of his many previous attempts to sell the A-4 as decisive. In any case, Hitler emerged from the meeting intoxicated by the missile and told Speer to push its production as fast as possible. The Minister was to modify the decree for the tank program and present it for the Führer's signature, which Speer did on July 25. Hitler also granted the Minister's request (originating from Dornberger?) that Wernher von Braun receive the prestigious title of Professor. The Führer was amazed at von Braun's youth and so impressed by his talent that he made a point of signing the document himself.[55]

The Peenemünders naturally felt that they had made a great breakthrough after three and a half years of problems. Dornberger reports that Hitler even made a point of coming to him afterward and saying: "I have had to apologize to only two people in my life. The first is Field Marshal von Brauchitsch. I did not listen to him when he told me again and again how important your research was. The second man is yourself. I never believed that your work would be successful."[56] As dishonest and self-serving as Dornberger's memoirs are, it is hard to believe that such an amazing statement was not burned indelibly into the general's memory. But if Hitler did say this, it contradicts many of his statements since August 1941. Such inconsistency was, however, typical of the Führer's previous behavior regarding the rocket program.

Whatever transpired on July 7, one thing is clear: Dornberger and his followers have greatly overrated the importance of this famous trip to the Wolfsschanze. Speer had already granted the program absolute first priority in early June, missile production had been greatly accelerated since the end of 1942, and A-4 development had been been out of top priority status for only limited periods. Beyond confirming the Führer's enthusiasm for the A-4 and making it easier for Peenemünde to acquire German skilled workers, the only practical effect of the audience was to fuel a missile mania in the Armaments Ministry that was counterproductive. After the "comparison shoot" Speer's deputy, Karl Otto Saur, had undergone a conversion experience from A-4 skeptic to missile fanatic. He now wished to put Hitler's demands for increased production into action. On July 16 he overruled the Degenkolb program and promulgated his own, with the support of the head of the A-4 Special Committee. It projected an output of 900 missiles in October, rising to 1,500 by January, and parts production rising to 1,800 by April 1944! A fourth production site would be created at Degenkolb's old firm, DEMAG in Berlin-Falkensee. Schubert noted resignedly in his chronicle: "This is the eighth production program in a short period of time."[57]

When Saur issued his decree, the assembly lines at Peenemünde and Friedrichshafen had yet to produce a single A-4, and Wiener Neustadt was months away from being finished. The engine calibration test stands at each site were also far from complete. Thiel estimated that they would not be put into full operation until November at Zeppelin and January at Rax. Von Braun also told Stahlknecht that the facilities at Peenemünde and Friedrichshafen were not built to accommodate 375 engines a month, as the new program specified. There were, of course, many additional difficulties. As Dornberger had told Hitler on July 7, ethanol production (for A-4 fuel) depended on the potato harvest and could not be significantly increased. None of those practical problems deterred Saur, however. At meetings he called in Berlin, starting on July 22, he humiliated industrialists who said his program was infeasible—notably Paul Heylandt, the liquid-oxygen equipment manufacturer who had supported Max Valier a dozen years before. Schubert's chronicle noted after the meetings: "Some managers will be removed from their posts, and some of those fired will be charged with sabotage." Missile mania had gotten completely out of hand.[58]

In the end it was Speer who called a halt to the proceedings. At the beginning of August he decided that the Saur plan would be "postponed for the moment," thereby reinstating the Degenkolb program of nine hundred A-4s a month at three sites by December. In the circumstances this overly ambitious scheme may now have looked at least reasonable to the Peenemünders. But in the meantime new conflicts with the A-4 Special Committee had erupted. Earlier in July someone in the Armaments Ministry had tried to have Schubert fired as head of the Peenemünde plant. Later in the month Degenkolb assigned four engineers to the center to investigate A-4 production, led by Albin Sawatzki, who had been decorated for Tiger tank production at the Henschel company. Under the impression that they would take over the factory, the four attempted to give orders, even though it was an Army facility. Sawatzki also put forth a new privatization plan for the plant. Dornberger was outraged and arranged for General Fromm to issue a rebuke to the A-4 Special Committee. The tense relations between Degenkolb and the Peenemünde Commander had thus not improved one iota over the preceding months.[59]

Despite those battles, the two had no choice but to work together. They shared a fanatical belief in Hitler and in the ballistic missile as the salvation of the Third Reich. On August 4 they met in Peenemünde to settle the outstanding issues. Each was accompanied by his deputy (Zanssen and Kunze), and a representative of Fromm also attended. This group decided to assign Sawatzki to plan production under Stahlknecht, while two of his colleagues were to be made the directors of production at Rax. All of them were to refrain from meddling directly in the internal organization of Peenemünde. Dornberger agreed that three hundred A-4s would roll off the Peenemünde assembly line in December, but only on the condition that the Special Committee guarantee the availability of a sufficient number of high-quality parts and subassemblies—a sure sign that Dornberger did not have much faith that this goal could be met. In addition, von Braun's deputy for the development shops, Eberhard Rees, was to be made the director of production in Peenemünde, effectively shunting Schubert aside. Rees had been on assignment since May directing the assembly of development A-4s in one of the Production Plant's buildings.[60]

The meeting's most important decision concerned the labor force. Hitler had asked on July 7 that only German workers be used in the A-4 program for security reasons, but like many problematic Führer orders, that demand had to be quietly ignored. Dornberger's minutes from August 4 read: "As a basic principle, production in all four assembly works will be carried out by convicts." (Dornberger used the word *Sträflinge* instead of the correct SS term, *Häftlinge,* meaning detainees or arrestees.) Peenemünde was to have 2,500 concentration camp prisoners, a total that included a "buffer for the other works," whereas Zeppelin, Rax, and Falkensee would have 1,500. (Speer's decision to postpone DEMAG as a fourth site apparently had not yet been made official.) Peenemünde was to receive more German skilled workers, but the ratio of prisoners to Germans was to be ten or fifteen to one at all locations. Any serious effort to recruit foreign labor from Sauckel's system was thereby abandoned. Rocket assembly would be done primarily by slave labor, a concept Dornberger fully accepted. In a draft of a letter to Saur that he wrote in advance of the meeting, he said: "*Production by convicts—no objections.*" To him they were merely factors of production.[61]

Thus, as the summer of 1943 waned, full-scale missile manufacture finally appeared to be within reach. The assembly line at Peenemünde was scheduled to start in late August, and more concentration camp prisoners were to arrive at all locations. An uncomfortable but mutually tolerable division of labor had been made between the Army and the Armaments Ministry, while the SS had been largely confined to the role of a supplier of labor power. Those relationships, however, were about to change. Allied air power would be the *deux ex machina* that would usher in a dramatic new phase in the history of the rocket program.

Chapter 7

The Move Underground

Shortly after 1:10 A.M., Wednesday, August 18, 1943, the sound of anti-aircraft artillery jolted General Dornberger awake. After a moment of confusion, he leaped out of bed and began to dress. A bomb blast rocked the Development Works guest house where he was staying. Attired in uniform breeches, pajama top, trench coat and bedroom slippers, he pushed his way past broken glass and doors blown off their hinges, only to stand "transfixed" in the garden. Artificial fog from smoke generators rolled across the complex; a full moon, searchlights, exploding shells, and descending British target-marking flares lit up the sky; flak, bomb blasts, and the "monotonous drone" of four-engine bombers "assaulted" his ears. The attack was not a complete surprise. There had been many alarms in the recent past, and a handful of bombs had previously fallen in the area. In the almost comic first raid in July 1940, a solitary and lost RAF aircraft had killed a cow and set a haystack on fire. But this time it was clear that the Allies knew about the rocket center and had set out to destroy it.[1]

The initial target was the Settlement, a residential community of three to four thousand with its own school, community buildings, and firehouse. The RAF had quite consciously decided to try to catch the leading engineering personnel in their beds. Because of a target-marking error, however, many bombers overshot their aiming points by 3 kilometers. Bombs rained down first on the Trassenheide construction labor camp, where more than three thousand foreign workers, mostly from Eastern Europe, were trapped inside barracks or behind barbed-wire fences. Hundreds were killed. The Settlement soon received hundreds of tons of bombs too, destroying at least three-quarters of the houses and apartment buildings. The residences near the beach housing young Labor Service women were particularly hard hit. Most of the leading personnel had made it safely to makeshift shelters and trenches. From the standpoint of the project, the only really irreplaceable loss was Dr. Walter Thiel, who was killed with his whole family by a direct hit on their shelter while their house remained relatively undamaged. Peenemünde's rocket engine development would suffer from his absence.[2]

Shortly after 1:30, the attack shifted to the Production Plant and then to the Development Works. But the tendency to overshoot continued, with the result that more bombs fell on the Settlement and Trassenheide. No devastating damage was inflicted on the important facilities farther north; the all-important wind tunnels and guidance and control buildings were almost untouched. Still, at least twenty-five buildings in the Development Works were set afire or damaged, including House 4, the headquarters building. After the raid ended at 2:07, Wernher von Braun and one of his secretaries risked their lives salvaging secret documents from the burning structure. Only scattered bombs fell on the test stands even farther up the island, suggesting that the British had not perceived their importance, and Peenemünde-West was untouched because the flying-bomb program was unknown to the Allies.

In the initial shock after the raid, Schubert estimated more than one thousand dead, but according to Dornberger's postwar accounts, 732 or 735 were killed, of whom about five to six hundred were foreign laborers. Among the concentration camp prisoners in the F1, Willi Steimel later reported eighteen dead and sixty injured, but the scattered bombs that fell on that enormous factory building did not damage it much. They exploded high overhead on the roof, while the heavy concrete floor of the main hall helped to protect the equipment and prisoners in the basement.[3]

Dubbed "Operation Hydra" by the RAF, that raid was the opening of what was soon called "Crossbow," the Allied campaign against the German secret weapons sites. Ever since the close of 1942 London had received an increasing number of reports about rocket development from inside Germany. The Propaganda Ministry's growing threats of "vengeance" and "wonder" weapons had encouraged careless talk by knowledgeable Germans and had frightened opponents of the regime. The greatly expanded foreign labor force at Peenemünde was a further source of information to the Allies. British intelligence had dismissed earlier reports as too fantastic or as German disinformation, but as evidence mounted in the spring of 1943 that Peenemünde really was the center of some kind of rocket work, high-altitude RAF reconnaissance planes had repeatedly photographed the complex.[4]

On the German side, the growing Allied air raids had made those responsible for the program increasingly concerned about the possibility of espionage and air attack. After discussions with the A-4 Special Committee, the Army had given its facility a cover name as of June 1, "Home Artillery Park 11," and use of the word "Peenemünde" was banned from documents. Karlshagen, a small hamlet near the Settlement, became the new postal address. On May 21 Himmler had ordered that SS guard posts be erected at checkpoints a few kilometers south of the base gate at Karlshagen, probably because of the approaching arrival of concentration camp prisoners. With the support of Speer and Milch, the Luftwaffe also reinforced the local flak batteries protecting the facility. Nonetheless, when the British attacked, the center was "woefully unprepared" for almost six hundred RAF bombers carrying 1.5 million kilograms of explosives. Except for the unfortunate foreign laborers, Peenemünde had been lucky. The center had been saved more than anything else by the technical difficulty of a precision night raid against a series of relatively small targets. It was fortunate also that the damage appeared so devastating the next day that the RAF canceled plans for follow-up raids by itself or the Americans.[5]

Inevitably, the attack set off a whirlwind of activity in Berlin, the Wolfsschanze, and elswhere. It was a powerful reminder of the A-4 program's increasing vulnerability to air attack. On June 21 at Friedrichshafen, and on August 13 at Wiener Neustadt, Allied bombers had unknowingly damaged the other two missile assembly sites in raids on neighboring facilities. Something obviously had to be done. What was done would result in fundamental changes to the character of Peenemünde and the roles of the various groups vying for control of the rocket program. The center's personnel would decrease in numbers and would be dispersed over a much wider area. More important,

A-4 production would depart from Peenemünde altogether and would move underground—to the benefit of the SS and at the cost of thousands of prisoner lives.

DISPERSAL, DECLINE, AND INHUMAN DEGRADATION

Not long after dawn Dornberger and von Braun took a light plane aloft to survey the destruction. Later that morning, Speer flew into Peenemünde, "where General *Dornberger,* still covered with dust and lacking sleep, reported on the damage." The Minister then took off to inspect the results of the first of the daring and costly American daylight raids on the ball bearing factories at Schweinfurt, which had occurred the previous afternoon. The next day, the nineteenth, Speer flew on to East Prussia to discuss the latest raids with Hitler.[6]

It was Heinrich Himmler, however, who first had the Führer's ear regarding the future of the rocket program. The Reichsführer-SS had arrived at his nearby Hochwald (High Forest) headquarters on August 15 and thus could offer his solution to Hitler when he went to Wolfsschanze on the eighteenth and nineteenth. Eager to gain as much control over the program as he could—in part because he considered the SS the most zealous and competent organization for carrying out the Führer's wishes—Himmler proposed that the A-4 be produced underground "with the increased use of workers from his concentration camps." Arguing that spies must have betrayed the location of Peenemünde, the Reichsführer-SS told an enthusiastic Hitler that using prisoners underground promised complete secrecy, because they could be cut off from the outside world. The SS could also send skilled workers and engineers to the factory from the jails and camps, thus easing the manpower problem. He further convinced the Führer that development work at Peenemünde should be moved to a Waffen-SS training area in Poland.[7]

At Hitler's request, Speer and Saur went to Hochwald on August 20 to discuss the details with Himmler, who had just received a surprise appointment as Minister of the Interior. Himmler followed up the visit with an arrogant missive to the Armaments Minister. It opened: "With this letter I hereby inform you that, in my capacity as Reichsführer-SS, I am taking over the production of the A-4 device in line with our conversation of yesterday." He thought the moment had come to seize a large piece of the ballistic missile program for the SS.[8]

In his discussions and letter, Himmler mentioned the man who would construct the new facilities: *Brigadeführer* (Brigadier General) Dr. Hans Kammler. An architect and civil engineer by training, the forty-two-year-old Kammler had been a party member since the end of 1931 and and an SS member since 1933. He had risen quickly as a construction expert in the Agriculture and Air ministries before being taken full-time into the SS leadership in 1941. There he played a prominent role in the building of the ultrasecret extermination camps and gas chambers of Auschwitz-Birkenau, Maidenek, and Belzec. In early 1942 he became head of Division C (Construction) of the newly created SS Economic and Administrative Main Office. To Dornberger and Speer, Kammler at first sight cut an impressively handsome, energetic, "Nordic" figure, but they would soon learn that his energy and decisiveness barely concealed overweening arrogance and absolute ruthlessness.[9]

While discussions proceeded at the highest level in East Prussia, Dornberger and his subordinates concentrated on restoring Peenemünde's ability to function. To convince the Allies that the center was wrecked, many burned-out buildings were left as they were, while others, like House 4, were repaired in a way that left them looking unusable in reconnaissance photographs. On Saturday, August 21, the first 333 victims were buried, provoking a conflict that brought into the open some of the tensions between conservatives and extremists in the National Socialist ruling elite. The Nazi Gauleiter for Pomerania tried to withdraw all soldiers from the funeral rites when the Catholic and Protestant pastors began to preach but was overruled by the Commanding Admiral for the Baltic. Dornberger later claimed credit as well for having permitted a church service to go ahead over the objections of the Gauleiter, who departed in a huff.[10]

With the initial postraid tasks finished, the leadership of the rocket program considered its next moves. On August 22 Dornberger merged the Production Plant into the Development Works and shunted Schubert into the repair of Peenemünde; it was clear that A-4 production would be evacuated to a less vulnerable location. The next day one of his staff officers sent a memorandum to General Fromm asking that

Dornberger be appointed "sole responsible leader *[Führer]*" of the A-4 program, thereby subordinating Degenkolb to the Army. Apparently Dornberger too thought that his moment had come.[11]

At a meeting on August 25 in Degenkolb's Berlin office, however, Dornberger first learned of Hitler's decisions about underground production. Later that day he phoned Peenemünde, precipitating a meeting chaired by von Braun. That gathering discussed the evacuation of production, along with the concentration camp prisoners and some German employees, to subterranean sites in the Saar area of western Germany. It was the first time that von Braun was involved in decision-making about the SS prisoners, although he certainly must have been aware of their role in the program. Prior to the raid decisions about the prisoners had been handled solely on the production side, through Rudolph to Schubert and Dornberger.[12]

It was only on the following day, August 26, that the new underground site was selected. Speer called Saur, Dornberger, Kammler, and Degenkolb into his office for "very secret negotiations." They settled on tunnels near the small city of Nordhausen in Thuringia as the location for the primary A-4 factory. That facility would soon be called the Mittelwerk ("Central Works"), a deliberately vague allusion to its geographical position in Germany.[13]

One of Degenkolb's production planners had first scouted the tunnels about mid-July, after it had become increasingly obvious that missile production might be in danger from the air. Ever since 1936 a government corporation with the innocuous name Economic Research Ltd. (German acronym: "Wifo") had been expanding an old and unprofitable gypsum mine near Nordhausen into a strategic oil and chemical weapons reserve. Two massive tunnels, "A" and "B," each big enough to accommodate dual railroad tracks, were driven into Kohnstein Mountain. Wifo also dug smaller cross tunnels for storage every few dozen meters, forming a ladderlike network. Plans were made to push the tunnel system all the way through from north to south, a distance of about 1.8 kilometers (a little over a mile). But when the Armaments Ministry seized the facility from Göring's Four-Year Plan organization at the end of August, only tunnel B had broken through to the south side, and not all of the cross tunnels were finished. One of the first tasks of the concentration camp prisoners, whom the SS trucked from Buchenwald with amazing speed (the first group arrived on August 28), would be to push the digging and blasting forward.[14]

The choice of the underground factory site was by no means the end of the postraid reorganization; in fact, it was only the beginning. On September 4 General Fromm took Dornberger out of Ordnance and made him an immediate subordinate with the opaque bureaucratic title of Army Commissioner for Special Tasks (BzbV Heer). In that position, Dornberger assumed responsibility for the creation and training of the operational rocket units but gave up direct supervision of Peenemünde-East. Fromm's order had two purposes: to strengthen Dornberger versus the Armaments Ministry and the SS, and to bolster the rocket chief's position *within* the Army, since Dornberger had very much wanted to be the operational commander of the world's first ballistic missile troops. (In the fall of 1943 the opening of the attack on Britain was still optimistically scheduled for the end of the year). But Dornberger's larger ambition to be the dictator of the A-4 program was not within Fromm's power to grant. As Kammler and Himmler also found out in short order, the central role of Speer and his organization could not be reduced by anyone but Hitler, and that was hardly likely at a time when the Minister's reputation for producing armaments miracles was at its height. After the August 18 air raid, the coalition of forces controlling the missile program had thus shifted to a new equilibrium: The SS had gained considerable influence, but it was still less powerful than either the Armaments Ministry or the Army.[15]

Dornberger's new position necessarily brought other changes. Colonel Leo Zanssen formally resumed his position as Commander of Peenemünde in early October, as had been planned after the failure of the SS plot against him in the spring. The return of Zanssen to Peenemünde (he had actually come back the day after the raid) created an uncomfortable situation, since his betrayer, Lieutenant Colonel Gerhard Stegmaier, was still on base. Dornberger designated Stegmaier as head of a new training school for the rocket troops at Köslin on the Baltic. Zanssen also assumed Dornberger's position as chief of Wa Prüf 11. That double burden proved to be too great, so the Ordnance rocket section was broken up in December 1943. Zanssen kept the liquid-fuel programs as the head of a new section, Wa Prüf 10, but the duties were soon given to another engineering officer, Brigadier General Josef Rossmann. The modest Zanssen received a promotion to general but retained only his Commander's post in mid-1944.[16]

In the meantime Hitler had given Dornberger command over military deployment on October 4, 1943, after the general had once more traveled to the Wolfsschanze. This visit had been occasioned by the destruction of the half-completed Watten firing bunker in two American air raids on August 27 and September 7. The Führer approved its replacement through a redesign of the projected rocket storage dump at nearby Wizernes, despite Dornberger's express opposition to fixed sites. But even in the case of operational command, the longtime rocket specialist was to meet frustration. In November and December Hitler ordered the creation of a special interservice corps, combining the A-4 with the Fi 103 (V-1) and long-range guns to be used against England. The corps commander, General Erich Heinemann, repeatedly clashed with Dornberger over the ballistic missile's immaturity as a weapons system. Heinemann finally succeeded in having him replaced as A-4 tactical commander at the end of December.[17]

Meanwhile, Dornberger was creating a staff headquarters at Schwedt on the Oder River, between Stettin and Berlin, and was moving most A-4 test and training operations to the Waffen-SS exercise area at Blizna. That heavily forested site near the confluence of the Vistula and San rivers, 150 kilometers northeast of Krakow, was a former Polish Army artillery range, which the SS had renamed *Heidelager* (Heather Camp). The plan, as outlined by Kammler, was to fire missiles at isolated regions to the north. In keeping with Nazi racial policy, the SS was concerned only about the accidental loss of German lives or property; the impact on Poles was considered irrelevant. On September 28 Himmler himself went to check on the progress of the launch site.[18]

Heidelager was not, however, destined to become the evacuation site for Peenemünde, as the Reichsführer-SS had planned. Kammler had suggested in the August 26 meeting in Speer's office that camp prisoners excavate an underground site in Austria for the development facility, probably because Blizna was too remote and was inappropriate for a subterranean factory. In mid-September two Peenemünde veterans who had become redundant, "Papa" Riedel and Ministerial Counselor Schubert, were sent off on a survey trip to the Austrian Alps, along with an SS officer. The location they chose, about 100 kilometers east of Salzburg, received its final code name from Kammler in December: *Zement* (Cement). Despite the opposition of Wernher von Braun, who thought the move would cause unnecessary disruption, in late October the Army and the SS gave approval for the complete withdrawal of guided missile development from Peenemünde to the Zement tunnels as early as the first months of 1944. The original rocket site was to become a mere artillery range, where some missiles would be launched for test purposes. By the beginning of 1944, however, the schedule for the project began to slip steadily because of high cost, difficult geological conditions, conflicts between the SS construction office and the Army, and lack of clarity as to what underground facilities would be required.[19]

As von Braun had pointed out in his October 2 memorandum against the Zement concept, Peenemünde had already taken effective measures against crippling damage from another attack. The center had dispersed a significant fraction of its development personnel to locations southeast along Usedom island. Beach hotels, restaurants, and other buildings were commandeered by the Army for the purpose, and the summer vacationers were sent home for good. The valve and materials testing laboratories were evacuated even farther, to an air base at Anklam on the mainland. In October Ordnance decided to move Rudolf Hermann's wind tunnels to Kochel in the Bavarian Alps, where a cooperative Luftwaffe Army hypersonic Mach 10 tunnel was planned. Hermann's group had never been a formal part of the Peenemünde Army center; rather, it had reported directly to Dornberger in Berlin. Now it was made into a independent government company with a misleading cover name.[20]

Combined with the damage left in place for camouflage purposes, this dispersal significantly altered the character of "my beautiful Peenemünde," as Dornberger called it in distress immediately after the bombing. No longer the "sleeping beauty," with almost lavish facilities isolated from the reality of the war (if only for the German personnel), living conditions became more difficult and efficiency suffered. As a result of the move, the wind tunnels were out of commission for nearly a year. Even the local travel needed for Peenemünde engineers and administrators to meet each other was an obstacle, particularly as gasoline shortages steadily worsened. On the other hand, Walter Wiesman, a young airman who worked as an administrator in the valve laboratory, thought the dispersal gave him more freedom to get around needless red tape. All in all, the Ordnance facility returned to being a fully functioning research and development center only one or two months after the raid, in spite of the loss of efficiency.[21]

The most fundamental change to Peenemünde was the evacuation of the Production Plant, which brought with it a decline in the center's size and, to a much lesser extent, its character as an "everything-under-one-roof" facility. At its peak before the raid, the Army center had at least 12,000 employees, foreign laborers, and prisoners. About half that number were in the Development Works; the Plant probably had about two thousand German blue-collar and white-collar workers. To that number must be added the staffs of the Commander and Construction Group Schlempp, at least three thousand foreign laborers and six hundred concentration camp inmates. After a postraid period of confusion, in early September evacuations of machinery and personnel to Thuringia began. Most employees of the Plant, as well as some administrators from the central procurement group, were shipped off to become the office personnel, foremen, and skilled workers of the Mittelwerk. On October 13 the SS prisoners boarded rail cars to the same place. By mid-November 1943 Peenemünde-East had shrunk to 7,278 employees spread over a much wider area, plus about 2,500 forced laborers. The latter's conditions must have been poor, because on October 19 Schubert reported an epidemic of typhus (a lice-borne disease) in the "Polish camp," putting 1,300 out of work. After the completion of repairs and a drastically reduced construction program, most of the foreign workers were withdrawn in the first half of 1944. By the summer the Army center was down to less than six thousand Germans and a few hundred POWs.[22]

Heading the evacuees to the Mittelwerk were Albin Sawatzki as chief planner and Arthur Rudolph as head of production. Since Detmar Stahlknecht's planning group had lost all its office files in the air raid and had never been well integrated into Degenkolb's organization, Stahlknecht left for other employment within the Armaments Ministry. Mittelwerk, which was called into being on September 24 as a government company financed by the Ministry, also received machinery and personnel from the other two A-4 assembly sites. At first there was some discussion of putting together missiles at Zeppelin and Rax, at least until Mittelwerk achieved its planned output of nine hundred a month. The threat and the actuality of further air raids against Friedrichshafen and Wiener Neustadt, however, quickly resulted in the abandonment of those ideas. Rax put together a few center sections (with tanks), but a November 2 raid then provoked the complete evacuation of the plant to Thuringia. Zeppelin remained as the primary center section and tank manufacturer.[23]

The reorganization of the assembly process also necessitated a restructuring of the engine-testing program. The Linke-Hoffmann company in Breslau, Silesia, manufactured the combustion chambers, but the plan had been to construct a liquid oxygen plant and engine-calibration test stands at each A-4 factory. Immediately after the August raids on Peenemünde and Rax, worried officials in Wiener Neustadt commandeered a brewery in central Austria that had underground facilities appropriate to a liquid oxygen plant. The Rax test equipment was moved there, and Kammler took over the construction of new test stands using concentration camp labor. Many Rax prisoners were sent to that facility, code-named *Schlier*, where they worked under terrible conditions twelve hours a day in construction. Prisoners built a similar area south of the Mittelwerk at a quarry near Lehesten on the Thuringian–Bavarian border. At Peenemünde, two test stands continued the mass production calibration of engines. Engine-testing at Zeppelin, however, was shut down shortly after going into operation in late 1943, probably because of the visibility of firings across Lake Constance in Switzerland.[24]

One of those who was intimately involved in these decisions was Wernher von Braun, chairman of the "Final Acceptance" subcommittee under Degenkolb. He flew to Austria twice in the fall of 1943, presumably to visit the Schlier and Zement sites, which were only 25 kilometers apart. He must also have driven to Lehesten on one of his longer stops in Nordhausen, the first of which was on August 30. Because of that responsibility, von Braun became implicated more deeply than ever before in the exploitation of slave labor. In an important November 1 meeting at Mittelwerk, Sawatzki demanded that more Peenemünde employees be transferred to Thuringia. Someone suggested using camp inmates at the test sites instead as a way to free up Germans for employment in the Mittelwerk. Von Braun, who was not present, seized on that idea and after some hectic calculations wrote Degenkolb that, because "you have now given the permission that . . . Schlier and Mitte [Lehesten] can be operated with detainees," only 120 of the 360 people to be employed there need be civilians. The prisoner-to-civilian ratio could not, however, rise above 2:1 "because of the difficulty of the testing processes to be carried out there." As two Austrian historians have noted, "considerations of a humanitarian nature never appear in the documents"—not in this case or in any other. Of course, the open advocacy of such considerations was ideologically unacceptable and even dangerous in the Nazi police state. Even so, it appears that the prisoners were only factors of production to von Braun and his engineers. The Peenemünders were under extreme pressure to produce the new "wonder weapon" quickly and many were believers in the regime anyway. Not surprisingly, they dismissed the treatment and fate of these unfortunates as someone else's problem.[25]

Nowhere, of course, were detainees needed more than in Mittelwerk, nor was any problem more critical to the restoration of the A-4 production program than the creation of the underground factory. As German workers, slave laborers, and machinery poured into the region, Sawatzki, Rudolph, and the SS construction staff struggled to turn a half-completed oil, gasoline, and poison-gas storage dump into a modern armaments factory. Workers and prisoners dismantled the huge petroleum tanks and loaded them on rail cars, while digging, blasting, and concrete-pouring operations continued deeper into the mountain. The first twenty cross tunnels, beginning with tunnel 0 near the north entrance, had cement floors, lighting, heating, and air conditioning. Farther in, conditions were primitive. For office space, the Mittelwerk company took over a school about 5 kilometers away, as well as numerous hotels, restaurants, and other buildings in surrounding towns.[26]

Completion of the legal arrangements for A-4 production lagged behind the extremely rushed *ad hoc* conversion of the factory. Along with Kammler, A-4 Special Committee representative Sawatzki was the real driving force behind putting Mittelwerk into operation, even though he was not even on the board of the company at first. The formal leadership comprised Kurt Kettler and Otto Bersch, two business executives on assignment from the national railroad and heavy industry, plus SS-Sturmbannführer (Major) Otto Förschner, the commandant of "Work Camp Dora" in the tunnels, which was a subcamp of Buchenwald. Förschner was on the Mittelwerk board as head of security but knew little about business management and concentrated instead on running his little empire of horror in overall conditions set by Kammler.[27]

In October and November the paperwork began to catch up with the reality. On October 19 Army Ordnance finally issued an A-4 assembly contract to the Mittelwerk company, but only after pressure from the Special Committee and from Heinrich Himmler, who had toured the tunnels four days previously. According to Kammler, Ordnance budget and counterintelligence bureaucrats had held up the contract because of its huge amount: 480 million marks for 12,000 missiles. The contract's unit price—40,000 marks per A-4, not including warhead, guidance equipment, and packing for shipment—was actually intended only as a guideline. Indeed, the company responded a month later with a more realistic estimate: 100,000 marks each for the first thousand A-4s, 90,000 each for the second thousand, and so forth down to a floor of 50,000 marks starting with the 5,001st missile. About the same time (late November), the Mittelwerk company finally concluded a contract to rent the facility from the Wifo for half a million marks a year, but a year later the formal construction contract for the factory still had not been completed by the Army. The Armaments Ministry lent at least 31 million marks for construction, but as the war went on the legal niceties seemed more and more irrelevant.[28]

The real cost of the Mittelwerk must be measured, however, in human lives and suffering. By October there were four thousand prisoners in the tunnels, all male and predominantly Russian, Polish, and French. (None were Jews; in line with its racial ideology, the SS lumped them into a separate category, regardless of nationality, and assigned none to Dora until the summer of 1944.) By the end of November there were perhaps eight thousand prisoners living underground. From the outset the SS had planned a regular barracks camp on the south side of the mountain, near the entrance to tunnel B. But Kammler gave that construction a lower priority, against the objections of the SS camp bureaucracy in Berlin. Armed with special powers from Himmler, he brushed those complaints aside, allegedly stating to his construction staff: "Pay no attention to the human cost. The work must go ahead, and in the shortest possible time."[29]

The results were horrifying. The SS walled off cross tunnels 43 to 46 at the southern, uncompleted end of the factory, and prisoners constructed wooden bunk beds four levels high. Until the bunks were ready, the prisoners had to sleep on straw or bare rock. The damp, dimly lit tunnels never got warmer than about 15° C. (59° F.) and the bunks, which were often shared by two or more individuals on the shift system, became full of lice and filth. Because of the proximity of the hectic, round-the-clock operations in the uncompleted tunnel A, explosions rocked the "sleeping tunnels" and dust filled the air. Jean Michel, a French resistance leader, later described his initiation to work at Dora on October 14, 1943:

> This first day is terrifying. The Kapos [prisoner bosses] and SS drive us on at an infernal speed, shouting and raining blows down on us, threatening us with execution; the demons! The noise bores into the brain and shears the nerves. The demented rhythm lasts for fifteen hours. Arriving at the dormitory . . . we do not even try to reach the bunks. Drunk with exhaustion, we collapse onto the rocks, onto the ground. Behind, the Kapos press us on. Those behind trample over their comrades. Soon, over a thousand despairing men, at the limit of their existence and racked with thirst, lie there hoping for sleep which never comes; for the shouts of the guards, the noise of the machines, the explosions and the ringing of the [locomotive] bell reach them even there.[30]

In bed or on the job, the miserable prisoners were subjected to the brutal whims of guards, Kapos, and block captains, while the ruthless struggle for existence the SS imposed on the prisoners resulted in widespread thievery, cruelty, and rivalry among nationalities. The hygienic conditions were simply catastrophic. The water supply was completely inadequate, and washing facilities were unavailable. There were no proper toilets for the prisoners, so oil drums were cut in half and boards placed over them. Chlorine was spread over the contents every hour, but the health of the whole workforce was endangered, and the stench was terrible. The doors to the main tunnels were therefore left

An A-4 is launched from Test Stand VII in Peenemünde, 1942 or 1943. After suffering numerous technical delays, the program came to depend on a successful flight. *(SI neg. no. 83-13847)*

General Fellgiebel (left), head of Army signals, congratulates Peenemünde-East Commander Colonel Leo Zanssen (center) after the first successful A-4 launch on October 3, 1942. Third from left is Dornberger, followed by von Braun. Second from right is Dr. Rudolf Hermann, head of the wind tunnels, and at far right is diploma engineer Gerhard Reisig, chief of the measurement section. *(SI neg. no. 87-5769)*

The propulsion and test group, headed by Dr. Walter Thiel (second from right), included at least two of the tiny handful of veterans at Peenemünde from the early rocket groups: Kurt Heinisch (second from left) and Helmut Zoike (third from right), both from the Raketenflugplatz. *(SI neg. no. 91-15663)*

After the first A-4 success, Armaments Minister Albert Speer (right, with armband) moved to take over missile production. Here he watches a launch with Propaganda Minister Josef Goebbels (center), who dubbed the missile "Vengeance Weapon 2" (V-2) in 1944. *(Bundesarchiv Koblenz)*

Reichsführer-SS Heinrich Himmler (center right) began to take an interest in the A-4 program in late 1942. During his June 1943 visit to Peenemünde, he is accompanied by General Dornberger (center left) and Development Works chief Lieutenant Colonel Gerhard Stegmaier (far left). Half hidden behind Himmler is a man in a black SS dress uniform—probably Wernher von Braun on the only day he is known to have worn it publicly. *(From V-2 by Walter Dornberger. Copyright © 1952 by Walter Dornberger. Used by permission of Viking Penguin, a division of Penguin Books USA Inc.)*

The Peenemünde housing settlement after the British air raid of August 18, 1943. Two coffins stand on the street corner. *(Deutsches Museum Munich)*

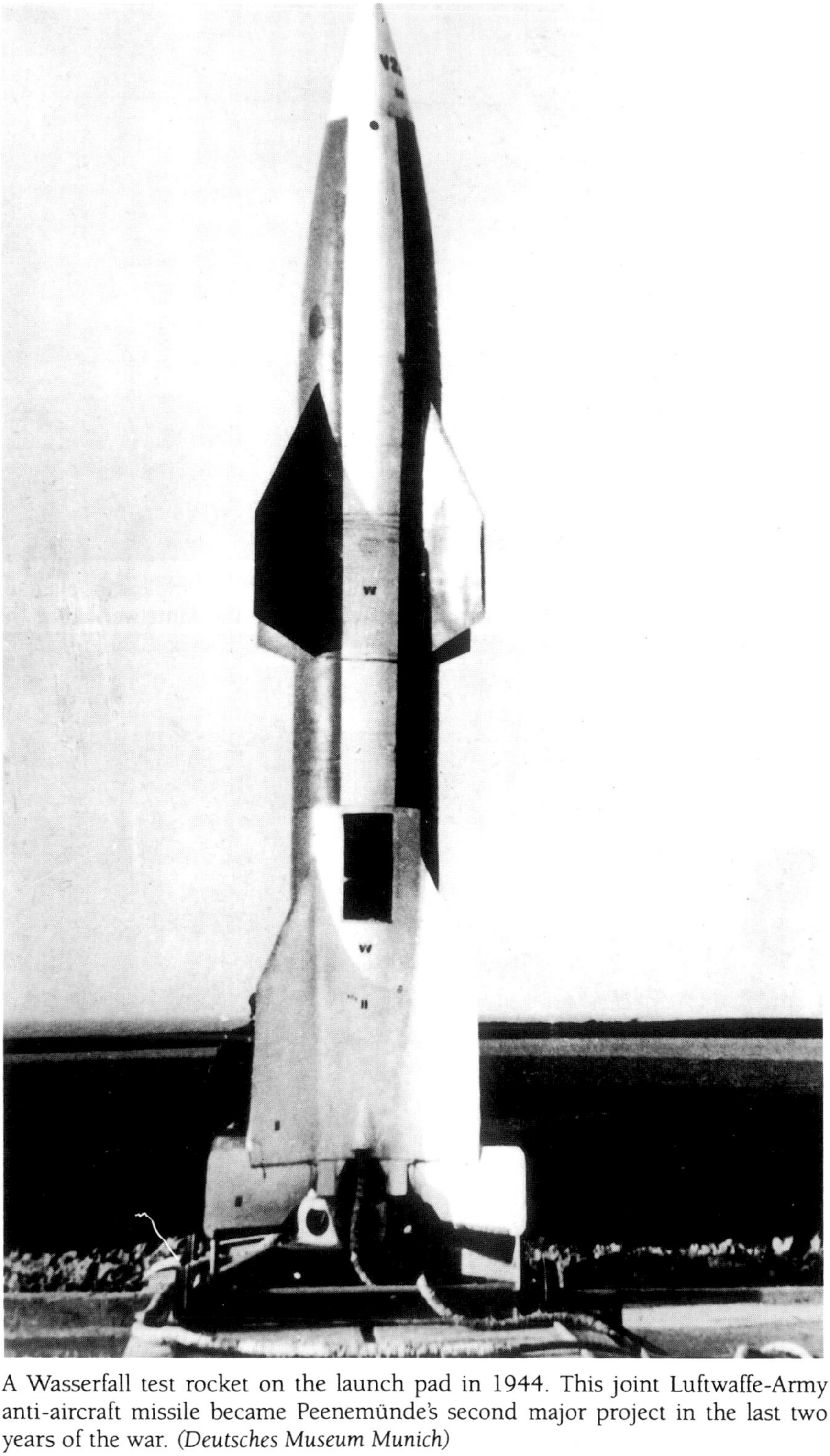

A Wasserfall test rocket on the launch pad in 1944. This joint Luftwaffe-Army anti-aircraft missile became Peenemünde's second major project in the last two years of the war. *(Deutsches Museum Munich)*

The ruthless builder of the Auschwitz gas chambers and the Mittelwerk, SS-General Hans Kammler, became the dominant personality in the Army rocket program after the assassination attempt against Hitler in July 1944. *(Archiv Jost W. Schneider)*

The first A-4b is prepared for launch in late December 1944. Originally called the A-9, the winged A-4 was revived late in the war. *(SI neg. no. 76-7772)*

A new start: Peenemünde's core personnel is reassembled at White Sands, New Mexico, in 1946. Seventh from right in the front row, with his hand in his pocket, is Wernher von Braun. Fourth from left in the front row, in the short white jacket, is Arthur Rudolph, the production manager at Mittelwerk who was forced to leave the United States in 1984 for his role in the use of slave labor. *(SI neg. no. 77-14246)*

In the New Mexico desert, a captured rocket is prepared for a mission to the upper atmosphere in December 1947. The A-4/V-2 became the foundation of guided-missile development by all the major powers after World War II. *(SI neg. no. 80-4734)*

open, producing terrible drafts. The only time most detainees saw the outside world was during the Sunday roll calls, when they often had to stand outside in the cold for hours, clothed only in thin uniforms that were little better than rags. Michel describes the result:

> Some deportees are too weak and collapse. They have dysentery. They foul their trousers. They no longer have the strength to sit over the barrels, even to get to them. The SS beat them. The blows are useless, they do not get up. They will suffer no more. Those who know in their hearts that they too are almost at the end of their tether watch in silence. Will they be the victims tomorrow? Soon? . . . Woe betide the man who is turned away, not ill enough for the infirmary Kapo! His days are numbered. The Kapos in the tunnel will work him until he drops. As if we are not being driven to our deaths already! Often there are no mess-tins. Food is scarce and they cut down the number of rations.[31]

By December, epidemics of dysentery, tuberculosis, and pneumonia were raging through the camp. Whereas Dora recorded 18 deaths in October and 172 in November, in December the figure was 670. From that month through March, an average of twenty to twenty-five prisoners died every day, for an official total of 2,882 in six months. Of these, 29 percent were Soviet, a quarter were French, 14 percent were Polish, 13 percent were German, and Italians, mostly POWs, made up 9 percent. At first the bodies were hauled back to Buchenwald for cremation, but a portable crematorium arrived in January, and a permanent one was opened in the new camp at the end of March. To the official death toll must be added an equal number who were "selected" for three transports in the first quarter of 1944. On January 6 the SS shipped one thousand sick and exhausted detainees to Lublin-Maidenek in Poland, where those who survived the trip undoubtedly died of disease or were gassed. Exactly a month later an identical number was sent to the same destination, and at the end of March, a thousand were entrained for the northwestern camp of Bergen-Belsen (the SS was obsessed with exact numbers). Only the coming of warmer weather, the end of mining, and the movement of the prisoners into the barracks brought a significant decline in the death rate. At the beginning of 1944, 4,500 out of ten thousand still slept in the tunnels. The final dismantling of the underground accommodations did not come until May.[32]

Long after the war Speer claimed credit for the improvement of conditions in the spring, even as he admitted, in his usual self-serving manner, responsibility for the horrors of Dora. His account revolves around his visit to Mittelwerk in December 1943 and subsequent efforts to upgrade prisoner health. Speer's office chronicle reads: "On the morning of December 10, the Minister traveled to see a new factory in the Harz [Mountains]. The accomplishment of this monumental task demanded the managers' last reserves of strength. Some were so affected, that they had to be forced to take vacations to rest their nerves." Speer's memoirs quote this passage to indicate that prisoner conditions so shocked the Minister's entourage that some had to be given leave. Actually, the chronicle clearly refers to officials on the spot who were exhausted by the overwhelming pace of the conversion. Only a week after the visit, Speer wrote to Kammler praising him for turning the tunnels into a factory in two months, an accomplishment "that far exceeds anything ever done in Europe and is unsurpassed even by American standards."[33]

Only on January 13 does his chronicle mention a meeting with the Ministry's chief physician, "who painted the health situation in the Mittelwerk in the blackest colors." The next day another of Speer's assistants began an investigation, which may indeed have led to minor improvements. In his memoirs and in his later badly organized and unreliable book about the SS, he asserted that he had been responsible for the construction of the barracks at Dora and the consequent lowering of the death rate. Yet the camp was already under construction at the time of his visit. Speer's dishonesty here is reminiscent of one of the more famous stories in his memoirs: that he received only a vague warning about horrendous activities at an unnamed "concentration camp in Upper Silesia" in mid-1944. Actually, by then he had long corresponded with Himmler about Auschwitz and had been active in the deportation of the Berlin Jews.[34]

Speer and the Armaments Ministry do not bear sole responsibility for the criminal enterprise that was Dora/Mittelwerk. Without a doubt, the greatest portion of the blame lies with Kammler, Himmler, and the SS for designing the concentration camp system and for forcing the pace of conversion. But Sawatzki too rates his share for pushing the construction of the factory so hard. He was also feared in the tunnels for personally beating and kicking prisoners he considered lazy—in fact one French survivor's memoir symbolically uses his name for the company. At a further remove, Degenkolb, Dornberger, and their organizations all drove the A-4 program forward without regard for the cost to the detainees of Dora, not to mention those in Schlier, Lehesten, Zement, Friedrichshafen, and other locations. It is all too easy to forget that this brutal exploitation was not confined to Mittelwerk. Once mining operations began at Zement in late November 1943, many of the same horrors were repeated on a smaller scale, although without the peculiar refinement of an underground camp.[35]

With the sole exception of the new Austrian development facility, however, by the end of December the enormous sacrifice of the prisoners had brought the reorganization of the rocket program nearly to conclusion. On New Year's Eve Arthur Rudolph, who was on the second level of Mittelwerk management as production director, was called out of a party to straighten out the loading on railcars of the first four or five A-4s. Sawatzki wanted to make a symbolic show of some production by year's end. Those missiles were actually so unsatisfactory that they were returned to the tunnels for repairs within days, but their roll-out signaled the accomplishment of a remarkable, if callously executed, feat: In only four months the Army, the Armaments Ministry, and the SS had evacuated A-4 assembly from its planned locations and had initiated it in the bowels of the earth.[36]

THE ARREST OF VON BRAUN

When Speer sent his words of praise to Kammler on December 17, he promised to repeat his sentiments to the Reichsführer-SS. He may or may not have done so, but five days later the Minister wrote to Himmler indicating that he had given further construction projects to Kammler "because in this manner I have been promised additional laborers through the provision of prisoners." The manner in which the letter was phrased was every bit as cold and calculating as Himmler's letter to Speer immediately after the Peenemünde raid. Speer transparently wished to signal that, in the case of the rocket construction projects,

Kammler worked for the Armaments Ministry, not the reverse. That message no doubt irked Himmler, whose designs on the A-4 program had once again been partially thwarted. In the first months of 1944 he spun a new web of conspiracy against Peenemünde that led to one of the most curious episodes in the entire history of the Army missile project: the arrest of Wernher von Braun, Klaus Riedel, and others by the Gestapo.[37]

Because Himmler's file on this matter appears not to have survived the war, even the dates of key events are unclear. The arrests must have occurred on or about March 22, but von Braun's crucial meeting with Himmler beforehand is extremely difficult to pin down. The earliest complete account of the incident comes from the rocket engineer's manuscript article of the late 1940s:

> One day in February, 1944, I received a phone call to report without delay to SS Chief Himmler's headquarters in East Prussia. I must confess that I felt a bit jittery when I was shown into his office, but he greeted me politely and conveyed rather the impression of a country grammar school teacher than that horrible man who was said to wade knee-deep in blood.
>
> "I trust you realize that your A-4 rocket has ceased to be an engineer's toy," he spoke up, "and that the German people are eagerly waiting for it. I can well imagine what a pitiful position you are in: a poor inventor enmeshed by Army bureaucracy! Why don't you come to us? You know that the Fuehrer's door is open to me at any time, don't you? I shall be in a much better position to help you lick the remaining difficulties than that clumsy Army machine!"
>
> I replied coolly that in General Dornberger I had the best chief I could wish to have, and that it was technical trouble and not red tape that was holding things up. I ventured to compare the A-4 with a little flower that needs sunshine, fertile soil and some gardener's tending—and said that by pouring a big jet of liquid manure on that little flower, in order to have it grow faster, he might kill it.

According to von Braun, he was soon "politely dismissed," although in a later, even more embroidered version, Himmler was said to have given a sardonic smile while looking angry under a veneer of politeness.[38]

Von Braun's manuscript states that the incident took place three weeks before the arrest, yet in his 1947 affidavit about his political record he specified eight weeks. Furthermore, the meeting could not have taken place on the evening of February 21, as two popular histories have asserted, because von Braun's pilot log shows him flying back from Nordhausen to Peenemünde precisely at the time he was supposed to be heading east. Himmler's desk calendar entries are in fact silent about such a meeting, and there is no unambiguous entry in von Braun's log for a flight to East Prussia after October 1943. In short, there is no evidence that the meeting even happened except for von Braun's word, but it would have been out of character for him to have wholly invented such a story. All one can say is that he probably saw Himmler during one of the SS leader's periods of residency in Hochwald: February 7–11 or 17–22.[39]

The motives and sequence of events behind the Reichsführer-SS's actions are easier to surmise but equally hard to document. Himmler clearly hoped to use von Braun to seize control over Peenemünde and Army missile development. It must not be forgotten that the young Technical Director was an SS officer; Himmler probably ordered von Braun to appear in that capacity, otherwise the request should have gone through the Army. The Reichsführer-SS doubtless hoped that von Braun would be amenable to a conspiracy to place the rocket center under the control of Kammler, whom Himmler had promoted to Gruppenführer (Major General) in January.[40]

The conditions were right for such a conspiracy by Himmler. The technical and production problems delaying the A 4's military use were angering the Führer and putting the already weakened Army in an uncomfortable position. Moreover, Speer had been hospitalized since January 18 with a severe knee infection that escalated into a life-threatening illness. The Minister had attempted to run his armaments empire from a hospital bed; on February 10–11 he suddenly collapsed from exhaustion, fever, and a pulmonary embolism. Himmler used the opportunity to plot with Göring and others to undermine Speer's influence in a number of sectors.[41]

How much faith the Reichsführer-SS placed in von Braun's willingness to cooperate in a conspiracy is questionable, however. Since at least mid-October 1943 the leading Peenemünde engineers had been under surveillance by agents of the SD (the SS intelligence service). According to notes taken on March 8, 1944, by General Alfred Jodl, Field Marshal Keitel's deputy in the OKW, von Braun was guilty of making statements that were treasonous—at least from the viewpoint of the Nazi police state. He and two friends, Klaus Riedel, head of ground equipment for A-4 deployment, and Helmut Gröttrup, Steinhoff's liaison to the Dornberger staff, had allegedly stated that their "main task was to create a spaceship." Moreover, Jodl claimed, Riedel had called the A-4 an "instrument of murder," and the three had made "comments about the war turning out badly." Considering that his source was the SD, those charges must be taken with a grain of salt, but they certainly must have perturbed Himmler. Thus it is likely that, when the Reichsführer-SS invited von Braun to his headquarters, he saw it as the rocket engineer's last chance for self-redemption before the trap was sprung.[42]

Once von Braun had refused to take the bait, Himmler moved to prepare the political ground for the arrest of the three. His chances in this matter were improved by the absence from headquarters of the A-4's strongest defender, Speer, and by a sense of crisis in the German leadership as a result of heavy American bomber raids in late February. On March 5 Hitler asked the Minister's deputy, Saur, to reexamine the A-4's drain on productive resources in view of the pressing need for more fighter aircraft. Also influencing the Führer's attitude was the continuing delay in the missile's deployment and the cheapness of the Luftwaffe's alternative (the V-1). Only days earlier Saur had been put in charge of a new Fighter Staff, which included Milch and Kammler as members, with the aim of greatly increasing production in that sector. For the SS and Kammler, the Fighter Staff meant a significant gain in influence through the expansion of underground factory construction by slave labor.[43]

It was in that context that the SD apprised General Jodl of the evidence against the rocket specialists on March 8, including claims that Klaus Riedel and Helmut Gröttrup had been members of leftist or pro-Soviet organizations before 1933. In the case of Riedel, there is some verification for this charge. Rolf Engel, an original member of the Raketenflugplatz and later an SS officer, remembers him as an idealistic fellow traveler of the Communist Party in the heyday of the Berlin group. The SD report also tied von Braun into what Jodl called a "refined Communist cell" by noting that the three were close friends and that the Technical Director was "very friendly with Mrs. Gröttrup." Himmler must then have discussed the charges with Hitler in one of their many conferences at Führer headquarters, which had been moved to Berchtesgaden on the Bavarian–Austrian border at the end of February. The Reichsführer-SS had earlier met Hans Kammler in East Prussia on February 18 and now saw him again on March 6. Moreover, on more than one occasion in mid-March he separately met his deputies Gottlob Berger, who had been involved in the Zanssen affair, and Ernst Kaltenbrunner, head of the Reich Security Main Office.[44]

Shortly thereafter Kaltenbrunner's secret police arm, the Gestapo, struck. At 2:00 A.M. one morning, three agents appeared at the door of von Braun's apartment and hauled him into "protective custody" in Stettin. The same day or a short time later, Riedel and Gröttrup were arrested, as was von Braun's younger brother Magnus, a chemical engineer and Luftwaffe pilot who had been called to Peenemünde in 1943 to work on the Wasserfall program. The exact sequence of events may never be clarified, but von Braun must have been arrested between March 19, the beginning of an eighteen-day gap in his pilot log, and March 23, his thirty-second birthday, which he claims to have spent in jail. The most likely date is the March 22, since Gröttrup attended a meeting with Dornberger the previous day at the general's headquarters in Schwedt.[45]

Early on the morning of the arrests, Dornberger asserts, he was awakened by a phone call from Führer headquarters asking him to come at once to an urgent conference with Field Marshal Keitel. He and his driver spent most of the day crossing the Reich from the northeast to the Bavarian Alps. Arriving in Berchtesgaden rather late, he finally learned what had happened but received no explanation. The next morning he went to Keitel's office. The chief of OKW allegedly told him: "The charges were so serious that arrest was bound to follow. The men are likely to lose their lives. How people in their position can indulge in such talk passes my understanding." By his own testimony, Dornberger put his career on the line to defend the reputation of the arrestees. Von Braun and Riedel in particular, he argued, were absolutely essential to the resolution of the technical problems that were holding up military deployment. Keitel refused to act, however, pathetically claiming that he could not appear less zealous than the SS because he was the last voice of the Army officer corps in Hitler's inner circle. At Dornberger's behest, the Field Marshal then called Himmler's headquarters nearby, but the Reichsführer-SS refused to see Dornberger, referring him to Kaltenbrunner in Berlin. The general went back to Schwedt in a rage.[46]

The following day he went to the bomb-damaged SS headquarters in the center of the capital, where he was referred to Gestapo chief Heinrich Müller. Kaltenbrunner was away (he was with Himmler again at midnight on March 23). According to Dornberger's memoirs, Müller was extremely cold and even threatened the general with the fat file the Gestapo had on his responsibility for holding up A-4 development. If that indeed happened, it is another indication of how dangerous the Third Reich was becoming as it approached its end. Dornberger left without any resolution, but working through Abwehr (Military Intelligence, which had recently been put under SD control), he was finally able to secure from Führer headquarters an order for von Braun's conditional release. It was undoubtedly helpful that Speer had interceded with Hitler when the Führer came to visit him on March 19 or 23—probably the latter. The Armaments Minister had been briefly quartered near Berchtesgaden while on his way to recuperate from his dangerous illness in an Italian mountain resort and had seen the Führer then.[47]

Von Braun, meanwhile, languished in jail for nearly two weeks without the slightest indication of the charges against him and with no contact with the others. Dornberger appeared, he later claimed, just as he was being interrogated by a group of Gestapo officers. They had accused von Braun of sabotaging the A-4 project by concentrating more effort on spaceflight than on his duties, as well as of having a plane ready to fly to England with the plans for the missile, which was impossible to prove or disprove. Armed with the order from Führer headquarters, Dornberger was able to free von Braun for a preliminary period of three months. After Speer's return from Italy, Hitler grumbled "about the trouble he had gone to" in this case but promised the Minister in mid-May that, "as long as [von Braun] is indispensable to me [Speer], he will be exempted from any punishment, however serious the resulting general consequences might be." The others were conditionally released not long after their boss. In July the orders were renewed; in early August, Klaus Riedel was killed in a car crash near Peenemünde. Gröttrup may have remained under a form of limited house arrest until the end of the war, but von Braun's case was eventually dropped altogether.[48]

Once again Himmler's attempt to take over the Army rocket program had failed, but the whole affair certainly had an intimidating effect. For von Braun, however, it proved to be one of the most fortunate things that ever happened to him in the Third Reich. After the war his defenders were able to credit him with an anti-Nazi record that he never had. Moreover, the nominal grounds for the arrest fitted perfectly with the image that he and his group wanted to project in the United States, especially after Sputnik. Von Braun's "team" were, so the mythology later ran, apolitical space enthusiasts from the Weimar rocket groups who were forced to make a detour through military development in order to reach the stars. It is certainly true that spaceflight played a role in the thinking of Peenemünde; von Braun and a few close friends toyed with the idea in their spare time, and Dornberger used it to promote the *élan* of the group as the founders of a radical new technology. The evidence also indicates that von Braun expressed real regret during the war that the rocket had to be developed first through military funding, a statement that was risky, as the arrests proved. But he was an opportunist who had no overriding moral qualms about building missiles for the Third Reich, even when slave labor became involved; the same goes for almost everyone else at Peenemünde, insofar as they had any choice in the matter, which most of them did not. As for the spaceflight origins of the group, the center's engineering leadership had been largely recruited or drafted after 1938 and had little or no previous exposure to rocketry. The motivations of those later recruits ranged from avoiding the Eastern Front to working in a technically exciting field to making a contribution to the war effort. Spaceflight was not central to their concerns, even if they later became fascinated by it.[49]

Although the arrests of von Braun, his brother, and his collaborators would later be used as evidence that they were just apolitical or anti-Nazi engineers punished for their enthusiasm for spaceflight, Himmler in fact had orchestrated the affair in order to strike back at von Braun and to install the SS as the supreme power in the rocket program. For the moment Himmler's ambitions had again been frustrated by the Army and Speer, but within a few months new opportunities would present themselves. In the meantime the Army rocket group had to continue wrestling with two difficult challenges: manufacturing and deploying a balky A-4 and completing the development of Wasserfall.

NEW NIGHTMARES

The start of A-4 manufacturing in the Mittelwerk had by no means signaled the end of the ballistic missile's production and reliability problems, as the political troubles of March indicated. In fact, a whole new set of difficulties had cropped up once A-4 launches began in Poland in November 1943. Of the eight Peenemünde-built missiles launched from Heidelager by early December, only one had successfully reached the target area, and it had broken up during the last phase of reentry into the atmosphere. That "airburst" phenomenon was dismissed as a consequence of rough transport conditions breaking screws on the fuselage structure, but it was actually the first hint of a technical problem that would plague Peenemünde for the next year. At the same time the production of both missiles and ground equipment remained stuck in a morass of difficulties that once again delayed deployment for months. The A-4 was still too technically immature to be a reliable weapons system.[50]

Among the nastiest problems found in Poland was a pattern of tail explosions and engine cutoffs shortly after launch. The missiles sometimes fell back on the mobile launch set, destroying hard to replace equipment and endangering the lives of the troops in training. Missiles also frequently went awry during launch because of components that were poorly made or were oversensitive to handling and transport in the field. By the spring of 1944, however, the launch problems had been mostly solved during tests at Blizna and at Peenemünde, which had begun firing again in late 1943.[51]

It was the spontaneous breakup of up to 70 percent of the incoming missiles a few thousand meters over the target that remained particularly vexing. Prior to the RAF raid, all A-4s had been fired along the

Baltic coast from Usedom; dye markers colored the sea where the missile impacted. There had been no obvious indication from those firings, or from the handful of A-4s that had strayed off course and hit land, that the missiles had not come down in one piece. In April, after five months of firing in Poland, the causes were still undetermined. Two reports, one by von Braun and one by an officer on Dornberger's staff, noted that the possible theories included the overpressurization of the fuel or oxidizer tanks, the ignition of leftover propellants within the tanks, the loss of metal skin off the fuselage through to reentry heating and aerodynamic forces, or the loss of the tail fins due to the same causes, leading to disintegration of the missile.[52]

It proved to be extremely difficult to find out what was going on. The telemetry system could send a only handful of measurements to the ground, a more advanced system was mired in development problems, and no telemetry was available at all in Heidelager. The fragments that fell to the ground were silent as to the cause. Visually observing the airbursts was exceedingly difficult, because the missiles arrived at a velocity of 2,400 kilometers per hour (1,500 mph), but desperation drove the rocket specialists to try to accomplish that feat anyway. Sometime in May or June 1944, Dornberger and von Braun set up camp at the very center of the target zone in Poland, on the theory that no missile was likely to be perfectly accurate. Peenemünde's Technical Director "was only 300 feet [90 meters] from the impact of one live [armed] missile. I was standing in an open field and, knowing the accurate launching time from a warning sign displayed from an observation tower, I beheld the rocket coming out of the blue sky. I threw myself down . . . but a moment later a terrific explosion hurled me high into the air. I landed in a ditch and noted with some amazement that I . . . had not suffered as much as a scratch." During those trials, Dornberger did witness one airburst through his binoculars. The warhead and instrument compartment came down in one piece, but what caused the disintegration of the main body remained as obscure as ever. In June Peenemünde began vertical launches from the Greifswalder Oie, so that the missiles could be tracked and visually observed on reentry. Those launches sent A-4s as high as 176 kilometers (109 miles) into space, but observations again proved inconclusive.[53]

In the absence of definitive evidence and under heavy political pressure to produce results, the rocket group tried many solutions. An October 1944 report, based on dozens of launches in Poland, listed eleven measures that had no statistically significant effect on the rate of airbursts, eight that appeared to make matters worse, and ten that had a positive outcome. Yet the author could only draw preliminary conclusions as to the cause, in part because no solution produced dramatic results and in part because of the difficulty of sorting out causation when two or more measures were tried simultaneously.[54]

Earlier in 1944 an excessive rise in the alcohol tank pressure had been the favorite explanation of von Braun's group. As a result the tank ventilation system was modified. Eventually a number of missiles were launched in such a way as to exhaust the alcohol supply completely, which showed that this tank was not the culprit. Around June, General Rossmann suggested glass wool insulation between the tanks and the fuselage skin to prevent overheating of leftover propellants. That measure seemed to have a dramatic effect in the early firings and was incorporated into missiles intended for the front. Over a long run, however, glass wool did not make much of a difference. The best results were eventually obtained through the reinforcement of the center section of the A-4, in particular through the welding of a steel collar around the forward end. It appears that heat-weakened skin panels were torn off by aerodynamic forces close to the ground, leading to the breakup of the missile. Because all wind tunnel measurements and structural calculations seemed to rule out that explanation, it had been too long discarded in favor of an internal fault.[55]

Another irony of the airburst problem was that the A-4's warhead often came down successfully, just as Dornberger had witnessed. Why then did the Army rocketeers devote so much effort to eliminating missile breakup? One reason is that they had so long emphasized the added explosive effect produced by the supersonic impact of the 4-ton rocket body that they were trapped by their own propaganda. While von Braun did indicate in early April 1944 that the best solution might be a separable nose cone, as became standard in postwar nuclear missiles, nothing ever came of his suggestion, in part because of earlier promises. But it is also true that it was rather late to introduce a major design change into the production process.[56]

Because the warhead could not be separated, the danger that airbursts would cause it to explode forced the employment of a less sensitive fuse on operational A-4s (V-2s), at least until the airburst problem could be reduced. As artillery expert Adolf Hitler had pointed out to Wernher von Braun in July 1943, the A-4 needed a very sensitive fuse or it would penetrate too deeply before triggering and would "throw up a lot of dirt"—that is, explosive energy would be wasted. In the wake of the airbursts, that scenario came true, further reducing the military effectiveness of a weapon that made little sense to begin with.[57]

How long the airburst problem actually delayed the operational debut of the A-4 is another question, but the delay does not appear to have been more than two or three extra months. The V-2 campaign began before Peenemünde had secured a significant reduction in the rate of airborne explosions; in any case, there were many other problems holding up deployment in the first half of 1944. The provision of a sufficient quantity of ground equipment for the mobile units proved to be a particular headache. After the Peenemünde raid, the modification and wiring of the special vehicles had been moved to unused railroad tunnels west of the Rhine. Because that facility was in a wine-producing region, it was code-named Rebstock (Grape Vine); because it was highly specialized and secret, it remained a branch of Peenemünde. By one account, a subcamp of the Natzweiler concentration camp was set up at Rebstock to supply slave labor for either construction or production. In any case, the special vehicles emerged more slowly than expected, probably because of manpower shortages, delays in finishing construction at Rebstock, and problems with parts production in the overstrained electrical sector. The result was further delays in the outfitting and training of the operational rocket units.[58]

Of greatest import for lagging deployment, however, were the delays in A-4 production. The air raids set final assembly back four months; the fifty missiles shipped from Mittelwerk in January 1944 about equaled the likely output of the Peenemünde Production Plant in September. But the problems did not end there. The raid on the rocket center had destroyed the test models of the electrical and guidance devices, producing even more disruption in that sector. Moreover, the first few dozen A-4s produced underground proved to be so riddled with leaks, bad welds, poor connections, and faulty parts that they had to completely overhauled in the Peenemünde shops or at the DEMAG company's Falkensee facility outside Berlin. DEMAG did the installation of electrical components and the final testing until Mittelwerk could take over those operations later in 1944.[59]

The poor quality of the first production A-4s inevitably set off yet another crisis in the program, coinciding as it did with the rash of launch failures and airbursts. The new difficulties had their roots in old problems: a missile prematurely forced into mass production and a development organization ill-prepared for large-scale manufacturing. Months before, in early September 1943, von Braun had promised the Long-Range Bombardment Commission that "the development of the A-4 is practically concluded." In November Degenkolb had made "bitter complaints" to Speer about statements of that sort from the "development people"—and not without reason. According to Rudolph, a complete parts list for the missile was unavailable until mid-1944. Moreover, the situation in component production remained confused well into that year. After the Peenemünde raid, all the principal sections of the missile were to be subcontracted to companies like Zeppelin, which would in turn manage the relevant component contracts. Whether that reorganization ever really functioned is doubtful, because it conflicted with the concentration of production at Mittelwerk as Allied air raids disrupted suppliers. But even at the outset of the move underground, shifting responsibility to subcontractors proved difficult, because it clashed with the "everything-under-one-roof" origins of the production program and the state of A-4 development. Von Braun's engineers were accustomed to dealing directly with manufacturers and were forced to do so, because the immature state of the A-4's technology and shortages of critical materials imposed constant alterations in the details of various components.[60]

The result was poor-quality parts—at least until mid-1944—and ever changing specifications from Peenemünde. By the end of the war the center had issued about 65,000 changes to the A-4 blueprints, a number that seems high but was consistent with aircraft production. Few of the modifications touched the basic missile design. Although a standard "Series B" missile had been set down even before the air raid, in contrast to the Peenemünde-built test rockets, now designated "Series A," the A-4's gross configuration remained essentially that of 1941. Only in the case of guidance were there significant alterations. The two-gyro "Vertikant system" was unchanged in principle, even when Siemens gradually supplanted Anschütz as the main manufacturer, but Peenemünde had to alter its original plan to put radio cutoff and guide beam equipment on every missile. Difficulties in producing sufficient quantities of both systems and making them transportable by road, plus a growing concern with Allied electronic jamming in 1943–44, resulted in the guide beam's being incorporated on only about 10 percent of Mittelwerk rockets, while the Müller-type gyro accelerometer became the dominant engine cutoff system. One by-product was a further reduction in missile accuracy and effectiveness. During the V-2 campaign, the average error was at least twenty times worse than Dornberger's unrealistic 1936 goal of a less than a kilometer.[61]

The blizzard of drawings and specifications changes therefore involved myriad details, but they were very important details if the A-4 was to ever to be a functional weapon. What the program needed, and what it got only after months of hard work in 1943–44, was a coherent process for incorporating design modifications into manufacturing. The first keystone was "Production Supervision," an organization formed in Peenemünde before the raid and headquartered near Mittelwerk after May 1944. It assigned center engineers to companies to supervise the incorporation of changes and proved essential in straightening out the confusion. Another important innovation was the creation of a "Drawings Change Service" at the Baltic coast center, but it was not functioning until at least the spring of 1944. Finally, the engineers at Peenemünde and Mittelwerk agreed on the incorporation of changes in blocks of missiles at roughly the same time, thereby minimizing the chaotic intervention that had prevailed as a result of Peenemünde's lack of manufacturing experience and the underdeveloped state of the missile.[62]

There was one other important factor in the A-4 manufacturing crisis of 1944: the prisoners. Their murderous conditions during the winter, compounded by their lack of skill and experience, undoubtedly contributed to terrible workmanship in the first Mittelwerk missiles. Conversely, the improvement of their physical condition in the summer helped to make the factory function more efficiently. While six thousand prisoners died or were transported to certain death in the first seven months of operation, from April to October 1944 the toll declined to about a thousand in Dora and the other subcamps that had been created, mostly to provide a labor supply for Kammler's Fighter Staff underground projects. Conditions at the new camps, notably Ellrich and Harzungen, approached Dora at its horrifying worst, while the original camp became much better. The walkways and roll call square were paved, the barracks were finished, and a number of amenities were completed, such as a cinema and a sports field. Those latter facilities were reserved, however, for the Kapos, block captains, and other privileged inmates. In August Dora recorded 52 deaths, less than 7 percent of the 767 dead in March. The lower rate was gained to some extent by dumping weak prisoners on the other camps, but it was mostly a product of the evacuation of the "sleeping tunnels," the end of mining operations in the main plant, the summer weather, and an improvement in the food supply.[63]

Even when conditions in Mittelwerk were at their best, however, they were fundamentally barbaric. On June 22, 1944, the company felt compelled to issue a secret decree to managers: "On the part of the camp doctor . . . it has been repeatedly determined that detainees who work in the offices or on the shop floor have been beaten by company employees because of this or that offense, or even have been stabbed with sharp instruments." The decree went on to remind employees that all infractions were to be reported to the SS, who would deal out the punishment. Attached was a copy of an earlier decree from camp commandant Förschner, forbidding any direct contact with prisoners except for work purposes, on pain of joining those unfortunates in a striped uniform. A handful of such imprisonments is known to have occurred, and a small number of employees risked everything to help the detainees, but the June 22 decree is evidence that Nazi ideology and the example of guards and Kapos had brutalized a significant minority of the German workforce. One can only speculate about the impact of such abuse on production quality, but it seems certain that terror ultimately does not beget competent workmanship, let alone willing cooperation.[64]

It is unlikely that the decree had much effect. Further efforts to improve conditions at Mittelwerk were similarly feeble and were motivated solely by the desire to extract more and better work out of the prisoners. During the war crimes trial held in 1947, Georg Rickhey, a DEMAG executive and the new "General Director" of Mittelwerk as of May 1944, argued in his own defense that he had procured shoes and medical supplies for the prisoners. During his tenure the company also instituted a premium wage system; skilled prisoners could earn small amounts of money that could be spent at the camp canteen. One of the originators of that system may have been Rudolph. The two men were also allegedly among those who discussed a more fundamental change: a move from twelve-hour to eight-hour shifts for the prisoners. That idea went nowhere, because the camp administration rejected it. More slave laborers would have had to be procured, and it would have complicated coordination with the civilian workers, who also worked an eleven-hour day and a six-day week. In any case, the attitude of Sawatzki, Kammler, and the SS was callous indifference to, or sadistic enjoyment of, the sufferings of the detainees. There was very little chance that they would ever have approved a three-shift system, even if anyone had dared to push the measure energetically, which no one did.[65]

Recently defenders of Rudolph, Dornberger, and von Braun have claimed that those three men did confront members of the SS with the need for better conditions if manufacturing quality was to improve, and, moreover, that they acted out of humanitarian concern. However, the three never once raised those claims immediately after the war, when they had to explain their records to the American authorities, nor did they ever mention them in their memoirs and public interviews. Their defenders' assertions must be regarded with the greatest skepticism, especially as there is not a single document to back them up. At most, Rudolph, Dornberger, and von Braun argued that missile quality was not going to improve if the labor supply was not in better shape.[66]

In addition, all three were present at a Mittelwerk meeting on May 6, 1944, in which the enslavement of more prisoners was discussed. The meeting had been called by Rickhey immediately after the Army and the Armaments Ministry had installed him as the nominal head of the firm. (The post of General Director had been created allegedly in order "to push back the influence of the SS" and to straighten out administrative confusion.) During the meeting, Sawatzki said that he would ask the SS to enslave 1,800 more skilled French workers in order to meet shortfalls in the Mittelwerk labor supply, a problem caused by the horrific toll of the winter. The meeting minutes indicate that, at a minimum, Dornberger, Rudolph, and von Braun said nothing. Objecting would have been risky, of course, and because von Braun had been conditionally released from a Gestapo jail just a month before, he was clearly in no position to object. Still, nothing in the past behavior of the other two indicates that they had any moral qualms about slave labor.[67]

Von Braun's post-arrest situation makes the evaluation of his responsibility more complex, but there is no doubt that he remained deeply involved with the concentration camps. On August 15, 1944, he wrote to Sawatzki regarding a special laboratory he wanted to set up in the tunnels to check out "ground vehicle test devices." The letter begins:

> During my last visit to the Mittelwerk, you proposed to me that we use the good technical education of detainees available to you and Buchenwald to tackle . . . additional development jobs. You mentioned in particular a detainee working until now in your mixing device quality control, who was a French physics professor and who is especially qualified for the technical direction of such a workshop.
>
> I immediately looked into your proposal by going to Buchenwald, together with Dr. Simon, to seek out more qualified detainees. I have arranged their transfer to the Mittelwerk with Standartenführer [Colonel] Pister [Buchenwald camp commandant], as per your proposal.

Some A-4 electrical parts production had been moved into a factory adjacent to Buchenwald, which is why it was a source of skilled labor. Von Braun ended the letter by asking if Förschner could grant special privileges for the French professor, including the right to wear civilian clothes, "so that his enthusiasm for independent work can thereby be increased." Humanitarian considerations may or may not have entered into that appeal, but it is clear that von Braun's visit to Buchenwald and its commandant further implicated him in the system of slave labor.[68]

In assessing the responsibility of the rocket engineers for the treatment of the prisoners and the impact of working conditions on production, one final issue must be considered: sabotage. On January 8, immediately after the first few missiles were produced, Mittelwerk directors Kettler and Förschner issued a warning. It alleged that "over and over again our installation has been consciously and maliciously damaged through *intrusion, destruction and theft.*" The document ordered the Technical Division (Sawatzki and Rudolph) and the Business Division (Bersch) to investigate measures for the prevention and detection of sabotage. Those two divisions were also responsible for reporting any incidents to the SD staff that Himmler had installed for the security of the ultrasecret works. The almost inevitable outcome of such reports was the hanging of the accused in camp in the most gruesome manner possible; the SS specialized in slow strangulation, which prolonged the agony for minutes.

Civilian workers and managers were denied all access to the prisoner section of the camp or the "sleeping tunnels" and were not forced to watch hangings inside the factory until March 1945, but they surely knew that executions were carried out in such cases, if not the manner of them.[69]

In the case of Rudolph in particular, Sawatzki's secretary testified after the war that sabotage reports did go through the production manager's hands. Although no conclusive documentary evidence has survived to prove that charge, it is consistent with the January 8 order. Another rocket engineer who went to the United States also stated that one of Rudolph's immediate subordinates passed such reports to the SD. (Prisoners have named other deputies of Rudolph as being among those who engaged in brutal beatings.) Such evidence makes it very difficult to believe Rudolph's later assertion that he never saw such a report or heard of actual sabotage in the works, as opposed to accusations of sabotage based on poor workmanship. Survivors have testified numerous times that sabotage, whether carried out by isolated individuals or by resistance groups, was a fact in the Mittelwerk. The difficult question is to what extent such heroic actions contributed to quality control problems. The small Communist-led resistance organization at Dora reached its high point in the fall of 1944, at a time when such problems had been brought under control. It is therefore likely that sabotage played only a secondary, but not a completely insignificant, role in missile failures.[70]

Although the creation of a truly cooperative and skilled workforce was impossible in the conditions of slavery, and many technical issues remained troublesome, by the end of the summer the production problems had largely been solved, and the missile was ready to be deployed. Six more months had been lost, however, including delays caused by the airbursts. In addition, the first half-year's production had been used up in testing, and consistent monthly output totals could not be sustained. After delivering 253 missiles to the Army in April and 437 in May, Mittelwerk shipped only 132 in June and 86 in July. That uneven record reflected not only stoppages because of the airburst question but also political interference due to the failure of the A-4's promoters to deliver on their promises. Kammler's position in the Fighter Staff enabled him to secure the use of the first twenty cross tunnels of the Mittelwerk for the production of aircraft engines by the Junkers company. The Mittelwerk had to compress all its facilities into tunnels 21–46 starting in May or June, disrupting production.[71]

A further problem was that Hitler became infatuated with the V-1, which, after months of delay, was first launched against London on June 13, a week after the Allies landed in Normandy. Although the "buzz-bomb" campaign got off to a weak start, the Luftwaffe soon began firing hundreds of cruise missiles a week across the Channel, with a significant short-term impact on the morale of a war-weary British population. Later that month the delighted Führer told Speer to cut back A-4 production to 150 a month in order to put more resources into V-1 manufacturing. By late August, however, it became clear that this weapon had failed to change the course of the war. The Führer put renewed emphasis and hope on the A-4 (V-2). Afterward, from September 1944 through February 1945, Mittelwerk hit its stride, assembling six hundred to seven hundred missiles a month, or more than twenty a day. In early September the first A-4s were fired in anger. The ballistic missile had finally become a weapon.[72]

WASSERFALL: THE END OF AN ILLUSION

While Peenemünde struggled to make the A-4 work, the center also became increasingly preoccupied with the anti-aircraft missile program. In August 1944, 1,116 employees were working on Wasserfall, nearly a quarter of those in development. Moreover, that figure may not have counted the many top people, starting with Wernher von Braun, who had been giving a significant fraction of their time to the interservice project since early 1943. Although Wasserfall is often treated as a footnote to the A-4 story, it was the second principal task of the Army rocket center in the last two years of the war. It was also a project that showed, perhaps even more clearly than the A-4, how the growing desperation of the Third Reich's leadership led to self-delusion and to the squandering of crucial resources on weapons that had no chance of altering the course of the war.[73]

After Göring's approval of anti-aircraft missile development in September 1942, Ordnance had begun to design a small solid-propellant rocket (the C-1) with a range and maximum altitude of 20 kilometers (65,000 feet), and a larger liquid-fueled one (the C-2) with the same ceiling but a 50-kilometer range. The C-1 was quickly dropped by the Luftwaffe, probably because it overlapped with the work of the Rheinmetall-Borsig firm on a two-stage, solid-propellant anti-aircraft missile called Rheintochter (Rhine Maiden). Peenemünde was left with the C-2, which received the name Wasserfall no later than March 1943. In order to save development time and wind tunnel testing, the Army had always specified an external shape based upon the A-4. In the spring and summer of 1943, the missile emerged from the drawing boards of Ludwig Roth's Projects Office, which was now simply a Wasserfall design bureau. The missile would be 8.9 meters (28.5 feet) high, with a maximum diameter of just under a meter, and it would be powered by an engine of 8 metric tons thrust, about one-third that of the A-4.[74]

There were only two obvious external differences between the Wasserfall and its precursor. First, in order to provide increased maneuverability—a quality unnecessary in a ballistic missile—the anti-aircraft missile received four wings in a cross-wing configuration. As originally laid out by Roth's section, the wings were straight rather than swept back, but Hermann's wind-tunnel group demonstrated by mid-1943 that this design was unsatisfactory: the aerodynamic center of pressure moved toward the tail at supersonic velocities, making the missile too stable and therefore too hard to maneuver. By a combination of perceptive guesswork and trial-and-error testing, the aerodynamicists were able to come up with an effective design before the evacuation of the wind tunnels began in October 1943. The new wings had a backward sweep along their leading edges and were set farther back on the body, with the impressive result that Wasserfall's center of pressure did not wander much in relation to its center of gravity, even as velocity increased from zero to Mach 3. The second noticeable difference between the A-4 and Wasserfall was the much enlarged air vanes on the bottom of the tail fins. Although Wasserfall also had jet vanes in the rocket exhaust, the air vanes would play a much larger role, because the missile would continue maneuvering toward its target even after engine burnout. A ballistic missile, by contrast, coasts unguided after cutoff.[75]

Internally, Wasserfall incorporated a number of lessons from the A-4 program. In order to simplify the missile, permit the long-term storage of propellants in the tanks, and eliminate the problems of manufacturing alcohol and handling liquid oxygen, the engine would use a combination of a nitric acid oxidizer and a hydrocarbon fuel called *Visol* (vinyl ethyl ether). With additives, the two propellants were hypergolic, that is, they ignited on contact, eliminating the need for a pyrotechnic igniter. Wasserfall's engine also had a single injector plate instead of multiple "pots." In addition, the missile's tanks were integral to the main body, avoiding the A-4's structural framework inside the skin. Peenemünde had chosen that design in 1939 out of engineering conservatism; not enough was known at the time about the forces that would be exerted on the ballistic missile's structure. That margin for error proved useful, however, in making the A-4 a weapon that could survive rough handling in the field.[76]

The stresses of military deployment were naturally a consideration for Wasserfall too, but its integral tank structure could more easily bear higher loads, because von Braun's engineers had decided to return to pressure-fed engines. As in the A-5 and earlier rockets, compressed nitrogen would force the propellants into the combustion chamber, making the missile simpler and more capable of prolonged storage in the field; the complicated turbopump/steam-generator system could be eliminated. The number of valves would be reduced as well by employing special bursting membranes that would be blown open when a charge was fired to open the nitrogen tank. In principle, the triggering of the charge would result in the automatic opening of the pipelines to the engine, followed by ignition and liftoff.

The preliminary design phase, through the year 1943, thus went fairly well. Pursuant to it request by von Braun and his engineers, who had learned a lesson from the troubled history of the A-4 and the Peenemünde Production Plant, the missile was also made easier to manufacture by bringing the advice of an experienced firm into the project. At the May 1943 "comparison shoot," Field Marshal Milch announced that Henschel, which had built rocket-assisted glide bombs and was designing its own anti-aircraft missile, the Schmetterling (Butterfly), would act as a consultant. By autumn Henschel designers were able to suggest useful modifications that would facilitate mass production.[77]

The one clear trouble spot was personnel. At the end of 1942 the air force had formed a special unit, the Flak Experimental Center Peenemünde (later Karlshagen), to staff the Wasserfall project and the new test stands being built for it. The center paralleled the Army's Northern Experimental Command in that it could pull Luftwaffe personnel from other units and assign them to Peenemünde-East. This unit, which was not formally connected to the air force's own base at Peenemünde-West, had another dimension as well: Its headquarters staff served as a sort of supervisory office for Wasserfall. The growth of the Flak Experimental Center was slow, however, because other Luftwaffe units obstructed transfers. Moreover, in the summer of 1943 Milch became furious when he found that Luftwaffe personnel assigned to the Army had been pulled into A-4 work. Because of the extreme pressure to produce results, Peenemünde had clearly exploited the interservice project to prop up the ballistic missile program.[78]

Those troubles aside, however, Wasserfall appeared to be proceeding satisfactorily until January 1944, when the missile's optimistic schedule collapsed. At the end of that month Ludwig Roth wrote a memorandum to his boss, design bureau chief Walther Riedel, indicating how unrealistic was the Air Ministry's requirement for the delivery of a complete set of Wasserfall production drawings by April 15. That deadline had been advanced a month on Milch's promise that more draftsmen and designers would be detailed to Roth's office; some indeed were. But even May 15 was utopian. Roth outlined a number of serious development problems, such as difficulties welding the interface between the wings and the tank structure, but the most fundamental were the choice of a fuel and the guidance system's lack of definition.[79]

The propellant problem arose from a lack of capacity in the chemical industry, caused, in all probability, by Allied air raids and the Armaments Ministry's lukewarm support for the anti-aircraft missile program. As a result, there was no adequate supply of Visol for the deployment of hundreds of Wasserfalls and other defensive missiles by late 1944, the expected date. The Visol would have to be mixed with other chemicals, but that implied the expenditure of further research time to find a combination with the same performance. A new fuel combination would also change the density of the propellant and would therefore alter the mixture ratio between it and the nitric acid oxidizer, which itself had been cut with sulfuric acid to lower combustion chamber temperature and prevent engine burnthroughs. An uncertain mixture ratio threw into doubt the relative size of Wasserfall's two main tanks, which necessarily brought into question the entire structural design of the missile. Only in the summer of 1944 was Peenemünde able to settle on a fuel combination that performed well and had the same density as pure Visol, thereby salvaging the existing tank design.[80]

The difficulties with guidance and control were even more fundamental. As Roth noted, even such a basic problem as the creation of a sufficiently powerful control system was unsolved. The need for the missile to maneuver imposed much more demanding requirements on the vane servomotors, which had to exert and resist greater aerodynamic forces than the same systems on the A-4. The program's attempt merely to modify the existing Luftwaffe mass-produced hydraulic servomotors proved to be a failure, and substitute electrical servos were slow in coming. Further development would be required for both types, with the result that it was impossible to say in early 1944 what the final design of the Wasserfall control system would look like, or even which principle the servomotors would use. A final layout of the missile's tail section was therefore impossible.[81]

The situation in guidance toward the target was even worse. From the outset, it was assumed that the final Wasserfall guidance system would be based upon a modification of one of the existing "Giant" radars into a guide beam that would slowly turn the vertically launched missile in the direction of the enemy aircraft and bring it into the target's vicinity. In the autumn of 1942 von Braun still believed, on the basis of information from Luftwaffe and industry experts, that the accuracy and discrimination of those radars would allow the guide beam to direct the missile to the target, whereupon a signal could be sent to trigger the warhead at its nearest approach. He and his informants significantly overestimated the capability of the radars, however, particularly in the case of a bomber stream, where they were unable to distinguish individual aircraft. Von Braun was nevertheless aware that a homing device in the missile would be desirable; by 1943 it became increasingly apparent that it would be a necessity if the missile was ever to come sufficiently close. Yet even the physical principle of a workable homing device was debatable in 1943–44, although infrared (heat-seeking) systems seemed the most promising.[82]

In addition, the insufficient discriminatory capability of radar meant that Wasserfall needed a proximity fuse to trigger the warhead in the likely event of a near miss. Yet no such fuse was available. The United States was able to deploy a such a fuse on anti-aircraft shells in 1944, but only after a massive program to develop a miniature radar set that could withstand the shock of being fired out of an artillery piece. German resources were spread too thin to permit a similar success. Proximity fuse projects, like those to build homing devices, were started too late and were allowed to proliferate without adequate control from the Air Ministry, with the result that none produced a workable device during the war. Even with better management, however, it is unlikely that the Germans could ever have deployed enough proximity fuses or homing devices to affect the military situation even marginally, given their inferiority to the Allies in research and development resources and electronic technology.[83]

Ironically, when Roth gave his pessimistic assessment of the state of guidance development in early 1944, he did not even assume that the preliminary version of Wasserfall would have a homing device, proximity fuse, or automatic guide beam, although all of those features would be needed if the missile was to be truly effective. At that time Roth did not even have sufficient information to lay out the interim guidance system, which was based on a joy-stick operated by a controller on the ground. This system was a modification of the transmitter and receiver developed to direct the Henschel Hs 293 glide bomb against ship targets. The bombardier maneuvered the projectile using a joy-stick in the attacking aircraft. The procedure had scored some significant successes against Allied shipping in the Mediterranean, but directing a supersonic missile against an airplane at distances scarcely visible to the naked eye was a somewhat more difficult problem. Various schemes were outlined in 1943–44, using a telescope or telescopes that would be radar-directed to follow the missile as it ascended and approached the target. The controller could maneuver the missile in his cross hairs, although obviously not at night or on a cloudy day. Some proposed variants also had a radar screen displaying the missile's position vis-à-vis the aircraft, but that scheme was subject to all the limitations of 1940s radar technology. In any case, the joy-stick system was in such an underdeveloped state that in January 1944 Roth did not even have the layout of the missile's various antennae.[84]

Wasserfall's problems did not end there. Earlier in January Milch had once again become infuriated that Luftwaffe personnel in Peenemünde-East were being employed on A-4 work. He demanded that they be returned full-time to Wasserfall within three months. Tension erupted as well over the arrangements for the mass production of the missile. Much to the disappointment of Roth and others at Peenemünde, Henschel always refused to accept the production contract because of its own burdens. The idea of using the Mittelwerk had also been considered, and the A-4 Special Committee had promised in May 1943 to organize the manufacturing process. But Degenkolb did not pay much attention to the project, which is not surprising in view of the A-4's problems. In the spring of 1944 Peenemünde and the Luftwaffe did succeed in interesting Linke-Hoffmann in the prime contract, but they received little help from the Armaments Ministry. Speer and Saur found no reason to support the production of a missile that was so far from deployment, and they soon had the formal power to obstruct such a decision. In June, on Hitler's order, Göring gave control of aircraft manufacturing to the Armaments Ministry. Milch found himself circumvented and resigned as a result.[85]

If the situation for Wasserfall production was bleak in the first half of 1944, the launch schedule was scarcely better. Von Braun had originally wanted to begin launches in late 1944, allowing adequate time for development, but the Air Ministry demanded in early 1943 that the schedule be greatly accelerated in view of the war situation. Before Wasserfall's wing design was even revised, Peenemünde committed itself to component contracts for two missiles to be launched by the end of 1943. Development problems delayed those attempts into early

1944, then the first vehicle was damaged in an engine test. Just as in the case of the A-4, the second test vehicle became the first to be launched. On February 29, 1944, it rose off its launch pad on the Greifswalder Oie. With no guidance except two gyros for stabilization, it had a simple trajectory, but the system nonetheless failed, and the missile tumbled out of control and fell into the sea. The next attempt, with the repaired first vehicle, was no more successful. Because of the inexperience of the launch crew, which the Flak Experimental Center insisted on staffing with its own people, the filming of the March 8 test was bungled; no conclusions could be drawn about the cause of the failure.[86]

Siemens's late delivery of the improved control system for the second Wasserfall configuration, compounded by severe manpower shortages at Peenemünde, postponed the next launch until May 12. The missile went off course after twenty-two seconds, in all probability because of the failure of a vane servomotor. On the fourth attempt, June 8, one of the explosive bolts holding the missile to the mobile launch table did not fire, resulting in a fiasco: the vehicle took off with the stand attached and crashed after nine seconds. During the fifth flight in July, the engine malfunctioned and the vehicle blew up in the air. Meanwhile, because attempts to work with the joy-stick on the May flight had not yielded much useful information, the Army center had launched an A-4 with the relevant equipment on June 13. It was successful for half a minute, then the joy-stick operator lost sight of the missile and it strayed to the left, ultimately ending in an airburst over southern Sweden. That embarrassing failure placed much new information about the A-4 in the hands of the Allies, although the peculiar guidance equipment on board proved to be misleading to intelligence analysts.[87]

So it went—a checkered launch record that is not surprising in light of Peenemünde's previous experience, nor that of later missile development elsewhere. But the project's original schedule and the assumptions behind Wasserfall's guidance development demonstrate how much wishful thinking, impelled by desperation about the war situation, had permeated the anti-aircraft missile program from the outset. In May 1944 the Flak Experimental Center still wanted to launch hundreds of missiles by the end of the year, notwithstanding the bleak short-run prospects. It was hard to admit the truth: that Wasserfall, like the other anti-aircraft missiles, was a promising long-term development project but was illusory as an answer to Allied air attacks during the war. There was no way it was going to be finished on time. Even if it had been, the joy-stick system would have been ineffective and susceptible to Allied electronic jamming. Nonetheless, the program was bureaucratically entrenched in the Army and the Luftwaffe, and the desperation felt by the German authorities ensured that Wasserfall would continue, whether it made any sense or not.[88]

Thus, as the summer of 1944 turned into fall, the anti-aircraft missile program struggled onward, while the ballistic missile troops finally began to move into firing positions in the west. In the meantime, the rapid decline of the Army's power in the Reich had ushered in the final phase in the history of the Ordnance liquid-fuel rocket program: one in which Peenemünde was a civilian corporation and Heinrich Himmler's SS had finally secured the dominance it had long sought.

Chapter 8

Rockets, Inc.

On July 20, 1944, the anti-Nazi resistance tried valiantly, but failed miserably, to overthrow the Third Reich. After the bomb planted by Colonel Claus Count von Stauffenberg narrowly missed killing Hitler in the briefing barracks at the Wolfsschanze, the attempted military coup in Berlin quickly sputtered and died. A little after midnight, the Chief of Army Armaments and Commander of the Replacement Army, General Fritz Fromm, directed the summary execution by firing squad of his chief of staff, Stauffenberg, and three other prominent subordinates. Although the opposition movement had a base among traditional elites such as the officer corps, the aristocracy, and the civil service, it had been centered in Fromm's office because his control over Army units inside the Reich provided the basis for a coup. Fromm himself had played an ambiguous and indecisive role before the uprising, neither denouncing the conspirators nor committing himself to their plans. His summary action had been designed to save his own skin. But it did him no good. Hitler had already named Himmler as Fromm's replacement that afternoon. Not long after the executions, the general was arrested. He languished in jail until March 1945, when the Führer, in a belated act of revenge, had him shot for "cowardice."[1]

The failure of the plot was a further devastating setback for the Army, even though the senior service was predominantly loyal to the Nazi regime. As a wave of arrests swept across Germany, hundreds of officers, often of impeccable Prussian aristocratic background, were caught up in the Gestapo's net. Many were subjected to grotesque show trials and gruesome executions. Among those arrested was the early rocket veteran General Erich Schneider, head of Ordnance Development and Testing since mid-1943. He was fortunate to be released after a month because of Speer's intervention and, one can presume, because of a lack of any evidence to implicate him in the conspiracy.[2]

The arrests were only part of the Army's troubles. Hitler's appointment of the Reichsführer-SS to a leading command position was "a calculated act of humiliation for the officer corps" that also signaled the ascendancy of the SS and the virtual dissolution of the Army leadership as a coherent body. Himmler soon detailed his new duties to Obergruppenführer (Lieutenant General) Hans Jüttner, the head of the SS Leadership Main Office, which acted as a sort of general staff for the Waffen-SS. But Himmler's latest title allowed him to achieve at least one long-cherished objective. He gave Kammler full powers on August 6 to accelerate the deployment of the A-4. It was all Kammler needed to consolidate ultimate authority over that program.[3]

Adjusting to the SS's new power over them was, however, only one of two transitions the rocket engineers had to make in August 1944 as a result of the Army's fall from grace. On the first day of the month the Armaments Ministry also gained further influence when von Braun's development group became a government-owned, civilian corporation, "Electromechanical Industries, Karlshagen, Pomerania." Once again, the balance of power in the program had shifted dramatically: The SS was now strongest and the Army weakest, with Speer's Ministry in a still powerful but eroding position.

THE ROCKET PROGRAM IS REORGANIZED—AGAIN

Peenemünde's conversion so soon after July 20 suggests a sudden improvisation to keep the facility out of the hands of the SS. In fact, the move had been planned since at least late May, although it was indeed a response to the threat posed by Himmler and Kammler. The wonder is that the conversion was accomplished with so little fuss. Not much more than a year previously, a heated confrontation over proposals to privatize the facility had erupted between Dornberger and Ordnance on one side and Degenkolb and associated industrialists on the other. By contrast, there is no evidence of a fight in 1944 over the conversion, or even much evidence of the discussions that must have preceded it.

Dornberger's memoirs only mention laconically that incorporation was "tolerated because the measure would prevent the seizure of Peenemünde by any military or semimilitary organization," by which he apparently meant the SS.[4]

Since Electromechanical Industries would in effect be owned by the Armaments Ministry, the tension between the Ministry and Ordnance clearly must have eased enough since 1943 to allow a fairly painless transition. There were two explanations for this. First, Degenkolb had delegated virtually all his authority to his deputy in the A-4 Special Committee, Heinz Kunze, after the staff was evacuated to Thuringia in early 1944. Thus the locomotive czar's abrasive personality and blatant ambition were no longer a factor. (Degenkolb disappeared altogether in the autumn after making insulting comments about various Nazi leaders. Because of his connections, he ended up in a mental clinic instead of in the hands of the Gestapo. He resurfaced in April 1945 as a Ministry liaison to Kammler.) A more important reason for the Ordnance–Armaments Ministry *rapprochement* was the common threat of the SS, which drove the two closer together. Of course, without the Army's loss of power even before July 20, that service would never have conceded ownership of the heart of its missile development capability.[5]

Besides the decline of the Army and the rise of the SS, one more factor shaped the third major reorganization of the rocket program in a year and a half: Dornberger's battle to salvage what was left of his declining influence. His appointment as a special commissioner under Fromm had marginalized him even within his own service instead of making him the "Führer" of the A-4 program. The commander of LXV Army Corps had succeeded in removing him from tactical control of the rocket batteries. Moreover, Dornberger began to clash with Ordnance, to which he had devoted fifteen years of his life. He was no longer in the chain of command for Peenemünde, yet he continued to exert influence through Zanssen, von Braun, and others. That caused discomfort for Ordnance, which pushed its own candidate, General Rossmann, who came to head the liquid-fuel rocket section, Wa Prüf 10. To add insult to injury, when Speer returned from his convalescence in early May, he issued an order spelling out the division of powers in the A-4 program. The document's primary purpose was to circumscribe the role of the SS, but Speer omitted any mention of Dornberger, either because he had forgotten that the general was independent of Army Ordnance or because he had heard that the continued existence of Dornberger's position (BzbV Heer) was in doubt. On May 31 Dornberger issued an ultimatum to Fromm that was both an indirect response to Speer and an expression of frustration at losing his grip on his life's work.[6]

Dornberger once again demanded that the organizational confusion in the A-4 program be overcome by making him its leader and military commander. This time, however, he ended by claiming that he would go over Fromm's head and appeal directly to Hitler if necessary. "Fromm summoned me. I was reprimanded, threatened with punishment, my honor was impugned by a charge of unsoldierly conduct and cowardly dereliction of duty, all with the objective of inducing me to modify my demands." In the end Fromm did nothing, and Dornberger's threat proved to be empty. Dornberger's memoirs further claim that in July, before the assassination attempt, Himmler pressed OKW chief Keitel to appoint Kammler the leader of the program, but the status quo once again held. But not for long: After July 20 Kammler rapidly acquired the position Dornberger had so long sought.[7]

In the midst of those battles, the corporate conversion of Peenemünde was finalized between Ordnance and the Armaments Ministry. Dornberger's May 31 document mentioned in passing "Director Storch as head of HAP 11 [Peenemünde-East]." Paul Storch, one of the top managers of the giant Siemens electrical engineering firm, indeed became the head of Electromechanical Industries when it came into being on August 1. One may conclude, therefore, that some such move had been under consideration in the late spring. Storch had headed the electrical and guidance equipment subcommittee of the A-4 Special Committee and was thus by no means "practically a stranger to our work," as Dornberger later asserted. On June 28 an unsigned Peenemünde document, "The Tasks of Electromechanical Industries, Ltd.," discussed the objectives of the proposed company. A week later von Braun's deputy for in-house manufacturing, Eberhard Rees, wrote to Storch about the budget. The company's monthly expenditures, including personnel, were projected to be 13 million marks, a sizable sum. In July 89 percent of the materials and procurement costs would be spent on the A-4 and only 11 percent on Wasserfall. Even in December the ballistic missile's continuing development needs were going to eat up half the procurement and personnel budget, a fact that Rees thought Storch might find "strange."[8]

Rees's estimates were based on a total employment of six thousand, but the company actually turned out to be smaller because of the way the facilities were divided with the Army. As of August 19 Electromechanical Industries had 4,863 German employees, plus 379 East European forced laborers and prisoners of war. (There were no concentration camp prisoners, but the Luftwaffe still had an SS camp for Peenemünde-West.) Of the Germans, 3,580 were civilians (618 of them women) and 1,283 (all male) were in the military, including 559 in the Northern Experimental Command and 411 in the Flak Experimental Center. The German staff included 264 graduate engineers and scientists and 590 engineers with lesser training, but well over half the staff were blue-collar workers; Peenemünde thus retained a significant in-house development capability in spite of the slow erosion of its "everything-under-one-roof" approach. Finally, 601 employees were assigned to Production Supervision, the Mittelwerk company, or elsewhere, leaving 4,262 at Karlshagen, a category that included dispersed facilities spread over the island and mainland. Those numbers indicate that roughly a thousand people in Peenemünde must have remained in the direct employ of the Army.[9]

This arrangement, probably the result of a compromise during the negotiations, did not promote the highest efficiency. Electromechanical Industries was only a tenant in Army-owned buildings, the motor pool and aircraft were to be shared between the two, and each retained its own launch equipment, all of which caused friction between the company and the Army. Base administration was left in the hands of the former Commander's office, which became the Karlshagen Test Range, headed by a colonel. The senior officer on base was General Rossmann, whose Wa Prüf 10 also shared the facilities. General Zanssen was sent off to command a solid-rocket brigade on the Western front, because Kammler had refused to work with him, and Ordnance no longer wished to fight that battle.[10]

There were one or two advantages to corporate conversion: Electromechanical Industries could rid itself of some civil service red tape and pay salaries competitive with the private sector. Indeed, von Braun and his chief subordinates received large pay raises as of August 1, although it did not do them much good; virtually everything was rationed anyway. The reorganization also made von Braun Storch's deputy, but the manufacturing shops and the test stands no longer reported to the young rocket engineer directly. Still, not much changed in the way the place operated, since von Braun continued to be the real leader of the group. Storch, for all his familiarity with the guidance and production dimensions of the A-4 program, was a product of a corporate culture very different from that of Peenemünde. His quiet, authoritarian style clashed with the outspoken group fostered by von Braun and Dornberger. Nonetheless, the arrangement worked reasonably well, in part because von Braun applied his usual tact and energy to keeping rocket development going no matter what the difficulties—or costs.[11]

While the leading engineers were carrying out the conversion, Dornberger was fighting to save his career. As soon as Kammler received his new powers from Himmler, he opened a campaign to isolate Dornberger and seize control of his staff. The SS general claimed the right to give direct orders to Dornberger's chief of staff and longtime deputy for liquid-fuel rocketry, Lieutenant Colonel Georg Thom. Dornberger was to be left with only a fraction of his former responsibilities. Kammler doubtless saw him as a rival, and the two had clashed in the spring and summer over the airbursts and other technical troubles holding up A-4 deployment.[12] Moreover, Dornberger was a protégé of the jailed and now universally despised Fromm. Still, Kammler could not make a political charge against him stick. Despite Dornberger's proximity to the July 20 conspirators, there is no evidence that they tried to recruit him, and for good reason. The plotters could not have trusted someone who had so openly declared his enthusiasm for the Third Reich.

As a result of Kammler's offensive, Dornberger was in despair. He drew up a letter asking for a transfer to other duties and was allegedly talked out of sending it only when Wernher von Braun and Ernst Steinhoff came for a Sunday afternoon visit. They argued that he must not abandon them. Dornberger's despair must have been increased by the fact that Thom cooperated with Kammler's attempts to circumvent him; his chief of staff had stabbed him in the back. Relations between the two completely broke down. Meanwhile, the unscrupulous and ruthless Kammler seized control of the operational A-4 batteries at the beginning of September, shut out the LXV Army Corps, and used the SS's power of intimidation to force the OKW's acquiescence. Thus the opening of the A-4 campaign—against London and newly liberated Paris on September 7 and 8—occurred under Kammler's command. Twelve days later the designated tactical commander, Major General Richard Metz, resigned his now meaningless position.[13]

The SS general did not succeed, however, in forcing out Dornberger. On Kammler's order, Thom went to report to Jüttner about the battles over control of the Dornberger staff and the operational rocket batteries. From Thom's account of the September 14 conversation, it appears that Jüttner was himself angered at the rapaciousness of his SS colleague. He thought it "intolerable" that Kammler had forbidden Dornberger any direct approach to Jüttner himself or to Himmler on pain of being shot! Jüttner, as Fromm's *de facto* replacement, refused to let Dornberger drop. Perhaps he saw the usefulness of the rocket general's technical expertise, or perhaps he saw him only as a counterweight to Kammler. In any case, Kammler was obliged to make a truce with Dornberger. Thom left to become chief of staff of the "Vengeance Division" set up by Kammler to control the missile batteries. Dornberger became the SS general's representative at home, responsible for the training of new units and A-4 supply and shipment up to the border of Germany. His position as BzbV Heer was thus effectively restored at the end of September. Kammler and Dornberger no doubt continued to detest each other, but they had struck a deal they could live with, particularly as both were committed to trying to save the Reich by firing as many V-2s (as the Propaganda Ministry soon called them) as possible.[14]

Kammler, with Himmler's backing, inevitably tried to seize control of A-4 production as well. Immediately after Kammler's appointment, Speer shot off a pair of letters addressed to Jüttner but transparently aimed at the Reichsführer-SS. The second dealt specifically with the ballistic missile. Its message—stay off my turf!—was ignored by Kammler, who had his own foothold in production through the SS's role in the Mittelwerk, through his slave-labor construction empire, and through his position on the Armaments Staff under Karl Otto Saur. That body had superseded the Fighter Staff on August 1 after aircraft production had been totally absorbed by the Armaments Ministry.

Speer's position, meanwhile, was in decline because of SS gains and because of the rise of Saur as a rival power center within the Ministry itself. Thus, whatever the formal arrangements, in the fall of 1944 Kammler felt free to intervene directly in A-4 production by giving orders to his ally, Albin Sawatzki, the real power within Mittelwerk. Further indication of Speer's declining influence is given by yet another document outlining the division of powers in the A-4 program. Drafted by Dornberger's staff in consultation with the other parties concerned, it was finally signed by Jüttner on December 31, 1944. Although organizations controlled by the Ministry were included in the negotiations—the Mittelwerk company, Electromechanical Industries, and the Special Committee—a copy was sent to Speer only as an afterthought.[15]

In comparison to the Mittelwerk or the A-4 batteries, Peenemünde felt Kammler's heavy hand much more indirectly. The fortunate coincidence that most of the facility had been incorporated just after July 20 had indeed given von Braun's development engineers greater institutional protection against an SS takeover. Kammler also knew that he needed Peenemünde's technical expertise if he was to make his weapon system work. Moreover, the fact that the A-4 finally was put into action relieved some of the intense political pressure on the program that had built up during the long months of technical difficulties in the first half of 1944.

Even so, it is clear that Wernher von Braun, for one, had still not returned to the good graces of the SS. After the opening of the V-2 campaign, Himmler proposed three names to Hitler for a high noncombat decoration, the Knight's Cross of the War Service Cross: Dornberger, Kunze, and Riedel (Walther Riedel, head of the Peenemünde design bureau). Himmler ignored von Braun, just as he never raised him in SS rank again after the early promotion of June 1943. Only Speer's insistence and his remaining influence with Hitler got von Braun the Knight's Cross. In the end, the young engineer and Georg Rickhey (the nominal General Director of Mittelwerk) were added to the list, while Riedel's name was dropped.[16]

Whether Himmler's distaste for von Braun was reflected in Kammler's behavior is unknown, but the SS general had earlier attacked the Peenemünde technical director as too young and arrogant for his job.

Von Braun quietly returned the contempt. In November Kammler asked for an immediate solution to the extremely difficult problem of determining the impact points on enemy soil of individual V-2s, with the obvious goal of improving accuracy. Von Braun wrote sarcastically in the margin of the document mentioning Kammler's demand: "Trivial! The day after tomorrow!"[17]

Another marginal comment from the same period indicates von Braun's growing skepticism about the war and about the regime's promises of a miraculous reversal of its course. On a Wasserfall report that asserted the importance of the anti-aircraft missile for "the overcoming of enemy air supremacy and therefore for the achievement of victory," he wrote: "Final victory, well, well!" Peenemünde colleagues have also said that he seemed depressed when the A-4 was finally used against people, although his defenders have probably exaggerated his reaction. While there is no solid evidence that slave labor disturbed him much, his arrest and the hopelessness of the war alienated him more and more from the Nazi system as it neared its end. Moreover, von Braun later claimed that his last meeting with Hitler in July 1943 had disillusioned him. The dictator was "suddenly revealed to me as an irreligious man, a man who did not have to answer to a higher power. . . . He was completely unscrupulous." In dealing with Kammler and other fanatics, however, von Braun had to be extremely careful to present a loyal face, because the political atmosphere was so paranoid. It is a wonder that he was not more careful in his marginal notations and in his comments to colleagues. Clearly, his indispensability and the protection of Speer and Dornberger gave him a certain latitude, however small.[18]

Just as the increasingly disastrous war made the political atmosphere at Peenemünde touchier, so too did it make daily life more of a struggle. During the summer of 1944 three successive daylight American heavy bomber raids (on July 18, August 2, and August 25) inflicted extensive damage on Test Stands VII (the original A-4 launch site) and XI (the former Production Plant test stand used for calibrating mass production engines). Peenemünde-West was also attacked for the first time. On the Army side, a few dozen people were killed, but the dispersion of the preceding year had worked; those raids failed to have any lasting impact on the way the facility operated. Nevetheless, the attacks, along with the repeated air raid alarms that occurred throughout the period, slowed development and testing.[19]

The state of the war created numerous other difficulties too. The Allied bomber offensive made gasoline hard to obtain; travel became problematic even between the dispersed facilities of Electromechanical Industries. Liquid oxygen and many other materials became increasingly difficult to acquire as well. For the staff of Peenemünde, rationing became ever tighter in the fall of 1944, while work hours became even longer. In September the sixty-hour week was officially introduced for all employees. Finally, the country's manpower crisis created relentless pressure from the draft authorities to cancel previous exemptions, plus pressure from the Mittelwerk to transfer more workers and managers, which eventually resulted in a nasty feud between the two companies. By January 1, 1945, Electromechanical Industries had shrunk to 4,325 Germans, a loss of more than 10 percent of its staff in less than five months.[20]

In circumstances of decreased efficiency and labor shortages, it is not surprising that the technical work of the center suffered. But development at Peenemünde was distorted even more by the demands of the increasingly catastrophic military situation.

DEVELOPMENT AND DESPERATION

In the first six months of Electromechanical Industries' brief existence, most of its development work fell into two categories: regular projects that suffered only limited eleventh-hour intervention and true desperation projects that expressed the Third Reich's growing flight from reality. But the best example of the distorting effects of the military emergency was a project that fell between the two: the A-9 or A-4b, as the missile was renamed in October to take advantage of the A-4's high-priority ratings. On the one hand, the revival of the winged A-4 was merely a continuation of research halted in October 1942 to concentrate resources on A-4 production and Wasserfall development. On the other hand, as the A-4b, the glider missile began to function primarily as a further justification for the existence of von Braun's engineering team, which was threatened by draft callups and pressure from Kammler for instant results. Improvisation and desperation marred work on the A-4b, which had to be rushed to the launch pad as soon as was feasible.

The A-9 project was first revived in mid-June 1944 on a very small scale. Probably as a result of Air Ministry pressure, Ludwig Roth was ordered to hand over his remaining Luftwaffe people to the Wasserfall detail design group run by the Flak Experimental Center. With the four staff members he had left under his direct supervision, he was to restart work on the A-9. Passing references in documents from the intervening years show that this missile had always been regarded as the next project in the series, presumably because it promised a relatively low-cost way to increase range. The fact that a decline in A-4 development work was on the horizon may also have been a factor in the A-9 decision. Roth immediately complained to von Braun that he could not make worthwile progress with so few people. To make matters worse, Hermann's aerodynamics group was unavailable until October, as it was completing the reconstruction of its wind tunnels in the Bavarian Alps. Nothing much could be done about those problems, however, except for writing a couple of contracts with universities to make trajectory calculations. As a result, the A-9 project moved very slowly in the summer of 1944.[21]

Only after the German position in France collapsed in August did Kammler begin to pressure Peenemünde to find urgent ways to extend the A-4's range. Fortunately for the missile batteries and unfortunately for London, the Allied advance slowed to a crawl in September, leaving German-occupied areas in western Holland that were still within 300 kilometers of the British capital. The heavily bombed bunker at Wizernes had been abandoned during the retreat, as were a number of prepared sites for the mobile batteries, but the rocket troops launched their missiles from completely unprepared areas. As Dornberger had foreseen, this method worked quite well. No one in the High Command could guarantee, however, that a further withdrawal from the Netherlands would not happen. Although more V-2s were eventually fired at the Belgian port of Antwerp than at London, it was important to Hitler to retain the capability to attack Britain directly. His desperate strategic concept hinged upon knocking the British out of the war by terrorizing the war-weary civilian population of the enemy capital. The glider missile could fill the bill, since its projected range of 500 kilometers (310 miles) would permit firing on London from northwestern Germany. Alternatively, if Holland was retained, the A-9 could be used to attack more distant British cities.[22]

From early September, von Braun and Roth therefore ordered shortcuts in the A-9 program to accelerate the launch date. The revival of the subscale A-7 (an A-5 with wings) was abandoned, as were any extensive improvements over the A-4. For the first test models, the swept wing favored in pre-1943 wind tunnel experiments would merely be grafted on to an A-4 fuselage. The biggest unsolved problem was creating air vanes and vane servomotors powerful enough to deal with the increased demands on the control system. As was true with Wasserfall, the enlarged air vanes were the only system available for stabilization and maneuvering after engine burnout. But in the case of the A-9, even launch was a problem, because wind forces on the large wings made roll control difficult. Notwithstanding that difficulty, on October 10 von Braun proposed the construction of a "Bastard" version of the A-4b, as the missile was now called, using an A-4 tail with slightly enlarged air vanes. According to his rationale, that would be a way to investigate problems of the launch and transition through the sound barrier. No attempt would be made to have the vehicle glide upon reentry.[23]

Werner Dahm, an aerodynamicist who worked under Ludwig Roth, interprets those actions politically. Von Braun's aim, he says, was to "keep the [Peenemünde] group together" and to signal to the authorities that "there is something coming." Von Braun was also responding to pressure from above to produce an increase in missile range as soon as possible in view of the war situation. The best way to demonstrate progress to Kammler and others was to launch quickly, even if that violated a more rational development process. No attempt would be made to glide because, as Dahm notes, that would only demonstrate that the "Bastard" A-4b could not sustain stable flight during its descent.[24]

When it came time to launch the improvised vehicles, which were among the last assembled by the Peenemünde shops, the results were not surprising. On December 27, 1944, the first "Bastard" A-4b crashed within seconds as a result of a roll that began at launch. The control system on the missile was simply too marginal to deal with winds or suboptimum performance by guidance equipment. After some emergency improvements, Peenemünde fired the second "Bastard" in late January. It came through the launch phase in good shape, but a wing broke off during reentry. Since gliding was not included in the test, that failure was unimportant, but it too demonstrated how far Peenemünde had to go before it could produce a workable glider missile, let alone one that was militarily useful.[25]

Meanwhile, the Peenemünde engineers, in consultation with Hermann's aerodynamicists, had struggled to find solutions to the serious guidance and control challenges of the final A-4b design. There were doubts as to the stability of the missile in all flight regimes, the best wing and air vane designs remained unclear, and there were questions about creating an accurate guide beam for a missile flying at low altitudes more 400 kilometers away. The low speed of the missile at the end of the glide also raised the possibility of its being shot down like the V-1. Someone in Ordnance suggested adding jet engines, but a much simpler answer was to add a terminal dive similar to the V-1's, at the cost of cutting off range at 450 kilometers instead of 500.[26]

When the evacuation of Peenemünde was ordered only days after the second launch, the program effectively came to an end. No further A-4bs were ever built or launched, nor did any postwar nation pursue the idea further. Although the vision of a winged V-2 did play an influential role among space artists and advocates in the 1950s, the idea was a dead end militarily. The extension in range was not worth the extra technical complexity, nor did it make sense to trade away the great advantage over the defense provided by a ballistic missile's high-speed reentry. Even if we put those questions aside, however, von Braun's engineers had too little time and too few resources to make the A-4b work. Desperation had marred the program, compromising many aspects of the original A-9 idea. But since the A-4b program was as much political as anything else, perhaps it did not matter.

Emergency improvisations had less of an impact on the A-4, because its relative technical maturity put limits on how much further improvement could be extracted in a short period of time. Throughout the summer, fall, and winter, a number of long-term projects continued, with the ultimate objective of producing a simplified and more accurate A-4 "Series C." Near the end of the war the propulsion engineers finally were able to make a workable injector plate for the 25-ton motor. Steinhoff's guidance and control division, in collaboration with university and corporate researchers, pursued a great number of projects, including better accelerometers, improvement of radio guidance systems, and the production and testing of Kreiselgeräte's three-gyro stabilized platforms, the Sg 66 and Sg 70, which were ultimately scheduled to replace the simple but less accurate "Vertikant system." That equipment was tested at both Peenemünde and Heidekraut (Heather), a new launch area southwest of Danzig (Gdansk), which was set up because the Red Army's approach had forced the abandonment of Heidelager in late July 1944. In addition, there were continuing efforts to solve the airbursts and other problems, often on the same missiles.[27]

Although a rational program of long-term improvements did predominate in late-war A-4 work, Kammler's demands for instant results nonetheless made themselves felt. Various methods to extend the missile's range were studied through test launches, ground experiments, and calculations, but most changes either delivered only marginal gains of a few kilometers or entailed development of new components that would take time. Kammler also revived the idea of the railroad-borne rocket battery. Launches were carried out at Peenemünde in late November and early December with road-mobile vehicles mounted on flatcars. Why he was interested in that idea is a mystery, as trains were inherently less flexible and more vulnerable to Allied air attack in any case. The energies of von Braun's engineers and Dornberger's soldiers would have been better spent solving technical problems for the rocket batteries and dealing with endless changes in A-4 production caused by materials shortages, the loss of parts suppliers, and so forth.[28]

Electromechanical Industries' other principal ongoing project, Wasserfall, enjoyed relative freedom from emergency intervention, in part because it was outside Kammler's jurisdiction until February 1945. But desperation and wishful thinking had marred the program from the outset, with the result that the anti-aircraft missile was too technically immature to permit a quick fix of it problems. Although the Flak Experimental Center had overcome its slow start and had begun to fire about half a dozen missiles a month in the fall, most of those tests failed in ways that showed how deeply troubled the program was. Even when missiles completed powered flight, they had an alarming tendency to blow up immediately after engine shutdown, because the self-igniting propellants were thrown forward through pressure-relief valves at cutoff, where they mixed and burned.[29]

Wasserfall's propulsion and tankage system displayed a number of other serious design problems. The injector plate was inefficient, and the bursting membranes in the lines failed to function as planned, causing explosions at ignition. When the engine did operate in flight, thrust was poor, and shutdown came after about thirty seconds (instead of forty-five) because maneuvering threw the propellants around, uncovering the tank drains. The result was that the engine ingested nitrogen pressurization gas along with the oxidizer and fuel, leading to substandard performance and early cutoffs. Obviously, the range and effectiveness of Wasserfall would be severely curtailed if that problem could not be fixed. The original solution of Roth's group did not work. According to the Luftwaffe officer appointed to oversee the Wasserfall schedule in November, the project leadership had been far too slow to react to the problem. A contest was eventually held to gather suggestions. It was won by Werner Dahm, but his idea apparently did not definitively solve the gas-ingestion problem either, so the situation was little improved at the end of January 1945.[30]

The guidance problems remained equally perplexing. Just as in the case of the A-4b, acquiring adequately powerful vane servomotors remained a severe difficulty. In other areas of Wasserfall guidance and control, most of the serious difficulties of early 1944 had not been overcome either. The situation in homing devices remained extraordinarily confused because of the proliferation of competing projects, while the final form of Wasserfall's gyro systems remained unclear. To make matters worse, Siemens's aircraft instruments company was late on all its contracts in the second half of 1944, in part because of air raids. There were only one or two bright spots: The joy-stick, which was merely a modification of an existing system, proved itself in launch testing; a proximity fuse had been chosen, and its design was approaching completion. In the long run, Wasserfall was still a promising project, but its deployment as an effective weapon was still at least a couple of years away—years the Third Reich obviously did not have.[31]

In light of its dismal outlook, the project came under little further pressure for quick results. Instead, its political support threatened to collapse altogether. Speer and Saur continued to refuse to order Wasserfall into quantity production, causing the directors of Linke-Hoffmann to slow production even of test missiles. Nor did Hermann Göring's first visit to Peenemünde go well. Accompanied by Speer, he arrived on October 30, 1944, resplendent in bright red riding boots, opossum-hair overcoat, off-white uniform, and jewel-encrusted finger rings. According to Dornberger, the obese Göring popped pills every few minutes and behaved erratically. Afterward the Reich Marshal and the Armaments Minister decided to reduce Wasserfall to a mere long-term development project. Emphasis was to be placed on Schmetterling, which was most advanced, as well as on Enzian (Gentian), a newer project for a mostly wooden, unmanned, scaled-down version of the Me 163 rocket fighter. It is a sign of Göring's irrelevance that the decision never took effect. It was eventually overruled by the rump Technical Office of the Air Ministry, probably on the basis of a bizarre order by Hitler for accelerated anti-aircraft development in view of the enemy's fear of the "hell of German flak fire." Wasserfall would thus continue to limp along until Peenemünde was evacuated.[32]

The obvious failure of the program to meet its original objectives nonetheless resulted in some true desperation projects. Among the ideas floated was fitting the Wasserfall with small A-4b-type wings to create a scaled-down glider missile. But the most important desperation project was initiated earlier by the Luftwaffe officer responsible for Wasserfall test stands, Lieutenant Klaus Scheufelen. Taifun (Typhoon), a small unguided anti-aircraft rocket, is first mentioned in the documents in August 1944. The pencil-thin missile had a diameter of about 10 cm (4 inches) and a height of approximately 2 meters (6 feet). In Scheufelen's first version, it would be powered by the same hypergolic propellants as Wasserfall and would be fired off a launch rail, burning out in only three seconds. Taifun could potentially reach altitude faster than an artillery shell and would save gunpowder production, but its warhead was very small (around a kilogram), and its accuracy would have been poor. Essentially, this project was an admission of defeat; the guided anti-aircraft missile was coming along too slowly to alter the overpowering air superiority of the Allies.[33]

In September the Luftwaffe gave Taifun a higher priority than

Wasserfall, but there are indications that Electromechanical Industries paid only lip service to the project. At the beginning of December von Braun told Rees and the shop managers to treat the new contract for ten thousand Taifuns as filler work, except for the first hundred test vehicles. It is unknown whether Scheufelen and the Flak Experimental Center were informed of that order, but Taifun continued to expand as a project, at least on paper. The first liquid-fueled missiles had already been launched in November, and by January a second, solid-fuel version was proposed. As of New Year's Day 1945, 135 people were assigned to Taifun, a total dwarfed by the 1,940 staff members of the A-4 project, the 1,220 of Wasserfall, and the 270 working on A-4b (660 other employees were listed as "general" or "administrative"). Although the Taifun project did not entail a substantial diversion of resources, it surely expressed the catastrophic character of the military situation.[34]

A project that even more clearly embodied the mood of desperation was "Test Stand XII." That code-name was applied in late November 1944 to the idea for a U-boat-towed launch canister for a V-2. It might generously be described as a forerunner of the ballistic-missile submarine, but in its own context "Test Stand XII" was merely ludicrous. The rationale was that the United States might be given pause by the bombardment of New York, although it is hard to see how a few such shots would have done anything but make Americans more determined to take revenge on German cities. In any case, the practical difficulties were overwhelming. To launch, the U-boat would have to surface, then the crew would have to erect the canister by flooding its ballast, fuel the missile, and send it on its way. Accuracy would have been terrible. Notwithstanding those difficulties, a contract was immediately given to Vulkan Docks of Stettin to build a test version by March or April. Only a handful of people worked on the project before it was stopped by the evacuation. Those who participated, including General Rossmann and Walther Riedel, seem to have taken it seriously, but there was scarcely a clearer expression in Peenemünde of the escape from reality produced by the impending collapse of the Third Reich. In one way or another, desperation had come to overshadow all work in the Army rocket program.[35]

EVACUATION, MASSACRE, FLIGHT

By January 1945 the situation in Peenemünde had become truly bleak. The sound of Russian guns could be heard in the distance; endless streams of grim and tattered refugees came marching across the island of Usedom on their way to the bridge at Wolgast and points farther west. They brought with them tales of terrible atrocities committed by Soviet troops, although the Germans all too easily forgot that such outrages were revenge for the millions killed in the East by the Nazi regime and its armed forces. In the center itself, military employees had to carry guns, while all able-bodied male civilians were obliged since October or November to spend time exercising with the Volkssturm (Home Guard), preparing for a last-ditch stand. On January 18 rumors of Soviet tanks only a few dozen kilometers away caused panic. Scientists from Erich Regener's stratospheric research institute in Friedrichshafen were sent away before they could finish preparing their instrument package for an A-4 launch, a project begun in mid-1942 to explore the upper atmosphere for guidance purposes as well as science. The launch never happened, because Kammler ordered the evacuation of Peenemünde to central Germany less than two weeks later.[36]

A move had of course long been planned, although the destination was not the one expected. Throughout 1944 the prisoner-dug Zement tunnels in the Austrian Alps had remained the intended location for the new development works. But as the date for the move stretched into 1945, Speer and Hitler became impatient. In early July 1944 the Führer accepted his Minister's suggestion that part of the complex be given over to tank drive-train production. That measure never went into effect, because American attacks on German oil production provoked a sudden decision by Saur to use the nearly complete "A" tunnels at Zement for a refinery. The oil refinery began to set up its equipment in early August. After a period of confusion, Electromechanical Industries received confirmation that it would move to Zement's smaller and much less complete "B" tunnels, when they were ready. In the fall studies were made of testing locations in Austria; one suggestion was to fire the A-4b from sites near Vienna at targets in the sparsely populated Tyrolean Alps. The steady deterioration of the military situation also provoked studies of an emergency evacuation to other locations, but until early December von Braun and the Peenemünde leadership took for granted a move into Zement—and in the relatively near future. Then Saur handed over the remaining tunnels to aircraft engine production on December 8, terminating a year of planning.[37]

Storch, von Braun, and other Peenemünders must have had an inkling that the only place left to go was the Mittelwerk area, which had become an SS-dominated region devoted to underground production and secret weapons. On January 26 Dornberger wired von Braun that his BzbV Heer organization was evacuating to Bad Sachsa, a town near the subterranean plant. Movement of related firms and research institutes to the same area would be discussed in Berlin the following day at the first meeting of "Working Staff Dornberger."[38]

That committee had been appointed by Speer on January 13 in a last attempt to consolidate authority over anti-aircraft missiles and other advanced weapons. Speer gave Dornberger special powers within the Armaments Ministry to make emergency decisions, but neither man anticipated how easily Kammler could counter the move. The SS general went to Göring on the January 26 and had himself appointed V-1 commander and commissioner for the "Breaking of the Air Terror." Ten days later, after receiving further backing from Himmler, Kammler canceled Enzian, Rheintochter, and a number of other projects and subordinated Working Staff Dornberger to his command. Speer's influence over missile development was at an end, while Kammler's meteoric rise continued. More and more, however, he was becoming the ruler over a shadow empire of skeleton organizations, false hopes, and self-delusion.[39]

Back at Peenemünde, rumors about the evacuation were flying. On January 30 and 31, Gen. Rossmann issued orders telling everyone to stay calm, remain in place, and await developments. Northern Experimental Command soldiers were to take training in antitank weapons. But later on January 31, "a cold and cloudy Wednesday," according to Huzel, Kammler's order finally arrived. Von Braun immediately called an emergency meeting of all available department heads to plan the transport to the Mittelraum (Central Region) of what could be salvaged of the facility. First priority was assigned to A-4 and Taifun personnel, followed by A-4b and Wasserfall people. The launch crews were to remain in place until further notice. The meeting unleashed feverish activity, because all departments had to choose whom to move first, how much equipment to take with them, and what form of transportation to use. Allotted to Electromechanical Industries were a few trains, a number of barges that could be mobilized to move heavier material, and vehicles from the Peenemünde motor pool and other organizations. Storch had already ordered in September that a central archive be assembled and prepared for shipment.[40]

Reflecting the complexity and ambiguity of the command structure over the rocket program, the company sought confirmation of Kammler's order from the chief of Army Ordnance. General Leeb gave his approval the next day, February 1, but there was little chance that he would have defied the SS and an even smaller chance that he would have done so successfully. That course of events demolishes the postwar myth that the rocket group had a choice as to whether to stay or go. Von Braun joked decades later that "I had ten orders on my desk. Five promised death by firing squad if we moved, and five said I'd be shot if we didn't move." He therefore decided to go along with Kammler, he claimed, because it suited the group's desire to head for the Americans. While it is almost certainly true that orders must have existed to stand and fight, probably from the Gauleiter of Pomerania as commander of the Volkssturm, von Braun had no power to make such a decision. Even within Peenemünde, Storch and Rossmann stood above him, and above them came Dornberger, Leeb, and Kammler. Moreover, there is little doubt that, if push came to shove, Himmler's protégé could have overruled the Gauleiter.[41]

Similarly, the realities of the evacuation order destroy the myth that the rocket group steered itself in the path of the American forces. It is certainly true that von Braun had conversations with close associates about the postwar situation and how to salvage Peenemünde's unique expertise, especially in view of his self-appointed historical mission to develop rocketry for spaceflight. The outcome of those discussions, which National Socialist fanatics would have seen as defeatism and high treason, was unanimous agreement to surrender to the United States, if possible. As one unnamed engineer put it immediately after the war: "We despise the French; we are mortally afraid of the Soviets; we do not believe the British can afford us, so that leaves the Americans." But neither von Braun nor his subordinates had the power to choose their destination. Although Kammler may already have been thinking of the group as a bargaining chip for separate peace negotiations with the West, the move to Thuringia made perfect sense on its own: It concentrated part of the Reich's remaining technical expertise near underground facilities in an SS-controlled region relatively far from the front lines.[42]

Not long after the evacuation order, von Braun traveled south to survey locations for the various divisions of the company, while in Peenemünde preparations went ahead day and night for the departure. The transportation coordinator, Erich Nimwegen, an entrepreurial character who operated on the thin edge of the law, proved his worth in mobilizing much needed material and transport. He also allegedly thought up a way to exploit a mixup in some newly printed forms: BzbV Heer had been garbled as VzbV, which he turned into a top-secret agency under the SS. Soon, Huzel says, those initials "began to appear in letters several feet high on boxes, trucks, and cars." For movements by road in the chaotic conditions of the last months of the war, any SS credential proved useful to get through the many roadblocks set up to catch deserters and to stop unauthorized travel. Von Braun himself admitted to the American authorities in 1947 that he had put his rank of SS-Sturmbannführer (Major) on transportation orders, although he claimed that it was "the only time" he had ever exploited the title.[43]

After von Braun had spent three weeks in Thuringia and had met the initial arrivals there (the first train left the center around February 17), he returned late in the month for his last visit to Peenemünde. After staying only a few days, he departed, never to see the place again. All launch activity had been shut down, and the mobile crews evacuated from Heidekraut in January were sent to northwest Germany. The last convoys, trains, and barges moved out in early March, even though travel was difficult because of roving Allied fighter-bombers. Peenemünde began to resemble a ghost town, with only a skeleton staff remaining. Ironically, it would remain that way until May 5, because not long after Kammler ordered the evacuation it was discovered that Soviet forces were farther away than believed. Rather than head north, the Russian winter offensive was going east—in the direction of Berlin.[44]

Among the last to leave the center was Dieter Huzel, a special assistant to von Braun since the autumn of 1944. When he arrived in the town of Bleicherode in the middle of March, he found an "[e]xtremely primitive headquarters . . . in a former agricultural school. There was not much sense of order. We couldn't just bodily lift a whole engineering plant, drop it two hundred miles away, and expect it to continue functioning without interruption." Although Kammler had made Electromechanical Industries the center of a development cooperative of advanced weapons firms, virtually everyone perceived the exercise as futile. The prospect of total defeat hung like a dark cloud over everything. Von Braun chaired meetings and investigated underground and above-ground sites for restarting work, but there is little doubt that he was merely putting on a good show for the ever watchful SS. By the time Huzel arrived, von Braun was hospitalized with a broken right arm, because his driver had fallen asleep at the wheel one night (the only time safe to travel) and the car had flown off the road and crashed into an embankment. A couple of days before his thirty-third birthday, the young technical director was released from the hospital wearing a massive cast. He rejoined Dornberger in useless planning, while others carried out design work on paper. In divisions like the valve and materials testing laboratories, virtually no work was done, because they were situated far from the oversight of the SS and the headquarters group.[45]

Not many kilometers away, the concentration camp prisoners were dying *en masse,* although there is little evidence that the Peenemünde evacuees confronted that suffering directly. Von Braun asserted in 1947 that he last visited Mittelwerk in February 1945 and that working conditions inside the plant were, if anything, better in the last months of the war. While his statement is true in a sense because of the completion of the heating, air-conditioning, and lighting systems, it is also rather callous, because the prisoners were starving and were subject to arbitrary terror even more intense than before. "Dora" had entered its third phase: the relatively mild period of summer–fall 1944 had ended, and the camp death rate once again skyrocketed.[46]

The situation had taken a noticeable turn for the worse in November. Shortly after the SS had made Dora the main camp of a new "Concentration Camp Mittelbau" (Central Construction), the local SD security organization had succeeded in penetrating and destroying the

Dora prisoners' underground organization. Many of its leaders were locked up in the "bunker," and some were horribly tortured. Most ended up on the camp gallows, with a consequent acceleration in the execution rate.[47]

Beginning in January 1945 conditions dramatically worsened. Germany's war economy and transportation system collapsed because of Allied bombing and the Soviet invasion of the Upper Silesian industrial region. The food supply for the whole population deteriorated, but in the Nazi system distribution was deliberately structured to favor the "Aryans." At the bottom of the hierarchy, unsurprisingly, were the concentration camp prisoners, who began to starve even more rapidly than before. The food supply and the health situation in the Mittelbau camp system, which now included about three dozen subcamps, was worsened further by transports full of exhausted prisoners evacuated from the east, most notably from Auschwitz. Dora's population grew from nearly 13,500 on November 1 to more than 19,000 in March, while the total of all Mittelbau prisoners went from about 26,000 to more than 40,000 in the same period. These numbers would have grown even faster but for disease and starvation; the bodies began to pile up at the crematoria faster than they could be burned. From December 24 to March 23, the camp administration counted 5,321 deaths, of which 1,090 were in Dora. The SS created a particularly horrendous situation at the Boelcke Kaserne, a former barracks in the city of Nordhausen, by using it as a dumping ground for hopeless cases, many from the transports.[48]

With the evacuated prisoners came numerous SS men, including Richard Baer, the last commandant of the original Auschwitz camp. (The primary extermination center for Jews had been the nearby Auschwitz II-Birkenau.) On February 1 Baer displaced Mittelbau's camp commandant, Otto Förschner, who was probably dismissed for letting the underground get out of hand. The new commandant raised the level of horror yet higher with a gruesome wave of hangings in late February and March that focused on the Soviet resistance organization, although a number of prominent German Communist prisoners were shot or beaten to death in the "bunker." The unprecedented 162 executions in March included 133 Russians, 25 Poles, 3 Czechs, and 1 Lithuanian. Most of the sentences were carried out in a few horrific mass executions: 16 on March 3, 57 on March 11, and 30 each on March 21 and 22.[49]

Two of the hangings took place for the first time inside the plant instead of in Dora. Although it is unlikely that any recent arrivals from Peenemünde were present, one or both executions were seen by Arthur Rudolph and others assigned to the Mittelwerk. Erich Ball, a shop foreman sent south after the 1943 air raid, described the first to an American investigator in 1947:

> A large wooden plank was brought into tunnel B and attached to the hooks in the overhead crane. . . . By this time all work in the factory had stopped. Everybody was ordered to watch the hanging. When the [12 to 16] Haeftlinge who were to be hung came in they had their hands tied behind their backs. They had wooden gags in their mouths. [A] German [prisoner] put a separate rope around each man's neck[;] the other end of the rope had previously been tied to the plank. The man who normally operated the crane was a French Haeftlinge [sic] and under my control. However, I called [him] down from the ladder. [Chief SD officer] Bischoff asked why . . . and I said, this man works for me and if he has to hang his friends he will be sick and will not be of use to me. . . . [H]e was scared and crying. The [executioner] went up the ladder[,] then Bischoff . . . read the order by Himmler, and gave the order to raise the crane.

A French prisoner, Yves Béon, and his compatriots were forced to file past the still hanging prisoners later that night:

> Most of their bodies have lost both trousers and shoes, and puddles of urine cover the floor. Since the ropes are long, the bodies swing gently about five feet above the floor, and you have to push them aside as you advance. . . . [Y]ou receive bumps from knees and tibia soaked in urine, and the corpses, pushed against each other, begin to spin around. . . . Here and there under the rolling bridge, truncheons in hand, the S.S. watch the changing of the shifts. They are laughing; its a big joke to these bastards.

This gruesome spectacle was repeated a few weeks later with about thirty condemned, who were once again surrounded by guards carrying "machine guns."[50]

Notwithstanding the increased terror and the extreme chaos in the war economy, Mittelwerk apparently kept churning out missiles right up to the end of March, although documentation exists only for 362 A-4s shipped up to the eighteenth. (Total verifiable Mittelwerk production is 5,789; Peenemünde built 150–200 more test-model A-4s, and Mittelwerk also assembled a few thousand much simpler V-1s, beginning about October 1944.) That deliveries continued at such a high rate can only be attributed to the consolidation of parts production in the tunnels and the high priority the ballistic missile retained to the very end. In the last days of the Mittelwerk, even Karl Otto Saur installed himself in the tunnel offices, but a bizarre "Twilight of the Gods" atmosphere hung over the place. Prisoners later reported that empty champagne bottles lay scattered next to the offices in the morning, and women in Saur's entourage sat around in nightgowns until ten, chatting with the prisoners and giving them bread and cigarettes.[51]

Toward the end of March the western front collapsed. After Allied armies crossed the Rhine in large numbers, the last V-2s were fired at Antwerp and London on March 27, and the last V-1 the following day. Kammler then ordered his units to fight as infantry. Around that time he found a large number of East European workers wandering in the chaos of collapsing Germany and personally ordered his soldiers to massacre at least 207 of them. Apparently he did not have enough blood on his hands already.[52]

Soon the rapidly moving American armored spearheads approached Thuringia, forcing Kammler's hand. Late in the day on April 1, he called Dornberger's chief of staff and commanded an evacuation to the Bavarian Alps of about five hundred key people. Von Braun, still burdened by his heavy cast, was driven to the designated site in Oberammergau, where some Messerschmitt people had already been collected. Dornberger's staff proceeded with its own convoy. On April 6 a sleeping-car train ironically dubbed the "Vengeance Express" (it had been used as a residence in Heidelager, Heidekraut, and Thuringia) departed with the rest of the group. Kammler's apparent motive was to hold them as a bargaining chip for peace talks with the West, although killing the lot of them if negotiations failed may well have crossed his mind. One thing we do know: The five hundred Peenemünders once again had no control over their destination. Indeed, it would have made more sense for them to stay put and wait for the Americans in Thuringia.[53]

For the prisoners, evacuation was presaged by yet another horrible tragedy: Two successive RAF night firebomb raids on Nordhausen killed 1,500 of them at the Boelcke Kaserne. Beginning on the morning after the second raid, April 4, about 25,000 to 30,000 inmates of the Mittelbau camp system were then forced into railroad cars and shipped to Bergen-Belsen. Their journeys took days and—like all the camp evacuations—an enormous toll in human lives due to starvation, disease, and wanton cruelty. Even more catastrophic was the fate of thousands from outlying Mittelbau subcamps, who were forced to leave on foot for lack of transports. Many collapsed and died by the side of the road or were shot for straggling. Nothing, however, compares in horror with the worst single massacre in the history of Mittelbau. At Gardelegen, SS guards herded into a barn 1,016 evacuees exhausted from marching and set the building on fire, burning them alive. Any who escaped were gunned down. When the 3d U.S. Armored Division liberated the Nordhausen area on April 11, all that was left were 600 extremely ill survivors in Dora and 405 living skeletons at the Boelcke Kaserne.[54]

By the best estimate, of the roughly 60,000 unfortunates who passed through the Mittelbau–Dora system, at least one-third did not survive. Perhaps half (10,000) of the deaths can be linked to A-4 production. In addition, more than 8,200 prisoners died at Zement, although most were the victims of mass starvation in the last few months of the war, when the camp was no longer part of the program. A smaller but unknown number of victims must be attributed to Schlier, Lehesten, Rebstock, Zeppelin, and other locations. Attempting an exact total can become a meaningless numbers game, but it is clear that the A-4 was a unique weapon: More people died producing it than died from being hit by it. In round numbers, 5,000 people were killed by the 3,200 V-2s that the Germans fired at English and Continental targets. (More V-weapons were launched at Belgium than at Britain, although one would hardly know that from the literature on the topic.) By that measure, at least two-thirds of all Allied victims of the ballistic missile came from the people who produced it, rather than from those who endured its descent.[55]

While the concentration camp prisoners suffered and expired in the final catastrophes, the Peenemünders languished in uncertainty in Bavaria and Thuringia. Events were largely out of their control, including the fate of the prisoners. In Oberammergau, von Braun saw Kammler for the last time early in April. From the Berlin bunker, Hitler had bestowed on the SS general his last, highest, and most absurd title, "Plenipotentiary of the Führer for Jet Aircraft." Kammler set off on a frantic tour around a rapidly shrinking Reich, trying singlehandedly to stave off defeat. According to Dornberger, if he could not sleep, he would "wake the slumbering officers of his suite with a burst from his tommy-gun." Kammler's final fate is uncertain, but the most plausible report is that he arranged to be shot by his adjutant in Prague around the time of the final German surrender, rather than be captured by Czech partisans.[56]

Kammler's absence from Bavaria proved a boon to the rocket veterans. They succeeded in persuading the local SS to allow a dispersion to various towns instead of a concentration in Oberammergau. A small group around Dornberger ended up in a mountain resort hotel high up on the former Austrian–German border. Von Braun joined them later in April, as did Dieter Huzel and Bernhard Tessmann, who had separately made dangerous odysseys across the Reich after burying Peenemünde's archive in a mine northwest of the Mittelwerk. The clear intent of that action was to create the group's own bargaining chip for use after the war. For the rest of the month there was little to do but stare at the sky, play cards, and worry about the omnipresent SS.[57]

By then many Peenemünders had been found by U.S. Army Ordnance intelligence officers in Thuringia, where the great majority of the rocket group had remained. Only the five hundred sent to Bavaria were left. A few were found by French troops, but most waited for an opportunity to surrender to the Americans—the only time during the entire evacuation when they had any control over their fate. Among the last to encounter the invaders were Dornberger and von Braun. On May 2, two days after Hitler's suicide, Wernher von Braun's English-speaking brother, Magnus, was sent down the mountain on a bicycle to find American troops. He encountered a rather surprised patrol, and soon the arrangement was made. After fifteen amazing and eventful years, the Army liquid-fuel rocket program was over.[58]

Epilogue

Peenemünde's Legacy

Peenemünde's death was followed quickly by its rebirth elsewhere. Even before the war was over, teams from the major Allied powers began searching for the spoils of the Baltic coast center and its revolutionary technology. In central Germany, U.S. Army Ordnance moved quickly to seize parts for one hundred A-4s, as intact missiles were nowhere to be found. Speed was of the essence. Thuringia was to be part of the Soviet zone of occupation, and the Red Army might move forward as early as the beginning of June. Ordnance's Special Mission V-2 also managed to ferret out the location of the Peenemünde archive from a former manager who had seen Huzel and Tessmann before they disappeared. Trucks whisked the 14 tons of paper out of the mine on May 27, allegedly just as the British began setting up roadblocks in what was to become their zone of occupation. After some delay, the Soviets occupied the Mittelwerk on July 5. Realizing what they had in their hands, a Soviet intelligence team that included Sergei Korolev, the Chief Designer of the Soviet space program in the 1950s and 1960s, was sent to investigate Peenemünde. What the Russians found was quite disappointing. The evacuation had stripped the center of much of its equipment, the defending forces had blown up many buildings, and the occupying Soviet units had carried off some of what was salvageable. Eventually the Soviets dynamited the rest.[1]

At the Bavarian ski resort of Garmisch-Partenkirchen, von Braun, Dornberger, and other Peenemünders were interrogated not only by U.S. Army Ordnance but also by numerous American and British intelligence teams, many of which showed themselves to be laughably ill-informed, in the Germans' opinion. Ordnance officers were well aware that the guided missile interested not only other Allied powers but also other American agencies, such as the Army Air Forces, which everyone expected to become a separate service in the near future. Although the incipient Cold War certainly played a role in Ordnance's motivations, it was no more important than a "denial policy" that applied to everybody. The American government sought to prevent a repetition of the Weimar Republic's secret rearmament in other countries, and American services and agencies sought to exploit German achievements to benefit the nation and themselves—not necessarily in that order. (The American tradition of interservice rivalry puts the Third Reich's internal battles in perspective.) Thus, when the leaders of the German rocket program sought to negotiate with Ordnance, they scarcely needed their document cache as a bargaining chip.[2]

The real questions became, how many Peenemünders would come to the United States, and on what basis would they be hired? Ordnance's interest in the German Army rocket program was influential in the creation of "Project Overcast" by the U.S. Joint Chiefs of Staff in July. The official rationale for Overcast was the temporary exploitation of 350 German specialists to help in the defeat of Japan, but the atomic bombing of Hiroshima and Nagasaki soon rendered that purpose moot. Overcast went ahead anyway, and Colonel Holger Toftoy, chief of Ordnance Rocket Branch since late June, moved to fill his quota of one hundred Peenemünders. Von Braun was chiefly responsible for drawing up a balanced list of people and then persuading them to sign on for what was, theoretically, only a six-month commitment. The former technical director decided that he could not do with fewer than 115 or so specialists, and Toftoy went ahead to acquire them. Like many of his colleagues, he did not let the wording of orders get in his way if he could avoid it.[3]

The Soviets were naturally disappointed not to get von Braun, Steinhoff, and other leading engineers of the program, although they did find low-ranking people and much equipment following their occupation of the Nordhausen area. They broadcast offers to Peenemünders to come over to their zone, where they would receive excellent positions at good pay. A few individuals were willing to accept, the most prominent being Helmut Gröttrup, who had been arrested with von Braun and had been Steinhoff's deputy in guidance and control at the end of the war. Although he was one of the most left-wing members of the rocket group, personal resentments more than political affinities seem to have caused him to cross the line after being evacuated from Thuringia by the Americans. He was not satisfied with the deal that Wernher von Braun was trying to strike with the United States. Besides, other Peenemünders falsely accused him of being the one who revealed the location of the documents. Gröttrup became the head of a rocket institute near the Mittelwerk after the Russians had begun to restore some of its manufacturing capability.[4]

Although the French also began to contact some rocket specialists, U.S. Army Ordnance's main competitor for leading Peenemünders turned out to be the British, and that only temporarily. In order to understand better how the A-4 worked, the British Army had created Operation Backfire. After some inter-Allied conflict, the two countries forged an agreement to lend some of the Germans earmarked for the United States to Backfire, which launched three missiles from the German North Sea coast in October 1945. A few other Peenemünders not wanted by von Braun and Toftoy were taken to Britain, most notably Walter "Papa" Riedel, who had been exiled to the Zement project in late 1943. The one person Ordnance could not get back from the British was Dornberger. According to a U.K. interrogator, the former rocket general had "extreme views on German domination, and wishes for a Third World War." Moreover, the British were determined to try Dornberger in Kammler's place for indiscriminate V-2 attacks on civilians. They kept him in a POW camp until 1947, but the hypocrisy of such a charge made a trial untenable—roughly 1 million Germans and Japanese had been killed by Allied bombing. Because of the narrow focus of war crimes investigations, the rocket general also avoided trial on the one charge that could have stuck: complicity in the exploitation of slave labor.[5]

While Dornberger sat in jail, U.S. Army Ordnance conveyed across the ocean nearly 120 selected Peenemünders, the essence of a development organization that had once employed six thousand people. Von Braun had already departed for the United States by airplane with six others in September 1945, followed in stages by the rest, who traveled by ship during the winter months. All were eventually assembled in the desert at Fort Bliss, near El Paso, Texas. One of their first tasks was to assist in scientific and military V-2 launches that began at White Sands Proving Ground, New Mexico, in April 1946. The chief role of the von Braun group was, however, to cooperate in planning future rocket development with Project Hermes, which Ordnance had contracted to General Electric in 1944 after the magnitude of the German rocket effort became clear. The Germans, however, were frustrated at their primitive laboratories and the postwar cutbacks that seemed to derail any hope for a return to accelerated work.[6]

Toftoy and the Ordnance Rocket Branch had to struggle to satisfy the Peenemünders in the face of limited budgets and the restrictive boundaries of their ambiguous status. (They ironically called themselves "prisoners of peace"; they were not legal immigrants, and their freedom of movement was limited.) In order to retain a few valued specialists, the Army, like the other services, also had to bend the rules regarding exclusion of individuals with dubious Nazi records. Under Project Paperclip, which had replaced Overcast in March 1946, the long-term use of former enemy scientists and engineers had been provided with a stronger legal basis. But security reports for a number of individuals, including von Braun, had to be revised or fudged to circumvent the restrictions that still existed. Some writers have seen those actions as evidence of a conspiracy in the Pentagon to violate a policy signed by President Harry Truman, but it really reflected a conscious choice by the U.S. government, approved up to the level of the Cabinet at least, to put expediency above principle. The Cold War provided ample opportunity after 1947 to rationalize that policy on anti-Communist grounds, but the circumvention of restrictions on Nazis and war criminals would have gone ahead at some level anyway, because the Germans' technical expertise was seen as indispensable.[7]

Thus when the Army's own investigators came looking for witnesses and evidence for the Mittelbau–Dora war crimes trial, which was held at Dachau in 1947, it is no surprise that Ordnance was none too cooperative in granting access to the Fort Bliss Germans. The whole story of Mittelwerk and its prisoners was to be obscured as much as possible, because it would besmirch Army rocket development. Indeed, from the very end of the war, if not before, the Peenemünders had divorced themselves from any responsibility for slave labor; the SS provided a convenient scapegoat for all the crimes associated with the program. It was a position that the American authorities found easy to accept.[8]

With that issue buried rather quickly, von Braun's group was free to continue to play a historic role in the rise of the guided missile and the space launch vehicle, particularly after the Cold War spurred heavier American investment in the technology. In 1950 the Army transferred the group to Redstone Arsenal in Huntsville, Alabama, where they became the premier rocket development group in the United States. Their arrival in the States had in fact changed the whole balance of Army rocket activities, since the Germans displaced the smaller groups that had begun to flourish in World War II, like the Jet Propulsion Laboratory in Pasadena, California.

At Huntsville, one of the keys to the Germans' success was the "everything-under-one-roof" approach developed at Kummersdorf and Peenemünde under the direction of Becker and Dornberger. It proved very compatible with the U.S. Army's "arsenal system" of in-house development. Under von Braun's leadership, the German-dominated group successfully developed the nuclear-tipped Redstone and Jupiter missiles in the 1950s. The Redstone—which was really just a much-improved A-4—then became the vehicle that put the first American satellite and first American man into space. Finally, under NASA aegis after 1960, the Peenemünders crowned their success with the phenomenally reliable Saturn vehicles, which launched Apollo spacecraft into orbit and put humans on the moon.[9]

The rebirth of Peenemünde in Huntsville was necessarily unique, because the center's engineering leadership had survived as a coherent group. But the Baltic coast center was also in some sense reborn in the many other postwar missile projects that sprang up in the United States and elsewhere. The transfer of Peenemünde's technology was crucial to the U.S. Army's work on anti-aircraft missiles and the Air Force's early research on intercontinental ballistic missiles (ICBMs), although personnel and ideas from the smaller German rocket programs had a role as well. In France and to a lesser extent in Britain, Peenemünders played an essential role in the development of missiles and space launchers. Even in the Middle East and China, the transfer of German rocket technology, often through indirect routes, was critical to guided missile proliferation.

But the Soviet Union was the most important heir to German rocket technology outside the United States. After the Soviets had built up Gröttrup's rocket institute, they rounded up the Germans at gunpoint on October 22, 1946, and shipped them off to Russia, along with thousands of other specialists from the eastern zone. Gröttrup and a few other leaders received reasonably comfortable accommodations near Moscow, while the majority of the rocket engineers and their families were dumped on a rather primitive island in a lake north of the capital. A year later, on October 30, 1947, the Russian military fired the first remanufactured V-2 (R-1) from a bleak semi-arid site in the south not too distant from Stalingrad.[10]

Gröttrup and some of his assistants were important advisers at those launches, but the Germans soon found, to their frustration, that they would not be fully integrated into the Soviet rocket program. They were set to work designing new advanced missiles that remained paper projects, because Stalin's military had decided to pump the Germans for their knowledge and then to toss them aside. The Gröttrup group was gradually cut off from contact with regular design bureaus. Beginning in 1951 its members were sent back to their native country. Communist paranoia and traditional xenophobia had prevented an effective integration of German talent such as had occurred in the United States. The Soviets also had many highly capable rocket engineers of their own. Impelled by their inferiority in nuclear weapons and long-range bombers, they pushed ballistic missile development more energetically than did the United States prior to 1954. The result was Korolev's Sputnik surprise of October 1957, a triumph that rested to no small extent on a rapid and effective absorption of Peenemünde's technological revolution.

The German Army rocket program clearly had a profound impact on science, engineering, and warfare in the second half of the twentieth century. But that inevitably raises a question that might be called the central paradox of Peenemünde: Why was the Army's guided missile technology such a bad investment for the Third Reich when it was so valuable to everyone else after the war?

To answer that question properly, a cost-benefit analysis is needed. A systematic accounting of the Nazi regime's expenditure on Army guided missile research, development, and production does not exist, but it is possible to make a rough estimate. According to May 1945 statements by Dornberger and von Braun, the Army facility at Peenemünde (including, in all likelihood, the ill-fated Production Plant) cost 300 million "gold marks" to build. The center's mid-1944 monthly expenditure of 13 million marks equaled an operating budget of about 150 million annually, although it would have been less earlier in the war, when Peenemünde was smaller. (Expenditures before 1939 were so small by comparison that they can be disregarded.) The largest single expense was A-4 production. By Mittelwerk's price list alone, the Reich paid the company approximately 450 million marks for nearly six thousand missiles, but this cost omitted the warhead and the guidance system. There is no reliable data on the total cost of mobile launch vehicles, troop training, construction of bunker sites and liquid oxygen plants, expenditures at Zeppelin, Schlier, Zement, and so forth. Thus 2 billion marks would appear to be a reasonable, even conservative, estimate. If those marks are converted according to the gold standard of that era (4.2 marks to the dollar—a problematic assumption), this amount equals about half a billion U.S. dollars of World War II vintage or about 5 billion current (early 1990s) dollars.[11]

By way of comparison, the Manhattan Project to build the atomic bomb cost the United States about 2 billion 1940s dollars, or four times as much. Since the German war economy was significantly smaller than the American one at its peak, the Army rocket program imposed a burden on the Third Reich roughly equivalent to that of Manhattan on the United States. Such a comparison makes it almost superfluous to explain why the German Army rocket program was, in military terms, a boondoggle. Even compared with Anglo American conventional strategic bombing, the V-2's results were pathetic. The total explosive load of all A-4s fired in anger was scarcely more than a single large RAF air raid! Moreover, the 5,000 Allied civilians killed by V-2 attacks (leaving the prisoners aside) were dwarfed by many tens of thousands of dead in single raids on Hamburg, Dresden, and Tokyo, not to mention Hiroshima and Nagasaki.[12]

The missile's psychological and material impact on the Allied war effort was equally unimpressive. Although no one should dismiss the terrible effects of individual V-2 hits, which at their worst killed hundreds of people, outside of East London and Antwerp the missile was little more than a nuisance. Only the onset of the V-1 campaign in June–July 1944 produced popular disquiet in Britain, and that was mastered by the end of the summer, when anti-aircraft defenses became efficient. One of the ironies of the Luftwaffe's "buzz bomb" (which cost a fraction of a V-2) was that not only did its noisiness create more terror, the fact that it could be shot down diverted much more Allied effort into stopping it. Since there was no defense against the ballistic missile, and the numbers launched were much smaller (3,200, as against 22,400), the Allies expended considerably fewer resources on A-4 countermeasures, mostly for attacks on launch and production sites. In the last analysis, the V-1 was no "wonder weapon" either, but the disjuncture between total expenditure and results was not quite so large in that case. Yet it is clear that the Reich's expenditure on both weapons—and on no less than four different anti-aircraft missiles—could have been much better directed elsewhere. By the estimate of the United States Strategic Bombing Survey, V-weapons production in 1944–45 alone cost the Third Reich the equivalent of 24,000 fighters at a time when the annual aircraft production was only 36,000. In short, German missile development shortened the war, just as its advocates said it would, but in favor of the Allies.[13]

Inaccuracy was one of the main reasons why German missiles were so ineffective. The V-2 could barely hit a giant city with any certainty, the V-1 was even worse, and Wasserfall and the other anti-aircraft missiles were never deployed because guidance problems, above all else, paralyzed their development. As V-weapons historian Dieter Hölsken has argued, World War II electronics and computers were too primitive for missile technology to be cost-effective. Thus the guided missile was not "too late" to change the course of the war, but rather was "too *early*" to have any significant effect on it—an important insight that also explains the central paradox of Peenemünde's missiles: complete short-term ineffectiveness *versus* profound long-term importance.[14]

But Hölsken underestimates a technology that stands even more clearly at the roots of that paradox: nuclear weapons. The ICBM and its twin, the SLBM (submarine-launched ballistic missile), were the truly revolutionary military offspring of Peenemünde—and Los Alamos. Although the exploration of space may in the long run be more important to the future of humankind, the early space race itself would have been impossible without massive investments in rocket technology by the major powers. Those investments made sense only because of the revolutionary strategic implications of the nuclear-tipped long-range ballistic missile. By contrast, the conventionally armed ballistic missile has remained militarily ineffective, in spite of minor propaganda successes in the Iran–Iraq and Gulf wars, and anti-aircraft, cruise, and other missiles with high-explosive warheads, however important, have not had anywhere near the transformative impact of the nuclear missile.

Thus the missile came "too early" because Peenemünde developed the technology before it had either the warhead or the electronics to make it effective. But that fact raises a second important question about the German Army rocket program: Why did the Third Reich invest so heavily in a technology that, in retrospect, had no hope of changing the course of the war? Part of the answer is obvious: The Germans clearly did convince themselves that the ballistic missile *could* change the course of the war. The question then becomes: What arguments were adduced to prove the war-winning capability of the ballistic missile, and what historical conditions allowed those arguments to prevail?

In the early years of the German Army rocket program, one man was primarily responsible for promoting the technology as decisive: Karl Becker. Without him, it is scarcely imaginable that the program would have gotten off the ground in the early 1930s. Becker was convinced that the surprise introduction of a radical new weapon could produce a stunning psychological blow against an enemy. As an artillery expert, he was also aware that the rocket-powered ballistic missile promised to break through all the technological limits of conventional gunnery. He probably dismissed the failure of the Paris Gun in 1918, a project on which he had worked, as the natural outcome for a weapon that was just not revolutionary enough to produce the desired effect. That was ironic, because the A-4 ended by becoming another, much more spectacular, Paris Gun. It shelled enemy cities with little political or military effect and was, like its spiritual ancestor, the product of a blinkered technological enthusiasm that displayed little insight into the psychology of the enemy.

There were a number of reasons why Becker's views gained so much support within the Army during the 1930s. The man himself was highly competent, and he surrounded himself with excellent technical officers like Dornberger. His political views and his enthusiasm for the Nazis also speeded his rapid rise to the top of Army Ordnance after 1933. The important role of artillery in the World War I trench stalemate gave officers of that branch an unusual prominence in the Army leadership of the Third Reich as well, which left Becker's rocket project with an even stronger constituency. For historical reasons the German Army officer corps was also noteworthy for combining high tactical and technical competence with disastrously short-sighted strategic thinking—precisely Becker's problem.

The rapid growth of the Army rocket program in the 1930s was fostered further by structural and political conditions that were inherent to the Third Reich and its armed forces. Although the Army's autonomy was steadily eroded by the Nazi regime, and strong ideological affinities existed between the officer corps and Nazism in any case, until the early years of the war the Army did have some room to administer its own affairs and to implement some of the armaments projects it wished to pursue, notably the ballistic missile, which the service's highest leadership had accepted as decisive. The rapidity of rearmament, the weakness of interservice coordination, and the "polycratic" character of the regime as a collection of competing bureaucratic empires further reinforced the Army's ability to go it alone on the ballistic missile, if it so desired. The interservice rocket alliance with Göring's politically potent Luftwaffe did, however, provide an essential boost to the program, enabling the construction of Peenemünde in 1936–37.

The real technical achievements of the rocket group assembled by Becker, Dornberger, and von Braun were a further necessary condition for the program's growth; increased investment in turn assured that more technological successes would be possible. Especially noteworthy is the revolutionary breakthroughs in key technologies that von Braun's engineers achieved between 1936 and 1941. By building up a massive, secret in-house development capability and by adding university and corporate research and development resources after the outbreak of the war in 1939, Peenemünde was able to prove that it could make the guided missile a reality.

The conflicting demands of the war and the accelerating decline of the Army's power in the Reich, however, brought the further growth of the rocket program into doubt after late 1939. Under the leadership of Army Commander-in-Chief von Brauchitsch, an acquaintance of Dornberger, the senior service therefore had to exploit what was left of its autonomy and political clout to protect its pet project. Despite innumerable and demoralizing priority crises, the Army leadership by and large succeeded in minimizing politically induced delays to A-4 development, if not to the construction of Dornberger's ill-advised Peenemünde Production Plant. But the approval and execution of missile mass production came to depend more and more on the new Armaments Ministry and, above all, on the whims of the Führer.

Fortunately for the program, Albert Speer was a Peenemünde enthusiast. Even more important, after August 1941, and especially after November 1942, Adolf Hitler became increasingly infatuated with the A-4's potential for rescuing Germany from the catastrophic strategic situation he had created. Believing that terror could be answered only with terror and that the regime needed to exact "vengeance" for the strategic bombing of Germany, Hitler readily accepted the arguments of Dornberger and other advisers that the British populace would easily succumb to psychological pressure. Army assertions that there was an international missile race that Germany could ill afford to lose—something that Becker and Dornberger had argued since the early 1930s—gave Hitler yet another reason to approve A-4 production in late 1942. Once that decision was made, no further arguments were really needed to justify the rocket program. Barring an insurmountable technical problem, the development, production, and deployment of the ballistic missile would become a fact, however illogical and "early" that might be.

The Führer's endorsement, in combination with the Army's decline, greatly strengthened the role of the Armaments Ministry, but it also brought another power into the program: the SS. Heinrich Himmler, in his relentless empire-building, tried and eventually succeeded in grabbing large pieces of the rocket program. But not all the impetus came from the SS. Problems finding a production labor force in the face of severe manpower shortages also drew the program's senior managers quite voluntarily into the exploitation of SS concentration camp labor. That fact raises a final key question about the rocket program: How "Nazi" was it or, less crudely put, how much influence did the ideology and practices of National Socialism have on Peenemünde and the program?

The most "Nazi" aspect of the Army rocket program was, in fact, the employment of slave labor in A-4 production, even though the decision to do so was a 1943 improvisation. The exploitation of concentration camp prisoners in the war economy was entirely typical of the Third Reich after 1942, and it was rooted in a Nazi racial hierarchy that many Germans took for granted. By contrast, nothing in the original conception of the program or its technology had identifiably National Socialist ideological roots; Becker's and Dornberger's military ideas originated in the thinking of the ultraconservative Weimar officer corps, and their technological enthusiasm for rocketry was bolstered by the spaceflight movement of that era. The rapid adoption of Peenemünde's technology and even its engineering management structures by foreign powers after World War II also suggests that slave labor was the one uniquely "Nazi" aspect of the rocket program.

Yet looking at the problem in this way would be misleading. The leap at such an early date from small-scale rocket research to a massive program would not have occurred without National Socialism; Peenemünde grew and flourished under Hitler because of the very nature of his regime. As a result, the rocket program built an institution and a weapon that made little sense, given the Reich's limited research resources and industrial capacity—a perfect symbol of the Nazis' pursuit of irrational goals with rational, technocratic means.

The top leaders of the program also compromised themselves thoroughly with National Socialism in order to achieve their technical goals. Immediately after the seizure of power, Becker and his subordinates quite willingly used the new police state to suppress the amateur rocket groups, with the aim of creating, in modern military parlance, a super-secret "black program." Another case was Wernher von Braun, who essentially made a pact with the devil in order to build large rockets. Although he became disillusioned toward the end of the regime, that did not alter his basic motivations; after the war he bore proudly the nominal reasons for his arrest—putting spaceflight before military missile work—but there is no evidence that he ever stuck his neck out for the concentration camp prisoners before his arrest, nor did he show any obvious pangs of conscience about their fate until the 1960s and

1970s, when protests by French prisoner survivors forced him to confront the issue more directly.

The German Army rocket program was thus greatly influenced by—and integrated into—the structures and practices of the Nazi regime, whatever its ideological and technological origins. The ease with which its military and civilian leadership became involved in mass slavery in order to achieve technical and military ends is certainly one of Peenemünde's most troublesome legacies to the world. But a much more ambiguous legacy was the big rocket itself. The A-4/V-2 was and is the grandfather of all modern guided missiles and space boosters. Some of its successors—the Redstone, the Saturn V, the R-7 *Semyorka,* and the Ariane—have put application satellites into space, scientific instruments on the planets, and humans on the face of our nearest celestial neighbor. At the same time, the A-4's successors have threatened us for fifty years with nearly instantaneous nuclear destruction, and will continue to do so, despite the end of the Cold War. Starting from unlikely, even utopian origins in the Weimar spaceflight movement, and ending even more strangely with ineffective weapons and emaciated slaves, the German Army rocket program and its Peenemünde center without a doubt changed the face of the twentieth century.

Appendix 1

The German Army Ordnance Liquid-Fuel Rocket Series

A-1 The first *Aggregat* was Wernher von Braun's initial attempt at building a liquid-oxygen/75% alcohol rocket vehicle. Powered by a nominal 300-kg-(660-lb)-thrust engine, it featured a large stabilization gyro in the nose. It was abandoned in favor of the A-2 in early 1934.

A-2 The redesign of the A-1 entailed moving the gyro to the middle of the rocket in order to increase stability in flight and to separate the two propellant tanks. Two A-2s, *Max* and *Moritz,* were successfully launched in December 1934. Their initial weight (107 kg, or 235 lb), length (1.61 m or 5.3 ft), and diameter (31.4 cm, or 1 ft) were about the same as the A-1.

A-3 Designed and built 1935–37, four A-3s were launched in December 1937, but all failed because of guidance system inadequacies. Powered by a nominal 1,500-kg-(3,300-lb)-thrust engine, they were 6.5 m (22 ft) long and 0.7 m (2.3 ft) in diameter, with a fueled weight of 750 kg (1,650 lb).

A-4 Better known by its 1944 Propaganda Ministry designation V-2 (*Vergeltungswaffe 2*), the A-4 was first proposed in 1936 on the basis of a projected 25 metric-ton-(56,000-lb)-thrust engine. It was designed in detail in 1939–41. The warhead was one ton (2,200 lb), of which three-quarters was explosives. Nominal range was 270 km (156 mi), although special test models flew as far as 385 km (239 mi) in late 1944. Length was 14 m (46 ft), maximum body diameter was 1.65 m (5.42 ft), and maximum fin diameter was 3.56 m (11.7 ft). For the production *(Baureihe B)* version, empty weight was 4 metric tons (8,800 lb), and fueled weight was about 12.8 tons (28,200 lb). First launched in June 1942, about 6,000 were built and about 3,200 fired in anger.

A-4b See A-9.

A-5 The A-3 was redesigned in 1938 with new tail fins and no high-altitude instrument package. Its function was solely to test guidance systems. First fired without guidance in October 1938, from October 1939 to mid-1943 about three dozen A-5s were launched with three different guidance systems: Kreiselgeräte's, Siemens's, and the so-called Rechlin or Möller system.

A-6 The improved version of the A-5, with a much shortened engine, was canceled in September 1939 in order to concentrate resources on the A-4. None were ever built. In one December 1939 document, Dornberger used the designations A-6 and A-7 to indicate successor vehicles to the A-4, but this usage was not standard.

A-7 This subscale A-9 was an A-5 with wings. In October 1942 the project was canceled before any powered versions could be completed. Two engineless models were drop-tested from aircraft in late 1942, but neither glide was notably successful.

A-8 On February 5, 1941, Thiel and von Braun inspected Helmut von Zborowski's nitric acid/diesel oil rocket-engine development at BMW in Berlin-Spandau. That summer Thiel began his own testing of this hypergolic (self-igniting) and noncryogenic propellant combination. The designation A-8, which probably had already been assigned to an improved A-4 concept, was reconceptualized as a simplified A-4 with a projected 30-ton-(66,000-lb)-thrust hypergolic engine of this type. Range could have been improved to 450 km (280 mi), but the concept fell out of favor in mid-1942 for engineering and political reasons.

A-9 This A-4 with wings was first dubbed "Glider A-4" and the "Winged Projectile" *(Flossengeschoss)* in 1939 before receiving the designation A-9 in 1940. It was studied as a missile of 500-km range and as the upper stage of an A-9/A-10 combination of more than 5,000-km (3,000-mi) range. Research was suspended from October 1942 to June 1944. In early October 1944, the revived A-9 received the designation A-4b in order to use the priorities of the A-4 program. Two "bastard" versions (using A-4 tails with enlarged air vanes) were launched in the winter of 1944–45.

A-10 No later than 1936 the Army rocket group planned an A-4 successor with a 100-metric-ton-(220,000-lb)-thrust liquid-oxygen/alcohol engine, a warhead of 4 tons (8,800 lb) and a range of more than 500 km (300 mi). Peenemünde facilities built from 1936 to 1939 were designed to accommodate it. Priority problems in 1939–40 pushed the concept into the background, but the A-10 had a second life as the projected booster for the A-9. In 1941 planned thrust was increased to 180 tons (400,000 lb) in order to reach the United States from Western Europe. The A-9/A-10 was never more than a drawing-board concept and was shelved in 1942.

A-11 During the heyday of the Project Office in 1940–41, the idea of a three-stage A-9/A-10/A-11 ICBM or satellite booster was bandied about but was never seriously studied.

A-12 Equally vague and hypothetical was an A-10/A-11/A-12 launch vehicle for heavy space payloads.

B-7 This was the designation for the prototype version of the 1,000-kg-(2,200-lb)-thrust liquid-oxygen/alcohol *Starthilfe*. (The American acronyms for such a system were JATO or RATO, for jet-assisted or rocket-assisted takeoff). Two egg-shaped pods of this thrust were to be strapped on to heavily loaded bombers. Development was initiated by the Luftwaffe in 1938, and testing took place in 1940–41. The designations B-1 to B-6 were apparently not assigned.

B-8 The Schmidding company in Bodenbach was to have manufactured this production version of the B-7, which, after further modifications, was called the B-8a. The Luftwaffe canceled the project in 1942.

C-1 This solid-fuel rocket was proposed in late 1942 as a part of Army Ordnance's contribution to the Luftwaffe anti-aircraft missile program. Its range and maximum altitude were to be 20 km (about 65,000 ft). The solid-fuel rocket side of Wa Prüf 11 was to design the propulsion system, while Peenemünde-East was responsible for the guidance. The project was canceled in the spring of 1943.

C-2 Better known as Wasserfall (Waterfall), this anti-aircraft missile used a scaled-down A-4 fuselage 8.9 m (28.5 ft) in length and 1 m (3.3 ft) in diameter, with a cross-wing and enlarged air vanes. The propulsion system had an 8-ton-(17,800-lb)-thrust engine using a 9:1 mixture of nitric and sulfuric acids as an oxidizer and Visol (vinyl ethyl ether), or Visol cut with other additives, as the fuel. Guidance was to have been through a joy-stick system, to be replaced by an ground-based guide beam and an on-board homing system. About forty test launches were made between February 1944 and February 1945.

Appendix 2

Organizational Structures of the Army Rocket Program

As is true of the Peenemünde story as a whole, much nonsense has been written about the rocket program's organizational nomenclature, owing to poor research and an uncritical attitude toward memoirs. The following summary is based on the close reading of original documents, plus Schubert's invaluable chronicle of the Production Plant (BA/MA, RH8/v.1206–10).

German Army rocket efforts had their origin in Karl Becker's *Heereswaffenamt Prüfwesen, Abteilung 1* (Army Ordnance Office Testing Division, Section 1), acronym Wa Prw 1. The sources often further specify subsection I (Wa Prw 1/I), which was probably under Major von Horstig before he became section head in 1933. In May 1935, rocket work was elevated to a separate subsection, Wa Prw 1/VII. Sometime before late November 1935, the expanded Kummersdorf rocket facilities received the name *Versuchsstelle West* (Experimental Center West).

Basic rocket research was also carried out in Erich Schumann's *Zentralstelle für Heeresphysik und Heereschemie* (Center for Army Physics and Army Chemistry) or Wa Prw 1/Z. Following Becker's appointment as chief of Testing Division in 1933, Schumann's research operation became section Wa Prw 11. After Becker's promotion to chief of Ordnance in 1938, Schumann's section was again directly subordinated to him as the *Forschungs-Abteilung* (Research Section), abbreviated Wa F.

In 1936, Wa Prw 1/VII became a special rocket section under Dornberger. It was called Wa Prw R for two months (July–August), but the name was quickly changed to Wa Prw D, presumably to obscure the section's purpose more effectively. No later than June 1937, it became Wa

Prw 13. In April 1938, after Becker's promotion, his renaming of his former division as *Entwicklungs- und Prüfwesen* (Development and Testing), and a change in acronyms, Dornberger's group became Wa Prüf 11.

The *Versuchsstelle Peenemünde* (Peenemünde Experimental Center), as it was first called, united the Army's *Werk Ost* (East Works) and the Luftwaffe's *Werk West* (West Works) under a *Kommandant* (Commandant). With the air force's withdrawal from the joint structure on April 1, 1938, its side became the *Versuchsstelle* (later *Erprobungsstelle*) *der Luftwaffe Peenemünde-West* (Peenemünde-West Air Force Experimental Center), while the East Works became the *Heeresversuchsstelle Peenemünde* (Peenemünde Army Experimental Center), or HVP. At some point after Leo Zanssen was named Commandant, his title was elevated to *Kommandeur* (Commander), with the rank equivalent to a regimental commander. Zanssen at first reported to Dornberger as subsection Wa Prüf 11/II, while Thiel's propulsion group at Kummersdorf was Wa Prüf 11/V, but those numbers later changed. Rudolf Hermann's wind tunnel group also reported directly to Dornberger, but its acronym is unknown.

In January 1939 the new factory planning group was formed as Wa Prüf 11/VI under *Ministerialrat* (Ministerial Counselor) G. Schubert in Berlin. The factory site itself was called the *Fertigungsstelle Peenemünde* (Peenemünde Production Plant)—acronym FSP. The two sides of the Army complex were combined as the *Heeresversuchsanstalt Peenemünde* (Peenemünde Army Experimental Establishment) on September 23, 1941. The abbreviation HVP remained unchanged, while the former HVP became the *Entwicklungswerk* (Development Works) or EW. The factory was renamed the *Versuchsserienwerk* (Pilot Production Plant) or VW. A January 7, 1942, Dornberger order shortened the center's name to *Heeresanstalt Peenemünde* (Peenemünde Army Establishment) or HAP, probably because it was now a production organization as well as a developmental one. Schubert's Berlin group was not officially resettled in Peenemünde until June 15, 1942, however.

Attempts to organize A-4 mass production began with the October 23, 1941, order forming the *Arbeitsstab Wa A-Vorhaben Peenemünde* (Ordnance Office-Peenemünde Project Working Staff). This committee was partially superseded in July 1942 by Detmar Stahlknecht's *Nachbaudirektion* (Production Planning Directorate), which was renamed the *Technische Direktion Serie* (TDS: Technical Directorate for Mass Production) in early 1943, paralleling Wernher von Braun's position as *Technische Direktor* (TD: Technical Director) for development. Both reported through Gerhard Stegmaier, the *Leiter* (Chief) of the Development Works (acronym EW/L), to the Commander (HAP/L or HAP/Kdr.). With Speer's order on December 22, 1942, to form the *Sonderausschuss A4* (A-4 Special Committee), the two preceding production organizations faded away by the autumn of 1943. The Committee was expanded to become *Sonderausschuss V-Waffen* (Special Committee for V-Weapons) in August 1944 but used cover names—usually *Sonderausschuss z.b.V.* Paul Storch's *Sonderausschuss elektrischer Zubehör für R-Geräte* (Special Committee for Electrical Equipment on R[ocket]-Devices) apparently led a separate existence from roughly mid-1943 to mid-1944.

A May 17, 1943, order from General Fromm, effective June 1, renamed the center *Heimat-Artillerie Park 11* (Home Artillery Park 11) or HAP 11, with Karlshagen replacing Peenemünde as the official location. After the August 18 air raid, the VW and EW were merged, preparatory to moving production underground. On September 4, Fromm named Dornberger *Beauftragter zur besonderen Verwendung Heer* (Army Commissioner for Special Tasks), or BzbV Heer. The rocket general had received the title *Der Beauftragte des Wa A für das A4-Programm* (The Ordnance Office Commissioner for the A-4 Program) in June, but now he was removed from Ordnance and Wa Prüf 11 altogether.

Dornberger also received the title *Artilleriekommandeur* (Artillery Commander) *191* or *Arko 191* after Hitler gave him tactical command of the A-4 on October 4, 1943. To that title was added the adjective *Höhere* (Higher), probably in December, making him *Harko 191,* but soon thereafter he lost the position. Meanwhile a December 1 Hitler order created *Generalkommando LXV. AK z.b.V.* (General Command, LXV Army Corps for Special Tasks) to control all V-weapons. After SS Major General Hans Kammler seized the A-4 mobile units at the end of August 1944, he reconstituted them as *Division z. V.* (*zur Vergeltung,* "for Vengeance"). About the same time Himmler named him *Sonderbeauftragter 2 des Reichsführers-SS* (Special Commissioner 2 of the Reichsführer-SS), acronym Sb 2.

On December 15, 1943, the liquid-fuel rocket program was removed from Wa Prüf 11 and constituted as Wa Prüf 10; solid-fuel rocketry remained under the old designation. Wa Prüf 10 became Wa Prüf (BuM) 10 in July 1944 or thereabouts, reflecting the separation of the Ordnance Development and Testing Division into two divisions, *Ballistik und Munition* (Ballistics and Munitions) and *Waffen und Gerät* (Weapons and Equipment), paralleling an earlier split of the *Industrielle Rüstung* (Industrial Armament, that is, Procurement) Division in 1940.

During the last two years of the war a number of limited-liability companies (*Gesellschaften mit beschränkter Haftung* or GmbH), technically private but government-owned, were created for the rocket program. The first was *Mittelwerk GmbH* (Central Works, Ltd.), on September 24, 1943, followed by the *Wasserbauversuchsanstalt GmbH* (Hydraulic Engineering Experimental Establishment, Ltd.) for the wind tunnel group that was evacuated to Kochel, Bavaria. A *Steinbruchverwertungs-Gesellschaft mbH* (Quarry Utilization, Ltd.) was created in late 1943 or early 1944 to run the *Vorwerke* (Engine Test Works) at Schlier and Lehesten, but it was later absorbed by the Mittelwerk. On August 1, 1944, the development sections of HAP 11 became the *Elektromechanische Werke GmbH* (Electromechanical Industries, Ltd.), acronym EW. The remaining Army base administration became the *Versuchsplatz Karlshagen* (Karlshagen Test Range).

On January 13, 1945, Speer ordered the formation of the *Arbeitsstab Dornberger* (Dornberger Working Staff) within the Armaments Ministry (which was called the *Reichsministerium für Bewaffnung und Munition* from March 1940 to September 1943 and the *Reichsministerium für Rüstung und Kriegsproduktion* thereafter). On February 6 Kammler subordinated the Working Staff Dornberger to himself. Göring had named him the head of a *Programm "Brechung des Luftterrors"* (Program for Breaking the Air Terror) on January 26, as well as V-1 commander. In the latter capacity, he folded the V-1 units into an upgraded *Armeekorps z.V.* (Army Corps "for Vengeance"). After the evacuation of Peenemünde to Thuringia, the EW became the core of the *Entwicklungsgemeinschaft Mittelbau* (Central Construction Development Cooperative). With the further evacuation of five hundred personnel to Bavaria, the Peenemünde organization can be said to have effectively dissolved.

Significant Abbreviations Used in the Notes

Arch.	Peenemünde Archive Report
Arm.Min.	Armaments Ministry
BA	Bundesarchiv Koblenz
BA/MA	Bundesarchiv/Militärarchiv Freiburg
BDC	Berlin Document Center
CTR	Chemisch-Technisches Reichsanstalt
DM	Deutsches Museum Munich
EW	Entwicklungswerk or Elektromechanische Werke
FE	Fort Eustis
FOIA	Freedom of Information Act
FSP	Fertigungsstelle Peenemünde
HAP	Heeresanstalt Peenemünde
HAP 11	Heimat-Artillerie-Park 11
HVP	Heeresversuchsstelle (or -versuchsanstalt) Peenemünde
INSCOM	U.S. Army Intelligence and Security Command
IWM	Imperial War Museum
KG	Kreiselgeräte GmbH
KL	Konzentrationslager
LC	Library of Congress
LZ	Luftschiffbau Zeppelin GmbH
NA	National Archives
NASM	National Air and Space Museum
OKH	Oberkommando des Heeres

OKW	Oberkommando der Wehrmacht
OHI	Oral history interview (deposited at NASM)
OSI	Office of Special Investigations, U.S. Justice Dept.
RLM	Reichsluftfahrtministerium
RSIC	Redstone Scientific Information Center
SRCH	Space and Rocket Center, Huntsville
VW	Versuchsserienwerk
WD	Walter Dornberger
WvB	Wernher von Braun

Notes

The abbreviated German date system has been used in the notes for clarity and brevity: day.month.year. Where possible, names have been given in primary document citations, although the originals may have been addressed to or from the acronym of a person's office. When the original author of a document is known and the document was signed by someone else, the actual author's name is given in brackets. Full citations of secondary works can be found in the Bibliography.

Prologue: Summer 1943

1. Huzel, *Peenemünde,* 44.
2. *Ibid.,* 30–38; Schubert chronicle, 17.6.43, 11.7.43, and 16.7.43, in BA/MA, RH8/v.1210.
3. Huzel, *Peenemünde,* 38–40.

1. The Birth of the Missile

1. Willy Ley, "Review in Retrospect" (ms., 1947), NASM, Willy Ley Collection, box 2701, folder 200.
2. Schneider, "Technik," 236; Hölsken, *V-Waffen,* 15; Müller, "World Power Status," 182; Ley, *Rockets,* 168–74.
3. WD, *V-2,* 19; WD, "Denkschrift," c. late 1943, DM (page 6 missing from copy in NASM, FE496); Ordway and Sharpe, *Rocket Team,* 17; Winter, *Prelude,* 51; Carroll, *Design,* 59–61; Homze, *Arming,* 8–10, 21–22, 39–41.
4. Oberth, *Rakete;* Neufeld, "Weimar Culture," 725–30; Winter, *Prelude,* 22. Unless otherwise noted, all translations are my own.

5. Neufeld, "Weimar Culture," 729–30; Winter, *Prelude,* 21–22; Goddard, "A Method"; Goddard, "Liquid-propellant Rocket Development" (1936).
6. Neufeld, "Weimar Culture," 731–41; Winter, *Prelude,* 35–37; Ley, *Rockets,* 105–23.
7. Neufeld, "Weimar Culture," 725–28, 742–52.
8. Oberth, *Wege,* 199–200.
9. WD, "Denkschrift," c. late 1943, 6–7, NASM, FE496, and DM; Schneider, "Technik," 236; WD, *V-2,* 19; WD, "European Rocketry," 249; Keilig, *Deutsche Heer,* 211/19, 211/68, 211/145, 211/302, 211/373.
10. Philipps, "Karl Becker," 293; Ebert and Rupieper, "Technische Wissenschaft," 469–70; Keilig, *Das Deutsche Heer,* 211/302, 211/373; Winter, *Prelude,* 52. That Becker had been aware of the spaceflight literature for some years is shown by Cranz's skeptical discussion of Oberth and Goddard in volume 2 (1926) of his textbook, Cranz, *Lehrbuch,* 2:402–19, 437. Becker had assisted in revisions to volume 1.
11. WD, *V-2,* 20; WD, "Lessons," 18; WD to Oberth, 12.6.64, in an unpublished Festschrift for Oberth in NASM file "Hermann Oberth"; Reisig OHI, 1989, 57–58; WvB, "Behind the Scenes," 8, 20, SRCH, WvB Papers. Whether Dornberger was actually hired first to look into liquid-fuel rocketry or only later was assigned to that task is unclear from the contradictory accounts in his book and memoir articles. In general, Dornberger's various memoirs are absolutely indispensable but not highly trustworthy. His memory was faulty on many details, and he suppressed or distorted damaging information, most notably regarding his dealings with Hitler, the SS, and concentration camp prisoners.
12. Neufeld, "Weimar Culture," 730–31. On the military-industrial complex and rocketry, see McDougall, *Heavens.*
13. Winkler, "Rückstoss-Arbeiten Winkler," 8.5.43, DM; Kunze, "Zusammenarbeit," 71–75. The powder rocket manufacturer Friedrich Sander, who had been involved in the Opel–Valier stunts, may have secretly built and launched a liquid-fuel rocket in 1929, but nothing further came of that work. Winter, *Prelude,* 37; Sänger-Bredt and Engel, "The Development," 221–22.
14. Franklin, *American,* 18–19; Rudolph OHI, 7–8; Essers, *Max Valier,* 247–65; Riedel, "A Chapter," 208–12; Riedel, "Raketenentwicklung mit flüssigen Treibstoffen" (ms., 1950), 9–10, IWM, German Misc. 148. See Winter and Neufeld, "Heylandt's Rocket Cars," for a fuller examination of this group.
15. Oberth to Dickhuth-Harrach, 11.1.34, in BA/MA, RH8/v.1226; Ley, *Rockets,* 124–27; Winter, *Prelude,* 38–39.
16. Ley, *Rockets,* 127–32; Barth, *Hermann Oberth,* 139–53; Neufeld, "Weimar Culture," 738–41.

17. Becker to Wimmer, 12.5.31, in BA/MA, RH8/v.1226; Nebel, *Narren,* 72–75; Oberth to Dickhuth-Harrach, 11.1.34, in BA/MA, RH8/v.1226. Nebel's memoirs are of questionable reliability, but some of his wilder stories are confirmed by other sources.
18. Magnus von Braun, *Weg,* 87–193; WD, *V-2,* 27; Ley, *Rockets,* 133–35; Ritter/CTR to Stucktay/Notgemein. d. Dt. Wissenschaft, 5.11.30, in IWM, MI 14/801(V).
19. Ley, *Rockets,* 136–39; WvB drawings of Mirak, 3.8.30, and Kegeldüse, 12.9.30, in SRCH, WvB Papers.
20. Becker to Wimmer/Wa Prw 8, 12.5.31, in BA/MA, RH8/v.1226 and IWM, MI 14/801(V). See also Wimmer to Becker, 6.5.31, in latter.
21. WvB, "Behind the Scenes of Rocket Development in Germany 1928 through 1945" (ms., late 1940s), 6, SRCH, WvB Papers. A heavily edited version was published as "Reminiscences of German Rocketry."
22. Marionoff, quoted in Winter, *Prelude,* 42; WvB, "Behind the Scenes," 6–7, SRCH, WvB Papers; Rolf Engel interview by Neufeld, 1991.
23. Ley, *Rockets,* 140–51; Winter, *Prelude,* 43.
24. Ley, *Rockets,* 140–54; WvB, "Behind the Scenes," 6–7; Winter, *Prelude,* 41–43; Nebel, *Narren,* 99–116; Sänger-Bredt and Engel, "Development," 218–21.
25. On the Paris Gun, see Klein, *Vom Geschoss,* 28–35, 56–57, 72–73; Ludwig, "Die 'Hockdruckpumpe'," 144; WD, *V-2,* 47–48; WD, "European Rocketry," 253–54; WD, "German V-2," 398. On surprise in Becker's thought, see Schumann, "Wehrmacht," 135. The solid-fuel rocket program's close connection to chemical warfare, leading to the battlefield *Nebelwerfer* of World War II, and Oberth's discussion of the ballistic missile's utility for long-range poison gas attacks make it certain that Becker and his subordinates also discussed the possibility of using liquid-fuel missiles for chemical warfare against civilians. But the only document that even hints that they advocated it is the war diary of General Franz Halder, Chief of the Army General Staff (1938–42). Halder records a tour of a chemical weapons plant with Becker on September 26, 1939. Poison gas and the use of the long-range rocket against London are mentioned but are not clearly linked; see Halder, *Kriegstagebuch,* 1:85. Müller, "World Power Status," 190, overinterprets this cryptic entry. If Ordnance believed in the 1930s that ballistic missiles would be decisive if equipped with chemical warheads, all documents mentioning this idea must later have disappeared.
26. Becker to AG f. Industriegasverwertung, 16.10.31, in NASM, FE724/a; Winter, *Prelude,* 52. By some accounts there was an important meeting on December 17, 1930, in which Colonel Karlewski, head of the Testing Division, backed increased funding for rocket development. But in the

absence of original documents such exactitude is suspicious; the chronology in the memoirs of this period is often demonstrably in error.

27. Riedel, "Raketenentwicklung," 11–14, IWM, German Misc. 148; AG f. Industriegasverwertung to von Horstig, 20.11.34, in NASM, FE737; Rudolph OHI, 6–15; Franklin, *American,* 24–27; Ley, *Rockets,* 146; Winter and Neufeld, "Heylandt's Rocket Cars."
28. Frey (for Hess and Hitler) to Nebel, 19.2.30, and Grundtmann (for Göring) to Nebel, 18.3.32, in NA, T-175/155/2685593–94; Nebel to Bodenschatz, 23.8.33, in IWM, MI 14/801(V); Nebel, *Narren,* 16–17; Nebel handbill in Winter, *Prelude,* 174; Horeis, *Rolf Engel,* 17–18; W. Kechmann article, *Berliner Zeitung,* 12.6.32, in WvB Papers, LC, box 53, scrapbook 1; Baumgarten-Crusius, *Rakete.*
29. Heylandt-Wa Prw 1 (Section 1) correspondence, 28.10.31–29.4.32, and von Horstig report, 2.5.32, in NASM, FE724/a; Reisig, "Peenemünder 'Aggregaten'," 46.
30. Heylandt-Wa Prw 1 correspondence, 1.10.32–21.11.32, in NASM, FE724/a; AG f. Industriegasverwertung to von Horstig, 20.11.34, in NASM, FE737; Wa Prw 1 documents on Belz, 21.12.31–25.6.32, in NASM, FE366/3.
31. Nebel, *Narren,* 133–35; WvB, "Behind the Scenes," 8, SRCH, WvB Papers. Ley, *Rockets,* 155–56, claims that the contact was initiated by Nebel, who wrote a "Confidential Memo on Long-Range Artillery," but it is possible that Ley confused the events of 1930 with those of 1932.
32. WvB, "Behind the Scenes," 8, SRCH, WvB Papers; Nebel, *Narren*, 135–37; Becker to Schiessplatzkommando Z, Kummersdorf, 6.6.32, Schneider report, 23(?).6.32, and Schumann (Wahmke) report, 1.7.32, in IWM, MI 14/801(V); Ebert and Rupieper, "Technische Wissenschaft," 471; Schumann files, BDC.
33. Schneider report, 23(?).6.32, in IWM, MI 14/801(V).
34. *Ibid.*
35. WvB, "Behind the Scenes," 8–9, SRCH, WvB Papers; WD, "Denkschrift," c. late 1943, in NASM, FE496; Ley, *Rockets,* 143–44.
36. Becker to Schumann *et al.*, 25.6.32, and attached drawing, 24.6.32 in NA, T-78/177/6116510–11; WD, *V-2,* 23–24; WvB, "Behind the Scenes," 11, SRCH, WvB Papers.
37. WvB, "Behind the Scenes," 9–10, SRCH, WvB Papers; Nebel, *Narren,* 138. Rolf Engel remembers von Braun making such a statement in the early 1930s. Engel interview.
38. WvB, "Behind the Scenes," 10, SRCH, WvB Papers. For the bland published version see WvB, "Reminiscences," 130. On von Braun's politics, see Ley, "Count von Braun," and Engel's impressions in Horeis, *Rolf Engel,* 24.

39. WD, *V-2,* 27. WvB's "Protokoll" of his Gestapo interview, 16.7.34, in IWM, MI 14/801(V), says December 1. The October 1 date in *V-2* is incorrect. According to the late 1943 anonymous typescript, "Werdegang . . . des Professors Dr. von Braun," in NASM, FE341, he took his *Vorprüfung* in mechanical engineering (roughly equivalent to a bachelor's degree) on November 3, 1932, and joined Ordnance the same day. Other secondary sources say November 1, but the 1934 document is closest to the original event.
40. Ebert and Rupieper, "Technische Wissenschaft," 471–72; Schumann, "Wehrmacht," 135–37; Ordnance-WvB contract, 4.4.33, in SRCH, WvB Papers.
41. Magnus von Braun, *Weg,* 234, 263; Hüttenberger, "Polykratie"; Kershaw, *Nazi Dictatorship,* 65–81.
42. WD, "Denkschrift," c. late 1943, NASM, FE496.
43. Horeis, *Rolf Engel,* 45–49; Engel interview; Päch, "Rolf Engel," 232.
44. Horeis, *Rolf Engel,* 47–49; Engel interview; Franz Mengering, "Die Magdeburger Pilotenrakete," *Corpsstudentische Monatsblätter* 41 (March 1933): 140–41, in IWM, MI 14/801(V); Winkler, "Rückstoss-Arbeiten Winkler," 8.5.43, DM.
45. Winter, *Prelude,* 44–46; Nebel, *Narren,* 125–28; Ley, *Rockets,* 154–59; *Raketenflug,* no. 8 (April 1933), and other clippings and publications in IWM, MI 14/801(V).
46. Winter, *Prelude,* 46–47; Raketenflugplatz publications and newspaper clippings, 1933–34, in IWM, MI 14/801(V).
47. Nebel to von Levetzow, 11.7.33, and Nebel to Bodenschatz, 23.8.33, and associated correspondence in IWM, MI 14/801(V).
48. Schneider Aktennotiz, 14.10.33, in IWM, MI 14/801(V).
49. Ley, *Rockets,* 157–58; von Dickhuth-Harrach to Nebel, 26.10.33, prosecutor's report, 30.11.33, and Fritz Beck/VfR to Wa Prw 1/I, 15.1.34, in IWM, MI 14/801(V).
50. Becker (Schneider) to V2, 22.12.33, and related documents in IWM, MI 14/801(V); Raabe OHI.
51. Horeis, *Rolf Engel,* 22–23, 50–51; WvB "Protokoll" of Gestapo interview, 16.7.34, in IWM, MI 14/801(V). Becker's rank was *Generalmajor* (Major General), but I shall translate ranks according to their American equivalents. In the German system of that time, *Generalmajor* was the lowest general officer's rank.
52. Winter, *Prelude,* 47–50; Becker (Schneider) to V2, 22.12.33, and von Horstig marginal notation on *Technik voran!* 15 (November 5, 1933) in IWM, MI 14/801(V); WD to W.A.(I), (?).7.34, and Arndt/RLM to Ordnance, 18.7.34, in NASM, FE366/3.
53. Horeis, *Rolf Engel,* 50; Nebel, *Narren,* 139; Polizei-Major Funcke/Ham-

burg to Bodenschatz, 27.7.34, in IWM, MI 14/801(V). When I interviewed Engel, he denied that he had met Seldte, as is asserted in Gartmann, *Men,* 96–97, so the contact may have come through Nebel. Any discussion of Engel's membership in Nazi organizations is drawn from his SS officer file in the BDC and *not* from the interview.

54. Wolfke/Warsaw to Nebel, 13.1.34, Nebel to Hitler, 20.2.34, Schneider Aktennotizen, 23.2.34 and 16.3.34, in IWM, MI 14/801(V).
55. Stud (Schneider) to Röhm through Wehrmachtsamt, mailed 10.3.34, and Becker (Schneider) to Wa Wi, 3.5.34, in IWM, MI 14/801(V).
56. Marginal notations of von Horstig and Schneider on Nebel to Reich Finance Minister, 20.2.34, Schneider Aktennotiz, 28.5.34, and Ohnesorge/Post Ministry to Seldte, 15.6.34, in IWM, MI 14/801(V).
57. Von Horstig (Schneider) to Abwehr, 4.6.34, Schneider Aktennotiz, 28.6.34, and Zwengauer (Dornberger) to Abwehr, 7.7.34, in IWM, MI 14/801(V); Nebel, *Narren,* 139; Gartmann, *Men,* 97.
58. Nebel, *Narren,* 139–40; Dornberger Aktennotiz, 10.7.34, in IWM, MI 14/801(V).
59. On Nebel, see IWM, MI 14/801(V), BA/MA, RH8/v.1226 and NA, T-175/155. Engel has made contradictory statements about his group's fate. In one version, Walter Thiel and Wernher von Braun, among others, pressured him to work for Ordnance. He was then forced to close the group when Hitler issued an order giving the Army a monopoly over rocketry. But no written evidence for such an order exists. Horeis, *Rolf Engel,* 51–52; Gartmann, *Men,* 97–98; and Päch, "Rolf Engel," 234. In the other version, he was threatened by a Gestapo representative, but surveillance stopped when a fellow student leader introduced him to Reinhard Heydrich, head of the SS Security Service, or SD, and administrator of the Gestapo. The group folded when he was transferred to the Nazi student leadership in Munich in 1935. Engel interview. Engel's "Lebenlauf" of 16.2.40 in his SS officer file, BDC, shows that he probably became an SD informant in 1936.
60. Wa Prw 1 material on Brügel, 22.6.34–30.6.35, in NASM, FE366/3.
61. Rudolph OHI, 17–28; Franklin, *American,* 38–43; Ordnance–Pietsch contract and statement, 15.5.33, in BA/MA, RH8/v.1225; WvB evaluation of Rudolph motor, 18.8.34, in NASM, FE727/c; Rudolph file, BDC.
62. Material on inventors is in BA/MA, RH8/v.1221–26, some of which is in NASM, FE366. The Oberth material is in RH8/v.1226. See also WD, "Denkschrift," c. late 1943, FE496.
63. WvB, "Behind the Scenes," 11, SRCH, WvB Papers.
64. WD, "German V-2," 395; AG f. Industriegasverwertung to Schneider, 10.11.32, in NASM, FE724/a.

65. Wa Prw I-Heylandt correspondence, 1.10.32–15.12.33, in NASM, FE724/a.
66. Riedel, "Raketenentwicklung," 24, IWM, German Misc. 148; WvB, "Behind the Scenes," 12–13, SRCH, WvB Papers. WD, *V-2,* 27, erroneously places Riedel at Kummersdorf in late 1932.
67. WvB, "Beiträge," 29–31; WD, *V-2,* 23–26; WvB, "Behind the Scenes," 11, SRCH, WvB Papers.
68. WvB memo, 14.12.33, in NASM, FE727/c; WvB, "Behind the Scenes," 11, SRCH, WvB Papers.
69. Wa Prw I correspondence and documents on aluminum firms, 13.4.33–15.9.33, in NASM, FE744; WvB, "Beiträge," 30.
70. WvB, "Beiträge," 31–35; WvB memo, 14.12.33, in NASM, FE727/c.
71. Chefkonstrukteur/Wa Prw 3 to von Horstig, 28.6.33, in NASM, FE74/b; WvB, "Beiträge," 35, 43; WD, *V-2,* 32–33.
72. WvB, "Behind the Scenes," 11–12, SRCH, WvB Papers; WvB, "Beiträge," 37, 43–44; Riedel, "Raketenentwicklung," 24–25, IWM, German Misc. 148. The oft-repeated assertion that the A-1 was abandoned because it was "nose heavy" is misleading.
73. WvB, "Beiträge," 26–28; Reisig, "Peenemünder 'Aggregaten'," 44–45.
74. WvB, "Beiträge,"; WvB doctoral exam records, 16.5.34–27.7.34, in Humboldt University archive, Phil. Fak. 759.
75. Keilig, *Das Deutsche Heer,* 211/19; Schumann, "Wehrmacht," 137; WD, *V-2,* 29.
76. Schneider order, 30.11.34, and Allgemeines Heeresamt to von Horstig, 11.12.34, in BA/MA, RH8/v.1945; WD denazification questionnaire, 14.7.47, in NA, RG 319, IRR files, WD dossier; Rudolph OHI, 51–52; WD, *V-2,* 36; Ordway and Sharpe, *Rocket Team,* 23–24.
77. WvB report on A-2 launches, 28.1.35, in BA/MA, RH8/v.1945.

2. *The Founding of Peenemünde*

1. Schneider order, 15.1.35, in BA/MA, RH8/v.1945; WvB, "Denkschrift," 18.1.35, in NASM, FE727/a.
2. WvB, "Denkschrift," 18.1.35, in NASM, FE727/a.
3. Von Horstig to Becker, 4.2.35, and related documents in BA/MA, RH8/v.1260; Riedel, "Raketenentwicklung," 32–33, and figs. 55, 56, 56a, IWM, German Misc. 148; WvB to von Horstig, 23.11.35, in NASM, FE727/a.
4. Zanssen note, 20.5.35, and marginal comment by "D." on von Horstig (Zanssen) to Wimmer/RLM Tech. Office, 22.5.35, in NASM, FE746; Baeumker to Becker, 10.10.34, in IWM, MI 14/801(V).

5. WvB, "Behind the Scenes," 16, SRCH, WvB Papers; von Richthofen to von Horstig, 6.2.35, Zanssen (WvB) report, 16.2.35, von Horstig (Zanssen) to RLM/LC II, 19.2.35, and Junkers report of 18.12.34 in BA/MA, RH8/v.1221.
6. Zanssen and WvB report, 26.3.35, on Schmidt visit, in DM, (FE722); Schmidt, "On the History. . . ," in Benecke and Quick, *History,* 375–84.
7. Walter–Ordnance correspondence, 15.10.34–29.11.34, in NASM, FE727/c; H. Walter, "Development of Hydrogen Peroxide Rockets in Germany," in Benecke and Quick, eds., *History,* 263, and I. Sänger-Bredt, "History. . . ," 326, *idem;* Becker (Zanssen) to RLM, 9.4.35, and WvB note to Zanssen, 30.6.35, in DM, (FE722).
8. Constant, *Origins,* 178–204; Zanssen note, 20.5.35, and Aktennotiz, 22.5.35, in NASM, FE746.
9. Von Horstig (Zanssen) to Wimmer, 22.5.35, in NASM, FE746.
10. WvB, "Stellungnahme," 27.6.35, and von Horstig minutes of 27.6.35 meeting in NASM, FE746.
11. WvB, "Stellungnahme," 27.6.35, NASM, FE746.
12. WvB NSDAP file card, BDC; WD denazification questionnaire, 14.7.47, NA, RG 319, IRR files, WD dossier; Nuss, "Einige Aspekte," 440–42. Dornberger in *V-2,* 20, and "German V-2," 394–95, asserts that when Ordnance had gone looking for contractors in the early 1930s, no one was competent or interested. That is at best a half-truth, as the fate of the Heylandt group demonstrates. See also Klein, *Vom Geschoss,* 94.
13. Von Horstig minutes of 27.6.35 meeting in NASM, FE746; von Richthofen minutes of 27.5.35 RLM/LC II meeting in NA, T-971/73 (no frame nos).
14. Draft agreement, 2.9.35, of Ordnance, RLM, Heinkel and Junkers, WvB to Lorenz/RLM/LC I, 25.10.35, Kirchhoff to WvB, 10.1.36, Kirchhoff to DVL, 10.1.36, and WvB to Lorenz, 24.4.36, in NASM, FE746; WvB to von Horstig, 23.11.35, in NASM, FE727/a.
15. WvB minutes of 16.10.35 meeting in NASM, FE746. Ernst Heinkel claims in his memoirs that he "met" von Braun in November and set up a rocket aircraft project as a private venture that had little or nothing to do with the Luftwaffe and Ordnance. The foregoing document shows this story to be nonsense. Heinkel, *Stürmisches Leben,* 448–50; Constant, *Origins,* 197–98.
16. Constant, *Origins,* 198; Heinkel works to WvB, 5.12.35, and WvB to Alpers/RLM/LC II, 14.12.35, in NASM, FE746.
17. WD, *V-2,* 38, 40.
18. WvB, "Behind the Scenes," 18, SRCH, WvB Papers; WD denazification questionnaire, 14.7.47, in NA, RG 319, IRR files, WD dossier; WD, *V-2,* 40.

19. WD, *V-2,* 40–41.
20. Von Richthofen minutes of 6.1.36 meeting, in NA, T-971/73 (no frame nos.); "Lageplan-Skizze" of 18.1.36 in BA/MA, RH8/v.1945; WD, *V-2,* 37–38; Rudolph OHI, 33–35.
21. WvB, "Reminiscences," 135. Unfortunately the corresponding page is missing from the manuscript version. This published version was heavily rewritten by an editor and no doubt overdramatizes the story even more than the original.
22. WD, "German V-2," 397; WD, *V-2,* 38–39; WvB, handwritten "Kostenaufstellung betr. Übungsplatz Peenemünde," c. 4.36, in BA/MA, RH8/v.1945. What Göring thought of the interservice rocket alliance is unknown, but he must have at least approved of it.
23. WD, "European Rocketry," 253–54. See also WD, *V-2,* 47–48, and WD, "German V-2," 398.
24. Ludwig, "Die 'Hockdruckpumpe'," 144; Klein, *Vom Geschoss,* 28–35, 56, 72–73.
25. David Irving in *Mare's Nest,* 18–19, has already made some of the same points. On the possibility that Ordnance was interested in using the ballistic missile for gas warfare, see note 25 of Chapter 1.
26. Murray, *German Military Effectiveness,* 1–38.
27. Schabel, "Wunderwaffen?" 19.
28. Genstbd.H.3.Abt.Att.Gr.I to Ordnance, 4.2.36, forwarding 4.1.36 "Science News Letter," in BA/MA, RH8/v.1945; von Horstig to Akimoff, 25.8.36, in NASM, FE366/4; Goddard, "Liquid-propellant Rocket Development"; WD, "German V-2," 400; WD Vortragsnotiz, 14.12.39, and WD to Leeb, 26.8.40, in NASM, FE349.
29. Germany, *Das Deutsche Reich,* 1:371–532.
30. WD, *V-2,* 39–41; von Richthofen report on 1.4.36 meeting with Kesselring in NA, T-971/73 (no frame nos).
31. Ebert and Rupieper, "Technische Wissenschaft"; Huzel, *Peenemünde,* 107; Reisig OHI, 1989, 24; WvB, "Behind the Scenes," 20, SRCH, WvB Papers; WD, *V-2,* 37–38.
32. Reisig OHI, 1989, 26–27; WD, *V-2,* 139; WD performance evaluation of WvB, 11.9.37, in SRCH, WvB Papers.
33. Becker order, 12.5.37, in NA, T-77/797/5528603–4; "Organisationsschema Werk Ost," c. 1937, in NASM, FE348; Ordway and Sharpe, *Rocket Team,* 31; Tessman OHI, 10–13; WD, *V-2,* 40–41.
34. WD, *V-2,* 48, 50–53.
35. Todt letter, 11.7.36, Riedel/Nebel patent of 3.8.36, Ordnance–Nebel–Riedel contract of 2.7.37 in NA, T-175/155/2685617–22 and 2685578–80; Ordnance correspondence on Nebel, 1936–37, in IWM, MI 14/801(V); WvB, "Behind the Scenes," 19–20, SRCH, WvB Papers;

and Reisig OHI, 1989, 55–56. The patent payment was probably based on the RM 72,000 paid to the Heylandt company for patents it received for its rocket car work. See Heylandt–Wa Prw 1 correspondence, 7.9.34–21.5.35, in NASM, FE749.

36. WvB, "Kostenaufstellung," c. 4.36, in BA/MA, RH8/v.1945; WvB to Lorenz/RLM/LC I, 24.4.36, in NASM, FE746.
37. Riedel, "Raketenentwicklung," 35, IWM, German Misc. 148; WvB German pilot logs and licenses in SRCH, WvB Papers; WvB SS officer file card, BDC; Tschirschwitz/DVL reports, 16.6.36 and 19.8.36, in NASM, FE746; WD, *V-2,* 124.
38. Riedel, "Raketenentwicklung," 35–36, IWM, German Misc. 148; Tessmann OHI, 38–39; WD, *V-2,* 125, states that the cabin was at the opposite end of the "carousel," but he is contradicted by Riedel and Tessman.
39. RLM/LC II reports, 18.2.37 and 1.3.37, on monthly meetings with Wa Prw D, and Pauls/RLM/LC II to WvB, 11.3.37, in NASM, FE746; Warsitz quoted in Ward, *Wernher von Braun,* 7–9.
40. Pauls, WvB, Künzel, and Warsitz report, 3.6.37, and LC II 2e report, 7.6.37, in NASM, FE746; WD, *V-2,* 125.
41. Ordnance–RLM and Ordnance–Heinkel correspondence, 1937–39, in NASM, FE746, especially Dellmeier report, 2.8.38; WvB to von Horstig, 23.11.35, in NASM, FE727/a; Heinkel, *Stürmisches Leben,* 456–64.
42. Stegmaier Starthilfe report, 3.2.42, NASM, FE692/e.
43. Marinewaffenamt document, 23.9.35, Kruse/RLM to von Horstig, 23.12.35, and Wa Prw 11 to von Horstig, 27.3.36, in NASM, FE724/b; Heinkel, *Stürmisches Leben,* 458–64.
44. Walter, "Development" in Benecke and Quick, eds., *History,* 263–80; Antz/RLM/LC 7III report, 14.10.38, in BA/MA, RL3/780; Schabel, "Wunderwaffen?" 183–84, 198; Stegmaier Starthilfe report, 3.2.42, NASM, FE692/e.
45. Gartmann, *Men,* 99–102; Sänger-Bredt, "Silver Bird," 203–6; Ordnance corr. on Sänger, 16.10.34–9.2.37, in BA/MA, RH8/v.1225; Sänger files, BDC.
46. Gartmann, *Men,* 103–6; Sänger-Bredt, "Silver Bird," 206–11; Sänger-Bredt and Engel, "Development," 237–44.
47. Rudolph OHI, 33–39; WvB to WD, 20(?).10.37, and WD "Notizen" for 3.11.37 meeting, 28.10.37, in NASM, FE348.
48. WD "Notizen" for 3.11. meeting, 28.10.37, in NASM, FE348; Becker orders, 11.7.38 and 15.11.38, in NA, T-77/797/5528609–12; WvB to RLM/LC II 2 e, 5.2.38, in NASM, FE746; "Abteilungsfest Wa Prüf 11," 16–17.7.38, in SRCH, WvB Papers; Zanssen biographical information courtesy of Mr. and Mrs. Gerhard Zanssen.

49. Riedel, "Raketenentwicklung," 24–25, 28, IWM, German Misc. 148; WvB, "Das Aggregat III," 29.11.37, DM, (FE1120).
50. WvB to von Horstig, 23.11.35, in NASM, FE727/a; Ordnance corr. with metalworking firms, 13.4.35–15.7.36, in NASM, FE744; IG Farben Bitterfeld to Kummersdorf, 23.11.36, and Zarges to Kummersdorf, 21.1.37, in NASM, FE752; WD to W Preispr.-Heer, 11.11.38, in NASM, FE745; Reisig, "Das kongeniale Vermächtnis," 106; Reisig OHI, 1989, 59–60; Rudolph OHI, 30–31.
51. Gievers, "Erinnerungen," 264–65; WvB, "Behind the Scenes," 14, SRCH, WvB Papers; WvB to KG, 11.7.34, in NASM, FE74/b; WD, WvB, Altvater, and (?), minutes of 9.11.37 meeting at Siemens, in NASM, FE119; Mackenzie, *Inventing,* 45–46.
52. WvB, "Behind the Scenes," 14, SRCH, WvB Papers; WvB draft contract documents for KG, 10.10.34, in NASM, FE74/b; Gievers, "Erinnerungen," 280–81.
53. WvB A-3 progress report, 28.4.36, in BA/MA, RH8/v.1260; KG description of A-3 guidance system, 11.36, NASM, FE23/a; WvB, "Das Aggregat III," 28–38, 29.11.37, DM, (FE1120); Mueller OHI, 13–18.
54. Patent, 1.5.36, on A-3 design, NASM, FE190; Ordnance, KG, Deutsche Glühfadenfabrik corr. and documents, 23.10.36–12.4.37; Goddard, "Liquid-Propellant Rocket Development."
55. WvB to Lorenz/RLM, 3.11.35, in NASM, FE746; Ordnance–TH Aachen correspondence, 15.1.36–19.1.37, in NASM, FE727/a; WvB, "Behind the Scenes," 18, SRCH, WvB Papers; WvB, "Das Aggregat III," 24–27, 29.11.37, DM, (FE1120).
56. WvB to Bassenge/RLM/LC II, 1.11.35, in NASM, FE746; WvB, "Das Aggregat III," 4 and 42–45, 29.11.37, DM, (FE1120).
57. WD, *V-2,* 42–44; WvB report on A-3 launches, 15.1.38, in NASM, FE747; "Operation Leuchtfeuer" documents, 1937, in NASM, FE367; Reisig OHI, 1989, 10–12.
58. WD, *V-2,* 44; WvB report on A-3 launches, 15.1.38, in NASM, FE747.
59. WD, *V-2,* 54–56; WvB A-3 launch summary, 7.38, appended as pages 57–69 of "Das Aggregat III," 29.11.37, DM, (FE1120).
60. WvB A-3 launch summary, 7.38, appended as pages 57–69 of "Das Aggregat III," 29.11.37, DM, (FE1120); WD Vortragsnotizen for Becker, 29.12.37, in NASM, FE367; WvB to Haas/KG, 17.12.37, in NASM, FE74/b.
61. WvB A-3 launch summary, 7.38, appended as pages 57–69 of "Das Aggregat III," 29.11.37, DM, (FE1120); Ordnance–KG meetings, 10.1., 15.1., 20.1.38, in NASM, FE119; Mueller OHI, 13–24.
62. Mueller OHI, 13–24; WvB report on A-2 launches, 28.1.35, in BA/MA,

RH8/v.1945; WD Vortragsnotizen for Becker, 29.12.37, in NASM, FE367; WD, *V-2*, 56.

3. *Breakthrough in Key Technologies*

1. "Werdegang . . . des Professors Dr. von Braun," anonymous typescript, late 1943, in NASM, FE341.
2. Thiel, "Empirische und theoretische Grundlagen zur Neuberechnung von Öfen und Versuchsdaten," c. 7.37, DM, GD634.190.8; Reisig, "Peenemünder 'Aggregaten'," 44–46. This chapter is greatly influenced by Reisig's article and my 1989 oral history interview with him.
3. WD, *V-2*, 50, 53, 149, 151; Minutes of meeting of Navy, Wa Prw.11, Walter, Electrochem. Werke München, and the CTR, 16.10.35, in NASM, FE724/b; WvB to Schumann, 31.1.36, in NASM, FE752; Wa Prw.11 (Thiel) to WD, 28.2.36, in NASM, FE727/a; Thiel–Seifert report on Seifert dissertation, 8.2.37, NASM, FE403.
4. Thiel to WD, 12.2.37, in NASM, FE746; Thiel(?), "Forschungsarbeiten über Materialfragen," etc., 8.4.37, in BA/MA, RH8/v.1260; Thiel to WD, Kummersdorf and Peenemünde, 1.10.37, in BA/MA, RH8/v.1352; WvB to Busemann, 4.4.35, in NASM, FE727/a; WvB to Vogelpohl, 4.6.35, and Thiel to Hase/TH Hannover, 11.12.36, in NASM, FE727/c; Ordnance documents on Hase, 6.8.37–29.10.38, in NASM, FE737; WD, "German V-2," 403.
5. WD, *V-2*, 51; Thiel to WD, 12.2.37, in NASM, FE746; Thiel, "Empirische und theoretische Grundlagen," c. 7.37, DM, GD634.190.8; Thiel to WD, 2.8.37, in BA/MA, RH8/v.1351; WvB, "Das Aggregat III," 15, 29.11.37, DM, (FE1120); Reisig, "Peenemünder 'Aggregaten'," 44–46.
6. Thiel, "Empirische und theoretische Grundlagen," c. 7.37, DM, GD634.190.8; Thiel to WD, 2.8.37, in BA/MA, RH8/v.1351; Reisig, "Peenemünder 'Aggregaten'," 45–46.
7. Thiel, "Empirische und theoretische Grundlagen," c. 7.37, DM, GD634.190.8; Thiel to WD, 2.8.37, in BA/MA, RH8/v.1351; Thiel to WD, Kummersdorf and Peenemünde, 1.10.37, in BA/MA, RH8/v.1352; Thiel to WD, 26.8.38, in BA/MA, RH8/v.1204; Casper minutes of meeting at Schmidding, with 25-ton engine drawings, 22.9.38, NASM, FE752; Reisig, "Peenemünder 'Aggregaten'," 44–46.
8. Thiel, "Empirische und theoretische Grundlagen," c. 7.37, DM, GD634.190.8; Thiel to WD, 2.8.37, in BA/MA, RH8/v.1351; Thiel to WD, Kummersdorf and Peenemünde, 1.10.37, in BA/MA, RH8/v.1352; Thiel to WD, 26.8.38, in BA/MA, RH8/v.1204; Thiel to WD, 16.1.39, in BA/MA, RH8/v.1205; Reisig, "Peenemünder 'Aggregaten'," 46.

9. Thiel–Schluricke patent application, 7.1.39 (English trans., 12.2.46), RSIC, FE196; Thiel to WD, 16.1.39, in BA/MA, RH8/v.1205; Schubert chronicle, 27–29.4.39, in BA/MA, RH8/v.1206; WD to HVP and Kummersdorf, 3.10.39, in SRCH, Tschinkel Papers; Reisig, "Peenemünder 'Aggregaten'," 45–46; WD, *V-2,* 52; Dannenberg OHI, 18–21.
10. Thiel film-cooling memo, 2.10.39, in SRCH, Tschinkel Papers; Thiel to Zanssen, 15.7.42, in NASM, FE171; WD, *V-2,* 52; Reisig, "Peenemünder 'Aggregaten,' 46–47; Dannenberg OHI, 21–27.
11. Dannenberg OHI, 21–28.
12. Hackh–Schluricke tungsten-molybdenum vane reports, 28.3.38 and 9.4.38, in NASM file "Peenemünde #2"; Thiel to WD, 26.8.38, in BA/MA, RH8/v.1204; WD, *V-2,* 58; WvB to von Horstig, 23.11.35, in NASM, FE727/a; M. Schilling, "The Development of the V-2 Rocket Engine," in Benecke and Quick, *History,* 289–91; Winter, *Rockets,* 49; Reisig, "Peenemünder 'Aggregaten'," 47.
13. Ordnance patent, 1.5.36, NASM, FE189, WvB to H. Walter, 12.3.36, in NASM, FE724/c; Ordnance–Walter corr., 2.4.36–30.9.36, in NASM, FE724/b; Schilling, "Development," in Benecke and Quick, *History,* 289–91; Reisig, "Peenemünder 'Aggregaten'," 47; WvB to LC II 2e, 5.2.38, and Dellmeier report, 2.8.38, in NASM, FE746; WvB development status report, 1.8.39, DM, (FE720).
14. WvB, "Behind the Scenes," 24a, SRCH, WvB Papers; WD, "German V-2," 403; WD, *V-2,* 63; von Brauchitsch (WD) order, 5.9.39, in NASM, FE342; WD to HVP and Kummersdorf, 3.10.39, in SRCH, Tschinkel Papers; Hölsken, *V-Waffen,* 20–21.
15. Ebert and Rupieper, "Technische Wissenschaft"; Ludwig, *Technik,* 216–31; Mehrtens, "Die Naturwissenschaften"; Trischler, *Luft- und Raumfahrtforschung,* 174–283.
16. Walker, *German National Socialism,* 17–20; Ludwig, *Technik,* 220, 230–31, 288–89; Trischler, *Luft- und Raumfahrtforschung,* 241–45.
17. Stegmaier minutes of 14.9.39 meeting at TH Dresden, and Thiel memo on 28.9.39 Kummersdorf meeting, in NASM file "Peenemünde #2"; WvB, "Behind the Scenes," 24a, SRCH, WvB Papers.
18. WD to Schumann, 19.9.39, in NASM file "Peenemünde #2."
19. Thiel to WD, 8.7.40, and list of university propulsion contracts, 1.2.41, in NASM, FE728/e; WD to Kummersdorf, 19.7.40, in NASM file "Peenemünde #2"; WvB estimate, 27.3.42, in NASM, FE692/f; Schilling, "Development," in Benecke and Quick, eds., *History,* 284–87; Reisig, "Peenemünder 'Aggregaten', 46–47, 73–77.
20. Thiel, "Empirische und theoretische Grundlagen," c. 7.37, DM, GD634.190.8; Thiel, "Unterlagen über die Grenzen der Leistungssteigerung durch Ofendruckerhöhung," 7.2.39, DM, GD634.19.2;

Thiel memo on 1939–40 program, 2.6.39., in NASM, FE171; WvB dev. status report, 24.4.41, in BA/MA, RH8/v.1260; Thiel, "Vorschläge zur Triebwerksvereinfachung des A 4," 4.9.41, in DM, (FE1068); final report on eighteen-pot engine dev., 15.9.41, NASM, Arch. 81/10.

21. H. H. Kurzweg, "The Aerodynamic Development of the V-2," in Benecke and Quick, *History,* 53; WD, "German V-2," 399; Reisig OHI, 1989, 16–17. American designers similarly used the .50 caliber machine gun bullet's shape for the fuselage of the X-1 rocket plane, which in 1947 became the first aircraft to break the sound barrier.
22. Constant, *Origins,* 108–9, 154–57; Hallion, *Supersonic Flight,* 11–12.
23. Zanssen and WvB report, 26.3.35, in DM, (FE722); WvB to Busemann, 4.4.35, and Ordnance-TH Aachen corr., 15.1.36–19.1.37, in NASM, FE727/a; WvB to Vogelpohl, 4.6.35, in NASM, FE727/c; WvB to Lorenz/RLM, 3.11.35, in NASM, FE746; Kurzweg, "Aerodynamic Development," in Benecke and Quick, *History,* 55; Hermann, "Supersonic Wind Tunnel," 435–36.
24. WD, *V-2,* 53; Trischler, *Luft- und Raumfahrtforschung,* 202–3, 262–64; Hermann *et al.,* "Denkschrift," 1.6.39, NASM, Arch. 66/11.
25. WD, *V-2,* 53–54; Hermann *et al.,* "Denkschrift," 1.6.39, NASM, Arch. 66/11.
26. WD, *V-2,* 114; H. Kurzweg FOIA release, INSCOM; Hermann, "Supersonic Wind Tunnel," 436–38; Smelt, "Critical Review," 903.
27. Hermann, "Supersonic Wind Tunnel," 436–38; Hermann *et al.,* "Denkschrift," 1.6.39, NASM, Arch. 66/11; WD, *V-2,* 115.
28. Hermann, "Supersonic Wind Tunnel," 438–40; Kurzweg, "Aerodynamic Development," in Benecke and Quick, *History,* 51–52, 58; Kurzweg interview by Neufeld, 1990; Aerodynamische Hauptabt., "Jahresbericht 1940," 10.1.41, NASM, FE346.
29. Kurzweg, "Aerodynamic Development," in Benecke and Quick, *History,* 53–56. It is often asserted that the A-5's design was based on the A-4, but in 1936–37 the A-4's shape and internal layout could only have been modeled on the A-3.
30. *Ibid.;* Kurzweg FOIA release, INSCOM; Kurzweg interview; WD, *V-2,* 56–57; Schröder to KWI f. Strömungsforschung/Göttingen and LZ, 10.3.38, in NASM, FE727/c; Zanssen (WvB) to WD, 13.10.39, in BA/MA, RH8/v.1260.
31. "Leuchtfeuer II" documents, 9–18.8.38, in NASM, FE747; Aerodyn. Abteilung, A-5 model report, 11.3.39, NASM, Arch. 66/6; Hermann *et al.,* "Denkschrift," 1.6.39, NASM, Arch. 66/11; WD, *V-2,* 58–59; Aerodyn. Hauptabt., "Jahresbericht 1940," 10.1.41, NASM, FE346.
32. Hermann drop test memo, 2.7.38, NASM, Arch. 66/3.
33. Hermann to WvB, 20.7.39, in NASM, FE737; Haeussermann OHI,

19–20; Hermann, "Supersonic Wind Tunnel," 442; Kurzweg, "Aerodynamic Development," in Benecke and Quick, *History,* 52–53, 62; Kurzweg interview. WD, *V-2,* 57, says the drops began in September 1938, which is quite unlikely.

34. Kurzweg, "Aerodynamic Development," in Benecke and Quick, *History,* 56–68; Hermann, "Supersonic Wind Tunnel," 441–46; Aerodyn. Hauptabt., "Jahresbericht 1940," 10.1.41, NASM, FE346.
35. Kurzweg, "Aerodynamic Development," in Benecke and Quick, *History,* 53–54; Aerodyn. Hauptabt., "Jahresbericht 1940," 10.1.41, NASM, FE346.
36. Patt proposal, 16.6.39, NASM, Arch. 71/1; WD Vortragsnotiz for Becker, 9.10.39, in NASM, FE746; "Entwicklungsstand der Gleiteraggregate am 1.1.1941," 21.1.41, in DM (FE1604); Schirmer/LZ A-9 reports, 1941–42, NASM, FE65/1-65/22; Hermann, "Supersonic Wind Tunnel," 443–44; WD memo, 12.7.43, in NASM, FE333.
37. Kurzweg, "Aerodynamic Development," in Benecke and Quick, *History,* 60–61; Hermann, "Supersonic Wind Tunnel," 443; Kurzweg interview; WD, *V-2,* 114–15; Aerodyn. Hauptabt., "Jahresbericht 1940," 10.1.41, NASM, FE346.
38. Fieber, "Zur Geschichte," 9; minutes of 9.11.37 meeting at Siemens in NASM, FE119; MacKenzie, *Inventing,* 46–47; Gievers, "Erinnerungen," 265.
39. Minutes of 9.11.37 and 24.1.38 meetings with Siemens, in NASM, FE119.
40. Minutes of meetings with KG, 10.1.38, 15.1.38, 20.1.38, in NASM, FE119; Mueller OHI, 25–27.
41. Fieber, "Zur Geschichte," 10; Karner, "Steuerung," 47–52; MacKenzie, *Inventing,* 37–44; Möller, "Die Rechliner Dreiachsensteuerung," 10.3.38, NASM, FE773.
42. Fieber, "Zur Geschichte," Vorwort, 2, 19–22; Karner, "Steuerung," 51–52; Otto Müller, "The Control System of the V-2," in Benecke and Quick, *History,* 80–84; MacKenzie, *Inventing,* 53–55.
43. Schröder, "Steuerung des Aggregats IV," 13.11.39, 32, NASM, FE303; minutes of meetings with KG, 15.1.38, 20.1.38, 14.2.38, and with Siemens, 24.1.38 and 5.4.38, in NASM, FE119; Steinhoff 1940 BSM report, 10.1.41, 4–5, NASM, FE769.
44. Fieber, "Zur Geschichte," 19–20; minutes of 3.2.39 meeting with Siemens in NASM, FE119; Mueller OHI, 38; Müller, "Control System," in Benecke and Quick, *History,* 82.
45. Fieber, "Zur Geschichte," 11, 24–26; Karner, "Steuerung," 52, 64n61; minutes of 5.4.38 meeting at P.-West, in NASM, FE119.
46. Würthner–Siemens D13 report, 15.4.40, NASM, FE461; Siemens minutes of 5.4.38 and 15–16.8.38 meetings with HVP, in NASM, FE119.

47. Heizer/KG to HVP, 22.10.38, and minutes of 24.1.38 meeting with Siemens, in NASM, FE119; RLM/LC 5 to W. Riedel, 23.5.39, in NASM, FE746; WD to Wa Prüf 12, 13.7.39, in BA/MA, RH8/v.1313; Würthner D13 report, 15.4.40, NASM, FE461.
48. Möller, "Die Rechliner Dreiachsensteuerung," 14.3.38, NASM, FE773; Askania-HVP corr. and meetings, 30.11.38–8.6.39, and TVA Eckernförde/Navy to HVP, 17.2.39, in NASM, FE119; WvB "Rundschreiben," 30.8.39, in NASM, FE750.
49. WD (WvB) to Lucht/RLM, 28.4.39, in NASM, FE119; WvB, "Terminlage LF III," 2.9.39, in NASM, FE750; Schröder, "Steuerung des Aggregats IV," 26–28, 13.11.39, NASM, FE303; WvB to Banse/RLM/LC 4, 9.1.40, in NASM, FE746.
50. Rudolph OHI, 72; Reisig OHI, 31; WD correspondence on Schröder, 1962, in DM, Luftfahrt/Persönlichkeiten/Männer, Dornberger files.
51. WvB, "Behind the Scenes," 24, SRCH, WvB Papers; Steinoff FOIA release, INSCOM; Steinhoff to Zanssen, 18.1.40, in DM, (FE 1821); Mueller OHI, 46–47; Haeussermann OHI, 18–22; Hoelzer OHI, 67–68.
52. "Nachweisung der am 1.1.40 für VP. eingesetzten Arbeitskräfte," in NASM, FE349; Haeussermann OHI, 5–6; Rees OHI, 5–6.
53. Stegmaier minutes of 14.9.39 meeting at TH Dresden, in NASM file "Peenemünde #2"; Steinhoff 1940 BSM report, 10.1.41, NASM, FE769; Reisig OHI, 1989, 33–34, 42–46.
54. WvB Aktenvermerk, 8.7.39, on meeting with Kreiselgeräte, in NASM, FE119; Mueller OHI, 54–67; Gievers, "Erinnerungen," 281, 284–85; Mueller, "A History."
55. Steinhoff 1940 BSM report, 10.1.41, NASM, FE769; Reisig OHI, 1989, 34–37; Steinhoff, "Development," 207–8; Haeussermann, "Developments," 226–28; Moore, "German Missile Accelerometers."
56. Steinhoff (?) to Banse/RLM/LC 4, 5.9.39, Günther/RLM/LC 4 memo, 24.9.39; WD to Peenemünde, 25.4.39, Wa Prüf 11 Stab Vortragsnotiz for Becker, 9.10.39, and WD to RLM, 15.11.39, re 21.10.39 meeting of Becker, Milch, and Udet, in NASM, FE746.
57. Hoelzer OHI, 8–10.
58. Steinhoff 1940 BSM report, 10.1.41, NASM, FE769; Hoelzer OHI, 14–21; Müller, "Control System," in Benecke and Quick, *History,* 88–93.
59. Steinhoff 1940 BSM report, 10.1.41, NASM, FE769; WD, *V-2,* 134–35; WvB dev. status report, 24.4.41, in BA/MA, RH8/v.1260; Ludewig memo on 4.12.41 meeting in NASM, FE727/a.
60. Steinhoff 1940 BSM report, 10.1.41, NASM, FE769; Hoelzer OHI, 28–39.
61. Hoelzer OHI, 28–39; Haeussermann, "Developments," 229–30; Tomayko, "Helmut Hoelzer's," 230–32.

62. Hoelzer OHI, 41–42; Tomayko, "Helmut Hoelzer's"; Haeussermann, "Developments," 229–32.
63. Steinhoff 1940 BSM report, 10.1.41, NASM, FE769; WvB dev. status report, 24.4.41, in BA/MA, RH8/v.1260; WvB minutes of 4.8.41 meeting, in NASM, FE733; WvB minutes of 20.10.41 meeting in BA/MA, RH8/v.1954; Steinhoff minutes of 11.12.41 meeting in NASM, FE728/b.
64. Fieber, "Zur Geschichte," 11, 29–30; Karner, "Steuerung," 53; sources in note 63.
65. Dannenberg OHI, 42–43.

4. Peenemünde's Time of Troubles

1. Rudolph OHI, 75–76; Franklin, *American,* 64.
2. Rudolph OHI, 76–78; WvB dev. report, 20.9.39, in BA/MA, RH8/v.1260; WD Vortragsnotiz, 14.12.39, in NASM, FE349.
3. Deutsch, *Hitler,* 220–30; Overy, *Goering,* 70–71.
4. Von Brauchitsch order to Becker, 21.11.38, in NASM, FE357; Reisig OHI, 1989, 64–65.
5. Schubert chronicle, 25.1.39, 20.2.39, and 20.5.39, and Schubert Vortrag, 7.6.39, BA/MA, RH8/v.1206; WD Vortragsnotiz, 14.12.39, in NASM, FE349; Rudolph OHI, 1989, 78, 81.
6. Schubert chronicle, 25.1.39, and Schubert Vortrag, 7.6.39, in BA/MA, RH8/v.1206; Schubert Aktennotiz, 18.10.41, in BA/MA, RH8/v.1208b.
7. WD Vortragsnotiz, 14.12.39, in NASM, FE349; Schubert chronicle, 25.1.39, in BA/MA, RH8/v.1206.
8. Nuss, "Einige Aspekte;" Leeb, *Aus der Rüstung,* 16–17.
9. Schubert Aktennotiz, 18.10.41, in BA/MA, RH8/v.1208b; Schubert chronicle of 10.2.39, and draft letter to Speer, 12.10.39, in BA/MA, RH8/v.1206.
10. Schubert chronicle, 25.1.39, and Schubert Vortrag, 7.6.39, in BA/MA, RH8/v.1206; WD draft Vortragsnotiz for Hitler, 5.7.41, in NASM, FE342; Overy, "Mobilization," 636–37; Overy, *Goering,* 158–60.
11. Rudolph Vortrag, 1.3.43, and Schubert to WD, 21.6.43, in BA/MA, RH8/v.1210; Rudolph OHI, 76–78; WD Vortragsnotiz, 14.12.39, in NASM, FE349.
12. Germany, *Das Deutsche Reich,* 1:470–72, 529–32; Overy, "Hitler's War," 274–75, 280–81; Schubert chronicle, 25.1. and 3.5.39, in BA/MA, RH8/v.1206; Heeresneubauleitung to Wa Stab, 28.3., and WD to W Rü, 31.3.39, in NASM, FE342.
13. WD, *V-2,* 64–66; Bergaust, *Wernher von Braun,* 55–60.
14. WD, *V-2,* 65–66; WD to W Rü, 31.3.39, in NASM, FE342; Schubert chronicle, 3–5.7.39, in BA/MA, RH8/v.1206.

15. WD, *V-2,* 63; Von Brauchitsch order, 5.9.39, and Burdach/Allgem. Heeresamt, "Zusatzbefehl," 8.9.39, in NASM, FE342; WvB, "Rundschreiben," 30.8.39, in NASM, FE750.
16. WD, *V-2,* 63; Schubert chronicle, 6.9.39, in BA/MA, RH8/v.1206.
17. Thomas to Funk and Todt, 15.9.39, in NASM, FE342.
18. Carroll, *Design,* 194–96; Overy, "Mobilization," 623, 637.
19. Draft order, 11.10.39, in NA, T-77/201/937537–38; Germany, *Das Deutsche Reich,* 5/1:370; Carroll, *Design,* 201; WD chronology, 5.7.41, in NASM, FE342; Schubert Aktennotiz, 16.11.39, in BA/MA, RH8/v.1206; Hölsken, *V-Waffen,* 21–22.
20. Udet to Jeschonnek, 12.9.39, in BA/MA, RL3/352; Schabel, "Wunderwaffen?" 151–52, 180–81; Heinkel, *Stürmisches Leben,* 464–73; Hüter Aktenvermerk, 15.6.39, Udet to Becker, 24.10.39, and WvB to WD, 22.6.40, in NASM, FE746; WvB dev. report, 20.9.39, in BA/MA, RH8/v.1260.
21. WD to Peenemünde and Kummersdorf, 3.10.39, in SRCH, Tschinkel Papers; Schubert Aktenvermerk, 16.11.39, in BA/MA, RH8/v.1206.
22. Becker Aktenvermerk, 21.11.39, on 20.11. meeting with Hitler, and WD to Wa I Rü 8, 17.1.40, both in NASM, FE342.
23. Quotes from WD, "The German V-2," 401; WD statement in Benecke and Quick, *History,* 296. See also WD, *V-2,* 69; WvB, "Reminiscences," 138, 140; Ordway and Sharpe, *Rocket Team,* 28–29; Irving, *Mare's Nest,* 18; McDougall, *Heavens,* 43. As result of Dornberger's memoirs, Hitler's role in the A-4 priority battle has been greatly distorted and mythologized. Only Hölsken's *V-Waffen* has given a more complete (if error-prone) account. For a fuller examination of the historiography, see Neufeld, "Hitler."
24. Hoffmann, *History,* 129.
25. Germany, *Das Deutsche Reich,* 5/1:409–26; Overy, "Hitler's War," 275–77; Murray, *German Military Effectiveness,* 229–43.
26. Schubert Aktennotiz, 1.12.39, in BA/MA, RH8/v.1206; Dornberger to Koch/Chef Wa Prüf, 2.12.39, in NASM, FE342; Dornberger Vortragsnotizen, 12.12. and 14.12.39, in NASM, FE349.
27. WD Vortragsnotiz, 14.12.39, and Schubert to WD, 6.1.40, in NASM, FE349; Becker to Koch, 20.12.39, in NASM, FE342.
28. WD to Zanssen, Schubert, and Heeresneubauamt, 8.1.40, and WD to Wa I Rü 8, 17.1.40, in NASM, FE342; Germany, *Das Deutsche Reich,* 5/1:423–46. Whether or not Hitler was involved in these discussions is unknown.
29. Dornberger Vortragsnotiz, 14.12.39, in NASM, FE349.
30. On decisiveness, see Dornberger, *V-2,* 104–6; on targeting, WD (WvB) to

Wa Prüf 1, 20.1.40, in BA/MA, RH8/v.1260; on accuracy, WD Vortragsnotiz, 14.12.39, in NASM, FE349.

31. Schubert chronicle, 28.2.39, in BA/MA, RH8/v.1206; Schubert chronicle, 7.2.40 and 8.3.40, in BA/MA, RH8/v.1207; Schubert Aktennotiz, 18.10.41, in BA/MA, RH8/v.1208b; Rudolph Vortrag, 1.3.43, and Schubert to WD, 21.6.43, in BA/MA, RH8/v.1210.
32. WD (WvB) to KG, 19.2.40, in NASM, FE119; Germany, *Das Deutsche Reich,* 5/1: 446; Schubert chronicle, 19.2. and 21.2.40, in BA/MA, RH8/v.1207.
33. Speer, *Inside,* 469; Schubert chronicle, 13.2.39 and 6.9.39, and minutes of 3–5.10.39 meetings, in BA/MA, RH8/v.1206; Zanssen program for 19.1.40 Speer visit in BA/MA, RH8/v.1207; OKW/WiRüAmt armaments inspectors meeting, 8.1.40, in NA, T-77/85/808733–34. On falsifications in Speer's memoirs, see Schmidt, *Albert Speer.* Speer's memories of the rocket program appear to have been influenced by Dornberger's account.
34. Germany, *Das Deutsche Reich,* 5/1:453–85; Schubert chronicle, 6.4.40, in BA/MA, RH8/v.1207.
35. Schubert chronicle, 8–9.4.40, in BA/MA, RH8/v.1207; Germany, *Das Deutsche Reich,* 5/1:474–75.
36. Leeb minutes, 19.6.40, meeting with Fromm, WD Aktenvermerk, 24.6.40, on 20.6. meeting with Leeb and Ordnance leaders, Haseloff to Leeb, 4.7.40, and WD to Wa Stab Ib *et al.,* 4.7.40, in NASM, FE342; WD Vortragsnotiz, 20.6.40, in NASM, FE349.
37. Germany, *Das Deutsche Reich,* 5/1:486–97, 502–22.
38. Hölsken, *V-Waffen,* 23–24; Göring priority order of 18.7.40, 26.7.40 entry in WD chronology of 24.9.40, Koch (WD) to Leeb, 3.8.40, Haseloff to Thomas, 5.8.40, Thomas order of 7.8.40, and Gschwender to WD, 14.8.40, in NASM, FE342; Stab II (Rüst)/Chef H Rüst u BdE to Leeb, 27.8.40, in NASM, FE349. For errors in Hölsken, see Neufeld, "Hitler." Whether the Führer played any role in this struggle is unknown; no information is available on his opinion of the rocket program between February and November 1940.
39. WD Aktenvermerk on 20.6.40 meeting, in NASM, FE342; WD to Leeb, Koch, Speer *et al.,* 6.8.40, in NASM, FE349. Hölsken, *V-Waffen,* 24, following WD's 1940/41 chronology of the priority battle (in NASM, FE342), dates the takeover as 15.8.40. Draft agreements in FE349 are inconsistent, but the 6.8.40 document is the best indication that the date was actually a month later.
40. Schubert chronicle, 3–5.7.39, and draft letter to Speer, 12.10.39, in BA/MA, RH8/v.1206; Schubert chronicle, 2–4.7.40, 24.7.40, and 27.7.40, in BA/MA, RH8/v.1207; Baugruppe Schlempp to WD, 21.6.40,

Todt to Dr. Timm/Labor Min., 18.7.40, and Abt. Rüstungsvorhaben/Arm. Min. to Aussenstelle Stettin/Generalinsp. d. deut. Strassenwesen, 3.8.40, in NASM, FE349; WD Aktennotiz, 18.7.40, on 17.7. meeting with Todt, in NASM, FE342.

41. Todt to Keitel, 3.9.40, regarding comments to Hitler, in NA, T-77/178/914227; minutes of meeting of Thomas, Fromm, Udet, and Witzell, 15.8.40, and OKW/WiRüAmt minutes of meeting with armaments inspectors on 13.9.40, in NA, T-77/85/809036 and 809075; Göring priority order, 20.9.40, in NASM, FE349.
42. Dornberger to Leeb, 26.8.40, and Löhr to Fromm, 28.8.40, in NASM, FE349; Schubert chronicle, 30.9.40, in BA/MA, RH8/v.1207; Lehmann, *Robert H. Goddard,* 298–340.
43. Demag-Zug to OKH, 4.10., Löhr to Fromm, 11.10., Wilson (Wa Prüf 11) note on 14.10. OKW phone call, Keitel to von Brauchitsch, 19.11., Löhr to Leeb, 9.12., and WD to Wa Prüf 11 sections, 16.12.40, in NASM, FE349; WD Vortragsnotiz, 8.10., and WD to Leeb, 13.11.40, in NASM, FE342; Schubert chronicle, 9., 16. and 30.10.40, in BA/MA, RH8/v.1207; Hölsken, *V-Waffen,* 25.
44. OKW Rüstungswirtsch. Abt. to Keitel, 30.10.40, in NA, T-77/17/728559; Maj. Kiesow lecture to armaments inspector trainees, 23.1.41, in NA, T-77/237/978433; Carroll, *Design,* 196.
45. Schubert chronicle, 24.1., 27.1., and 5.2.41, in BA/MA, RH8/v.1208a; Fromm phone call in WD chronology of 5.7.41 in NASM, FE342.
46. Schubert chronicle, 4.4., 5.4., 10.4., 22.4.41; WD Vortragsnotiz for Leeb, 28.4., and chronology of 5.7.41, in NASM, FE342; WD to Leeb, 8.5.41, regarding Leeb–von Brauchitsch meeting of 29.4. in NASM, FE728/e.
47. Speer quoted in Overy, *Goering,* 138.
48. WD to Leeb, 3.8.40, in NASM, FE342; Löhr to Leeb, 9.12.40, in NASM, FE349; Schubert chronicle, 9.10.40, in BA/MA, RH8/v.1207.
49. WD Vortragsnotiz, 20.6.40, in NASM, FE349.
50. Steinhoff 1940 BSM report, 10.1.41, NASM, FE769; Dornberger Vortragsnotiz, 8.10.40, in NASM, FE342.

5. Hitler Embraces the Rocket

1. Keitel to von Brauchitsch, Göring, and Raeder, 16.6.41, in NA, T-77/537/1711038-39; Todt to Thomas, Fromm, Witzell, and Udet, 11.7.41, in NA, T-77/194/929154; Germany, *Das Deutsche Reich,* 5/1:556–80; Overy, "Mobilization," 631–32.
2. Schubert chronicle, 23.6., 25.6., 27.6., and 3.7.41, in BA/MA, RH8/v.1208; Löhr (WD) Vortragsnotiz, 27.6.41, Kipping to Koch,

28.6.41, and Thom/Wa Prüf 11 to Koch and Wa Prüf 12, 20.8.41, in NASM, FE342.

3. Löhr (WD) Vortragsnotiz, 27.6.41, in NASM, FE342; WD, *V-2,* 71, "The German V-2," 402–3, and "European Rocketry," 256–59.
4. Löhr (WD) Vortragsnotiz, 27.6.41, Kipper to Koch, 28.6.41, Leeb (WD) Vortragsnotiz, 5.7.41, for adjutants of Hitler and von Brauchitsch, Löhr to Fromm, 5.7.41, and Rühne (Fromm's liaison to von Brauchitsch) to Fromm's chief of staff, 17.7.41, in NASM, FE342.
5. Todt to Fromm, and Todt to WD, 30.7.41, in NASM, FE342; Todt order, 2.7.41, Schubert chronicle, 12.7.41, Schubert to WD, 18.10.41, in BA/MA, RH8/v.1208.
6. Rühne to Fromm's chief of staff, 17.7.41, and WD Vortragsnotiz, 31.7.41, in NASM, FE342.
7. Graupe, "Berechnung. . . ," 29.7.40, Roth?, handwritten calc., 9.9.41, Roth, *"Zweistufenaggregat (180/30) A10 A9,"* c. 9.41, all in DM, (FE1135/1?); TA/Proj., "Entwicklungsstand der Gleiteraggregate am 1.1.1941," 21.1.41, in DM, (FE1604); WD Vortragsnotiz, 31.7.41, in NASM, FE342; Roth, "Die Flugbahn eines Zweistufengleiters," 20.10.41, NASM, Arch. 68/21; Dahm OHI, 8–9; Overy, "From 'Uralbomber' to 'Amerikabomber'."
8. WD Aktennotiz, 21.8.41, in NASM, FE341; movie script, c. 10.40, in NASM, FE338. Dornberger appears to have omitted this crucial meeting from his memoirs because it did not fit his self-serving view that Hitler was converted to the A-4 only in July 1943.
9. Hölsken, *V-Waffen,* 29; Walker, *German National Socialism,* 46–60; Speer, *Inside,* 309. For occasional contacts between Peenemünders and the nuclear project, see WvB minutes of 15.2.41 meeting on cyclotron in NASM, FE728/f; Stegmaier (Thiel) to Schumann, 31.8.41, in BA/MA, RH8/v.1260; WD, *V-2,* 105.
10. WD Vortragsnotiz, 21.8.41, and Keitel to von Brauchitsch, 20.8.41, in NASM, FE341; Stegmaier minutes of 29.8.41 meeting in NASM, FE357; biographical information courtesy of Mr. and Mrs. Gerhard Zanssen.
11. WD, *V-2,* 69, and *idem* "The German V-2," 401, indicate von Brauchitsch's role but are chronologically misleading. See Schubert chronicle, 22.9.41 and 13.10.41, in BA/MA, RH8/v.1208; Stegmaier to Stellv. Generalkommando II. AK/Stettin, 15.11.41, in NASM, FE728/f; WvB(?), report on A-4 and A-8, c. 10.12.41, BA/MA, RH8/v.1955; Hoelzer OHI, 55–56; Heimburg OHI, 18–19; Schubert minutes, 29.7.42, in BA/MA, RH8/v.1209.
12. Todt to WD, 13.9.41 (with Leeb's marginal comment) and 20.9.41, WD to Todt, 19.9.41, Todt to Leeb, 10.10.41, and Leeb to Todt, 28.10.41, in

NASM, FE342; WD Aktennotiz, 21.8.41, Thomas priority order, 15.9.41, and WD Vortragsnotiz, 8.10.41, in NASM, FE341; Schubert chronicle, 16.10.41, in BA/MA, RH8/v.1208.

13. WD to Leeb, 26.8.40, in NASM, FE349; Thiel, "Vorschläge zur Triebwerksvereinfachung des A4," 4.9.41, and Thiel minutes of 17.9.41 meeting, in DM (FE1068); Thiel Aktenvermerk, 5.9.41, in NASM, FE728/d; Broszat Aktennotiz of 8.9.41, in BA/MA, RH8/v.1260; OKW/WiRüAmt documents, 12–23.9.41, in NA, T-77/37/750312–19; WD to Peenemünde, 13.11.41, in NASM, FE342.
14. Thiel monthly reports, 6–11.41, in NASM, FE728/e; Thiel minutes of 17.9.41 meeting in DM, (FE1068); WD to Peenemünde, 13.11.41, in NASM, FE342; WvB(?), report on A-4 and A-8, c. 10.12.41, in BA/MA, RH8/v.1255; minutes of 20.3.42 meeting at OKW/WiRüAmt, in NASM, FE341.
15. Löhr order, 23.10.41, in BA/MA, RH8/v.1260; WD to Arbeitsstab members, 13.11.41, and related documents in NASM, FE357; WvB(?), report on A-4 and A-8, c. 10.12.41, in BA/MA, RH8/v.1955; Germany, *Das Deutsche Reich,* 5/1:541–42, 558–59.
16. WvB(?), report on A-4 and A-8, c. 10.12.41, in BA/MA, RH8/v.1955.
17. Stegmaier minutes of 3–4.9.41 meetings in NASM, FE728/b; WD draft Vortragsnotiz for Leeb, 18.9.41, in NASM, FE342; WvB minutes of 16.12.41 meeting with Hertel/RLM, in NASM, FE728/f; Rees OHI, 23.
18. Germany, *Das Deutsche Reich,* 5/1:607–24; Overy, "Mobilization," 632–33; Schubert chronicle, 19.12.41, in BA/MA, RH8/v.1208; Bartov, *Hitler's Army.*
19. Schubert chronicle, 1–3.10.41 and 8.12.41, in BA/MA, RH8/v.1208; same, 16.1.42, 21.1.42 (with WD marginal comment), and 28.1.42, in BA/MA, RH8/v.1209; Thom to Wa Z, 2.12.41, in NASM, FE728/f.
20. Germany, *Das Deutsche Reich,* 5/1:610–77; Overy, *Goering,* 205–8; Seidler, *Fritz Todt,* 256–60.
21. Schmidt, *Albert Speer,* 60–63; Speer, *Inside,* 260–66; Germany, *Das Deutsche Reich,* 5/1:677–78.
22. Speer, *Inside,* 262–65; Germany, *Das Deutsche Reich,* 5/1:678–79; Carroll, *Design,* 234–40; Overy, *Goering,* 208–12.
23. Speer, *Inside,* 467–69; Speer interrogations, 21.5.45, USSBS report 5, in NA, T-73/193/3406764–81.
24. Engel, *Heeresadjutant,* 121; Speer, *Inside,* 469; Speer minutes, 5–6.3.42 and 19.3.42, in Boelcke, *Deutschlands Rüstung,* 71, 74. See also minutes of 20.3.42 meeting at OKW/WiRüAmt, NASM, FE341.
25. Hölsken, *V-Waffen,* 32; WD to Thomas, 24.3.42, in NASM, FE333. See also WD intro., 21.3.42, to HAP/EW, "Vorschläge fur den Einsatz der Fernrakete A4. . . ," DM, GD639.1.4.

26. Speer quoted in Hölsken, *V-Waffen,* 35; WvB minutes of 16.12.41 meeting with Hertel/RLM, WD minutes of 11.3.42 meeting with Milch, and related documents in NASM, FE728/f. See also NASM, FE357.
27. WD marginal comment on Busemann/DFL Braunschweig to OKH, 21.3.38, and WD (WvB) to Banse/RLM, 25.4.40, in DM (FE722); P. Schmidt, "On the History. . . ," and F. Gosslau, "Development of the V-1 Pulse Jet," in Benecke and Quick, *History,* 375–418.
28. Werrell, *Evolution,* 7–42; Hölsken, *V-Waffen,* 33–35. For older accounts, see Irving, *Mare's Nest,* 23–24; Young, *Flying Bomb,* 10–12.
29. Minutes of meeting with LGW (Siemens), 27.11.41, in NASM, FE119; Steinhoff minutes of 11.12.41 meeting, in NASM, FE728/b.
30. Schubert Aktennotiz, 15.4.41, in BA/MA, RH8/v.1208; Reisig OHI, 1989, 67–69, 71.
31. Dellmeier Starthilfe report, 12.3.40, NASM, Arch. 70/1; Cerny/Peenemünde-West Aktenvermerk, 7.9.40, in NASM, FE746; Stegmaier Starthilfe report, 3.2.42, NASM, FE692/e; Starthilfe report of 6.41, with changes, 15.4.42, in NASM, FE692/f.
32. Overy, *Goering,* 155, 181–92.
33. WD (Dellmeier) to Beck/RLM, 24.11.41, in NASM, FE752; Stegmaier Starthilfe report, 3.2.42, NASM, FE692/e; WD, *V-2,* 126.
34. Stegmaier Aktenvermerk, 7.5.41, in BA/MA, RH8/v.1260.
35. Schabel, "Wunderwaffen?" 302–5; Klein, *Vom Geschoss,* 111–16; Halder(?), "Entwicklung und Planung in der deutschen Luftwaffe, Teil 3," Heft 9, c. 1945, NA, T-971/27/317–18.
36. WvB to Leeb, 13.5.41, in BA/MA, RH8/v.1260; WvB, rocket-fighter proposal, 6.7.39 (English translation, 29.4.46), RSIC, FE595; WvB, "Einsatzmöglichkeiten des Strahlantriebs. . . ," 27.5.41, NASM, Arch. 58/1; Schabel, "Wunderwaffen?" 197–99; Späte, *Top Secret Bird;* H. von Zborowski, "BMW-Developments," in Benecke and Quick, *History,* 297–324.
37. Steinhoff minutes of 12 11.6.41 meetings with L Flak and WvB report, 18.7.41, on Lucht visit, in NASM, FE746.
38. Udet to Leeb, 18.8.41, and WvB to Pauls/Peenemünde-West, 4.12.41, in NASM, FE746; Steinhoff minutes of 25.10.41 meeting with Brée/RLM, in DM, (FE1068); Fieseler-Werke, "Voruntersuchung für den Höhenjäger Fi 166," c. 11.41, NASM, FE639; Vorwald/RLM minutes of 6.12.41 meeting in NASM, FE368; WvB marginal comment on Holl/Arbeitsstab report, 24.1.42, in NASM, FE357.
39. Halder/Genst.6.Abt. (IVB) to Chef 6.Abt., "Ablauf der Flakentwicklung und Vorschläge," 5.41, in DM.
40. Revised drafts of Halder document, 5–6.42, in DM; v. Axthelm, "Übersicht über den Entwicklungsstand. . . ," 22.6.42, Göring order, 1.9.42,

and v. Axthelm cover memo, 18.9.42, in NASM, FE738/4; Speer minutes, 13/14.10.42, in Boelcke, *Deutschlands Rüstung,* 194; Schabel, "Wunderwaffen?" 304–6.

41. Boog, *Die deutsche Luftwaffenführung,* 204–8, 592; Speer, *Inside,* 364–65.
42. Speer, *Inside,* 468–69; Ludwig, "Die deutschen Flakraketen," 89–90.
43. WD Vortragsnotiz, 31.7.41, in NASM, FE342; WD to Koch, "Beantwortung von Fragen Gen St d H/Org.Abtlg.," 11.10.41, in NASM, FE341; minutes of 1.11.41 meeting in BA/MA, RH8/v.1260; Thiel minutes of 6.1.42 in NASM file "Peenemünde #2."
44. Thiel, "Vortragsnotiz über Stand der Entwicklung. . . ," 15.9.41, in BA/MA, RH8/v.1260; WvB minutes of 4.8.41 meeting in NASM, FE733.
45. Schubert chronicle, 21–23.10.41 and 4–6.11.41, in BA/MA, RH8/v.1208; WD to HVP, 7.11.41, in BA/MA, RH8/v.1260.
46. WD, *V-2,* 139; manned A-9 drawing, c. 1941, in DM, (FE1604); A-9/A-10 as cited in note 7; WvB minutes of 15.2.41 meeting in NASM, FE728/f; Stegmaier (Thiel) to Schumann, 31.8.41 (with WD marginal comment), in BA/MA, RH8/v.1260; Thiel to Oberpostrat Gerwig, 16.9.42, in NASM, FE692/f; contract with Reichspostministerium, 15.10.42, in NASM, FE331.
47. WD to HVP, 13.11.41, in NASM, FE342; Thiel report, 27.11.41, in NASM, FE728/e; Roth to Riedel I *et al.,* 25.2.42, in NASM, FE750.
48. WD to HVP, 23.12.42, in NASM, FE728/e; WD Vortragsnotiz, 21.8.41, in NASM, FE341; WvB to Stegmaier, 3.10.41, in BA/MA, RH8/v.1260; WD to HVP, 8.12.41, in NASM, FE728/b.
49. WD to HAP, 5.2.42, and WvB minutes of 31.1.42 meeting, in NASM file "Peenemünde #2."
50. WD to HAP, 5.2.42, in NASM file "Peenemünde #2."
51. *Ibid.* Other copies of this document exist in the BA/MA and in the FE microfilm, but Thiel's original is in NASM.
52. *Ibid.;* WD to HVP, 8.12.41, in NASM, FE728/b; WD minutes of 3.3.42 Arbeitsstab meeting, in NASM, FE357; minutes of 29.10.42 meeting, in NASM, FE750; WD, *V-2,* 72; Dannenberg OHI, 36–38; Reisig OHI, 1989, 73–74; Rudolph OHI, 50–51; Tessmann OHI, 29–30.
53. WD to HAP, 5.2.42, in NASM file, "Peenemünde #2;" Schubert chronicle, 26.5.41, in BA/MA, RH8/v.1208; Schubert to WD, 21.6.43, in BA/MA, RH8/v.1210.
54. Thiel, "*Diensteinteilung*. . . ," 18.3.42, in NASM file "Peenemünde #2;" Thiel minutes of 14.3.42 meeting in BA/MA, RH8/v.1954; Thiel A-4/V1 accident report, 20.3.42, in NASM, FE692/f.
55. "Terminauszug. . . ," 20.4.42, in NASM file "Peenemünde #2"; WvB minutes of 15.2.42 meeting, in NASM, FE692/f; WvB to Stegmaier, 29.7.42, in BA/MA, RH8/v.1954.

56. WvB minutes of 23.4.42 meeting in NASM, FE692/f.
57. Stegmaier order, 10.6.42, in BA/MA, RH8/v.1209; WD minutes of 13.6.42 meeting, in NASM, FE342; Steinhoff(?), BSM report, 31.7.42, NASM, Arch. 96/24.
58. WD minutes of 13.6.42 meeting in NASM, FE342.
59. WD draft lecture, 9.6.42, in NASM, FE358; WvB minutes of 20.4.42 meeting, in NASM, FE750; WD order to HAP, 6.6.42, in BA/MA, RH8/v.1959; Wa Prüf 11 note of Arm. Min. phone call, 8.7.42, in NASM, FE728/f; WD, *V-2*, 72; Dannenberg OHI, 36–38. In Peenemünde documents, "Papa" was Riedel I, Klaus (from the Raketenflugplatz) was Riedel II, and the second design bureau chief was Riedel III.
60. WD Aktenvermerk, 30.4.42, WD order to HAP, 6.6.42, and Leeb to Rüstungsamt/Arm. Min., (?).7.42, in BA/MA, RH8/v.1959; WD to HAP, 17.9.42, NASM, FE692/f.
61. Steinhoff(?), BSM report, 31.7.42, NASM, Arch. 96/24; WvB minutes of 17.6.42 meeting, in NASM file "Peenemünde #2"; WvB to Stegmaier, 29.7.42, in BA/MA, RH8/v.1954.
62. Thiel minutes of 21.8.42 meeting in NASM, FE358.
63. WD to HAP, 29.9.42, in NASM, FE342. On V7, WvB to Stegmaier, 29.7.42, in BA/MA, RH8/v.1954.
64. WD, *V-2*, 3–17; WD, "A4: Stand der Entwicklung," 8.10.42, in BA/MA, RH8/v.1228; Riedel III minutes of 4.11.42 meeting, in BA/MA, RH8/v.1954.
65. WvB, "Behind the Scenes," 25, SRCH, WvB Papers. For Dornberger rhetoric, see his draft lecture, 9.6.42, in NASM, FE358, and "Denkschrift," c. late 1943, NASM, FE496. For his political opinions, see entries of 31.10.38, 12.5.43 and 4.6.43, in WD notebook #1, DM.

6. *Speer, Himmler, and Slave Labor*

1. Speer office chronicle, 4–6.10.42, in BA, R3/1736; WD to Gen. Hartmann/Arm. Min., 6.10.42, in NASM, FE342; WD, "A4: Stand der Entwicklung," 8.10.42, in BA/MA, RH8/v.1228; HAP/EW/TDZ, "Versuchsschiessen A4," 7.43, pictured in WD, *Peenemünde*.
2. WD to Stegmaier, 9.10.42, in NASM, FE728/f; WvB to WD, 16.9.42, and WD to Leeb, 22.10.42, in DM, (FE1106). Dornberger must have forgotten Milch's announcement of "Cherry Stone" at the first A-4 launch attempt in June.
3. Stegmaier order and WvB "Rundschreiben," 10.10.42, in NASM, "Peenemünde #2"; A-7 drop test report, 10.11.42, in DM, GD630.0.10; WD, *V-2*, 139; Dahm OHI, 9–10, 13; Steinhoff minutes of 16.12.42 meeting

in NASM, FE753/a; Renz/RLM/L Flak to Milch, 13.1.43, in BA/MA, RH8/v.1235.

4. WvB, "Behind the Scenes," 26, SRCH, WvB Papers.
5. Speer minutes, 23.6.42 and 13–14.10.42, in Boelcke, *Deutschlands Rüstung,* 138, 194; Speer to Leeb, Milch, Fromm, and Zeitzler, 24.10.42, and Leeb (WD) to Fromm, 29.10.42, in NASM, FE342; WvB, "Einsatzplanung A4," 18.12.42, in NASM, FE333.
6. Speer minutes, 22.11.42, in Hillgruber, *Kriegstagebuch,* 2/2:1312; Hölsken, *V-Waffen,* 37, 83–89.
7. Speer minutes, 3/4/5.1.43, in Boelcke, *Deutschlands Rüstung,* 214; WD (WvB) to Abwehr/OKW, 21.11.42, on a 13.10.42 intelligence report, in NASM, FE741. For fanciful German reports on U.S. rocket development in 1943–44, see Abwehr to Wa Stab *et al.,* 10.3.43, in DM, GD629.1.2, "Bericht zur USA-Rakete 'Comet'," DM, GD620.0.9, and corr. in NA, T-78/663/347–53.
8. Speer minutes, 22.11.42, in Hillgruber, *Kriegstagebuch,* 2/2:1312; WD to Thiele/OKW/Ag WNV, 23.11.42, in NASM, FE692/f; Löhr to Wa Prüf, 24.11.42, in BA/MA, RH8/v.1230.
9. WD Vortragsnotiz, 14.12.39, in NASM, FE349; HAP/EW, "Vorschläge für den Einsatz," 21.3.42, in DM, GD639.1.4; WD to Zanssen, 10.11.42, and WvB, "Einsatzplanung A4," 18.12.42, in NASM, FE333; Reisig OHI, 55–58; WvB, "Die befestigte Abschussstelle für A4," 27.11.42, in DM, (FE1714); WvB, "Behind the Scenes," 28, SRCH, WvB Papers; WD, *V-2,* 103–4.
10. Stahlknecht minutes of 5.12.42 meeting, in BA/MA, RH8/v.1959.
11. WD, *V-2,* 75, 97–98; Freund and Perz, *Das Kz,* 35–38.
12. WD minutes of 22.12.42 meeting, in NASM, FE355; WD, *V-2,* 73–79; Hautefeuille, *Constructions speciales,* 20–27; Hölsken, *V-Waffen,* 38; Speer, *Inside,* 471, 694. Hitler could not have signed it on December 12 or December 22, as Speer inconsistently asserts. He was, however, at Führer headquarters December 15–17. See Speer office chronicle, 15–17.12.42, BA, R3/1736.
13. Schubert chronicle, 6.1.43 and 11.2.43, and Rudolph Aktennotiz, 9.2.43, in BA/MA, RH8/v.1210; Degenkolb to WvB, 29.1.43, and WvB to Degenkolb, 11.2.43, in NASM, FE732; WD lecture, 3.3.43, in NASM, FE358.
14. WD, *V-2,* 75–85; Schubert chronicle, 23–26.2.43 and 10–12.3.43, in BA/MA, RH8/v.1210.
15. Schubert chronicle, 20.12.41, in BA/MA, RH8/v.1208; minutes of 19.11.42 meeting in BA/MA, RH8/v.1209; Schubert chronicle, 16–19.3., 29.3., 15–17.4., 30.4., and 8.5.43 in BA/MA, RH8/v.1210; WD order, 8.6.43, in NASM, FE732; Freund, *Arbeitslager Zement,* 37–38. Speer's

opinion of the aggressive moves made by Degenkolb and his associates is unknown, but he must have at least tolerated them.

16. Stahlknecht to WvB, 24.2.43, in NASM, FE358; WD, "A4: Stand der Entwicklung," 8.10.42, BA/MA, RH8/v.1228; WD minutes of 22.12.42 meeting in NASM, FE355; Schubert chronicle, 31.3.43, in BA/MA, RH8/v.1210; Freund and Perz, *Das Kz,* 49, 62–63; Degenkolb circular, 2.4.43, in NASM, FE732.
17. WD, *V-2,* 89–90; WvB circular, 30.4.43, in NASM, FE732; Thom, "Notizen," 9.5.42, in NASM, FE728/f; WD order, 6.6.42, in BA/MA, RH8/v.1959; WD to Pleiger, 16.1.43, in NASM, FE356.
18. Leeb (WD) to Chef H Stab/Chef OKW, 8.4.43, in NASM, FE341; Storch/Siemens to Steinhoff, 20.3.43, in DM, (FE1224/1); WvB circular, 30.4.43, in NASM, FE732; Rudolph OHI, 88, 92.
19. Thiel to WvB, 16.3.43, in NASM, FE692/f.
20. Von Zborowski SS-officer file, BDC; SS corr. on v. Zborowski, in BA, NS19/3711; Schubert chronicle, 11.12.42, in BA/MA, RH8/v.1209; Himmler's daily schedule, 11.12.42, in NA, T-581/40A (no frame nos.); HAP/EW/TDZ, "Versuchsschiessen A4," 7.43, pictured in WD, *Peenemünde.* WD, *V-2,* 180, misdates this visit as April 1943.
21. Berger to Himmler, 16.12.42, NA, T-175/117/2642360.
22. Stegmaier to Berger, 26.1.43, Berger to Himmler, 1.2.43, and Himmler to Berger, 8.2.43, in NA, T-175/124/2599320–22; Himmler notes on Hitler meetings, 23.1.43 and 10.2.43, in BA, NS19/1474; WD lecture, 3.3.43, in NASM, FE358.
23. WD, *V-2,* 74, 88, 205–6; Himmler to Milch, 3.2.43, in BA, NS19/1197.
24. Speer minutes, 3/4/5.1.43, in Boelcke, *Deutschlands Rüstung,* 214; WD to Wa Prüf 1, 18.1.43, in NASM, FE692/f.
25. Hölsken, *V-Waffen,* 37; WD, *V-2,* 180–211.
26. WvB, "Affidavit of Membership in NSDAP. . . ," 18.6.47, El Paso, Texas, in FOIA release, INSCOM; WvB SS-officer file card, BDC.
27. WvB, "Affidavit," 18.6.47, in FOIA release, INSCOM; Reisig OHI, 1989, 86–87; SS-Stammrollenblatt for WvB, 28.2.34, from BA, Zwischenarchiv Dallwitz-Hoppegarten (Stasi archive).
28. WvB, "Affidavit," 18.6.47, in FOIA release, INSCOM; survey of membership in Nazi organizations from BDC records, INSCOM FOIA releases, Project Paperclip and IRR files in NA, RG 319 and 330. The sample is not truly random; Party members may be overrepresented because of the influence of publicized cases on the selection process. See Hunt, *Secret Agenda;* Hunt, "U.S. Coverup"; and Bower, *Paperclip Conspiracy.* The one ambiguous case is Gerhard Reisig, who says he was asked to join the Party, but no record of his membership exists in the BDC. Reisig OHI, 1985, 25–26. Four individuals, including Rudolph and two SS members,

were also in the SA at one time or another. As a general principle, names will be mentioned only if they have already been published by others.

29. Heimburg OHI, 50–51; interview with unnamed Huntsville engineer, 23.5.61, in WvB FOIA release by FBI; WvB to SS-RuSHA, 25.3. and 5.4.43, and marriage application, 5.4.43, in BDC; Himmler to Milch, 3.2.43, in BA, NS19/1197; v. Zborowski–Himmler correspondence, 1943, in BA, NS19/3711.
30. Mazuw telegram to Himmler, 27.3.43, Himmler to Berger, 29.3.43, and Berger to Himmler, 5.4.43, in NA, T-175/124/2599315–19. For earlier accounts, see Speer, *Infiltration,* 204; Irving, *Geheimwaffen,* 35–41.
31. Reisig OHI, 1989, 80–81; Rees OHI, 13; Heimburg OHI, 67–69; Rudolph OHI, 86–86; interviews with Mr. and Mrs. Manfred Schubert, Dr. and Mrs. Hans Geipel, and Mr. and Mrs. Gerhard Zanssen, by Neufeld, 1989 and 1990; Mrs. Leo Zanssen, biographical fragment, courtesy of same; Fromm to Himmler, 10.6.43, and Mazuw to Himmler, 27.3.43, in NA, T-175/124/2599307 and 2599319.
32. Himmler to Berger, 10.4.43, Himmler to Schmundt, 24.4.43, and Kaltenbrunner to Brandt/Pers.Stab RF-SS, 12.7.43, in NA, T-175/124/2599300-02, 2599314, 2599323; WD, *V-2,* 182–83; Schubert chronicle, 7.5.43, in BA/MA, RH8/v.1210.
33. Interview with Mr. and Mrs. Manfred Schubert, Dr. and Mrs. Hans Geipel, and Mr. and Mrs. Gerhard Zanssen, 1990; WD, *V-2,* 182–83; Kaltenbrunner to Brandt, 12.7.43, in NA, T-175/124/2599300–302.
34. Schubert chronicle, 11.5.43, in BA/MA, RH8/v.1210; WD draft letter (for Fromm?), c. 15–24.5.43, and WD draft letter to Leeb, c. 4–7.6.43, in WD notebook #1, DM. Dornberger, in *V-2,* 183–85, says he does not know who betrayed Zanssen, but his own notebooks show that to be false.
35. Draft of talk, 12.5.43, in WD notebook #1, DM; see also his draft talk, 4.6.43, in *ibid.*
36. Fromm to Himmler, 10.6.43, Brandt to Berger, 18.6.43, and Berger to Brandt, 22.6.43, in NA, T-175/124/2599305–9.
37. WD, *V-2,* 186–95; Schubert chronicle, 28–29.6.43, and Stegmaier program for visit, 28.6.43, in BA/MA, RH8/v.1210; Himmler Kalendernotizen, 28–29.6.43, in BA, NS19/1444; WvB SS-officer file card, BDC.
38. Schubert chronicle, 10–12.3.43, and Schubert to WD, 21.6.43, in BA/MA, RH8/v.1210; WD to Hartmann/Arm.Min., 6.10.42, in NASM, FE342, and WD, "A4: Stand der Entwicklung," 8.10.42, in BA/MA, RH8/v.1228; Homze, *Foreign Labor,* 40–41, 168–73, 271–77; Herbert, *Fremdarbeiter,* 76–78; Middlebrook, *Peenemünde Raid,* 29–32.
39. Schubert chronicle, 24.7.40, and Schubert Aktennotiz, 6.12.40, in BA/MA, RH8/v.1207; Schubert chronicle, 10.4., 16.4, 16.10., and

12.11.41, and Schubert Aktennotizen, 1.4. and 29.4.41, in BA/MA, RH8/v.1208; Schubert Aktennotizen, 9.4.42 and 17.4.42, in BA/MA, RH8/v.1209; Schubert chronicle, 13.8.43, in BA/MA, RH8/v.1210; minutes of RLM dev. meeting, 27.4.43, in NA, T-321/143/726–27; Garlinski, *Hitler's Last Weapons,* 81–84.

40. Rudolph Aktennotizen, 9.2.43, in BA/MA, RH8/v.1210.
41. Homze, *Foreign Labor,* 83; Herbert, *Fremdarbeiter,* 133–49; Herbert, "Labour," 151–53, 166–71.
42. Rudolph lecture, 1.3.43, and Schubert chronicle, 8–9.4.43, in BA/MA, RH8/v.1210; Freund, *Arbeitslager Zement,* 14–16, 39–43; Herbert, "Labour," 172–73; Krausnick and Broszat, *Anatomy,* 227–30.
43. Rudolph Aktennotiz, 16.4.43, in BA/MA, RH8/v.1210.
44. WvB, "Reminiscences," 140–41; Irving, *Mare's Nest,* 122. The chronology and agency of the decision is confused in Ordway and Sharpe, *Rocket Team,* 62–63, and Hölsken, *V-Waffen,* 51. Freund and Perz, *Das Kz,* 67–68, first revealed the document but missed its significance for the Rudolph case. For opposing views of the case see Hunt, *Secret Agenda,* and Franklin, *An American.*
45. WD Aktennotiz, 24.4.43, in BA/MA, RH8/v.1959; Burger, "Zeppelin," 56; Weinmann, *Das nationalsozialistische Lagersystem,* 629.
46. WD Aktennotiz, 24.4.43, in BA/MA, RH8/v.1959; Schubert chronicle, 13–14.5.43, Stegmaier program for Sauckel visit, 11.5.43, Rudolph Aktenvermerk on 2.6.43 meeting, and Schubert to WD, 21.6.43, in BA/MA, RH8/v.1210; Irving, *Mare's Nest,* 56–58.
47. Rudolph lecture, 1.3.43, Rudolph Aktenvermerk on 2.6.43 meeting, and Schubert chronicle, 17.6.43, in BA/MA, RH8/v.1210; Freund and Perz, *Das Kz,* 71.
48. Steimel statement, 26.5.47, in NA, M-1079/4/893–99; Krausnick and Broszat, *Anatomy,* 236–37.
49. Schubert chronicle, 11.7.43 and 16.7.43, in BA/MA, RH8/v.1210; Freund, *Arbeitslager Zement,* 47.
50. WD Aktennotiz, 29.6.43, in BA/MA, RH8/v.1954; WD, *V-2,* 93–98; Schubert chronicle, 26.5.43, in BA/MA, RH8/v.1210; Hölsken, *V-Waffen,* 40, 44.
51. WD, *V-2,* 95–96.
52. *Ibid.;* WD Aktennotiz, 29.5.43, in BA/MA, RH8/v.1954. In its final form, the "DE" rating ended on 31.12.43, rather than after a certain number of missiles.
53. Waeger/Arm.Min. order, 2.6.43, in NASM file "Peenemünde #2"; Waeger Aktenvermerk, 11.6.43, in DM, (FE 1224/1); Schubert chronicle, 9.6., 12.6, 15–16.6., 24.6., 28–29.6., and 9.7.43, in BA/MA, RH8/v.1210; WD, *V-2,* 99–100; Himmler notes on meeting with Hitler, 10.7.43, in BA,

NS19/1474; Himmler to Engel, 14.7.43, in NA, T-175/33/2542135. Krauch wrote a report critical of the A-4, but it quickly sank into oblivion because of the prevailing euphoria. See "Bemerkungen zum R.-Programm," 29.6.43, in NA, T-77/360/1201244–51.

54. WD, *V-2,* 100–104; Speer, *Inside,* 471; film script, 17.10.42, in NASM, FE338; Thom to AHA/In 10, 23.7.43, in NASM, FE355; Irving, *Mare's Nest,* 27; Hautefeuille, *Constructions speciales,* 25–45. Henshall, *Hitler's Rocket Sites,* describes the ruins today but is grossly in error on aspects of their history.
55. WD, *V-2,* 104–8; Speer, *Inside,* 471–72; Speer minutes, 8.7.43 and 25–26.7.43, in Boelcke, *Deutschlands Rüstung,* 280, 286.
56. WD, *V-2,* 107.
57. Stahlknecht order, 18.7.43 and Kunze Rundschreiben, 19.7.43, in NASM, FE750; WD, *V-2,* 98; Schubert chronicle, 16.7.43 and 29.7.43, in BA/MA, RH8/v.1210.
58. Thiel to Simon, 27.3.43, in NASM, FE692/f; WvB to Stahlknecht, 27.7.43, in NASM, FE750; WD, *V-2,* 104–5, 110–13; Schubert chronicle, 22–26.7.43, and Rudolph Aktennotiz, 27.7.43, in BA/MA, RH8/v.1210; Rudolph OHI, 93.
59. Stahlknecht to Arbeitsausschüsse, 3.8.43, and Kunze to WD, 9.8.43, in NASM, FE750; Schubert chronicle, 11.7., 23.7., 30.7., 3.8., and 4.8.43, in BA/MA, RH8/v.1210; WD, *V-2,* 99; WD draft letters to Leeb, Degenkolb, and Saur, c. 22–29.7.43, in WD notebook #1, DM.
60. WD minutes of 4.8.43 meeting and Zanssen "Sonderbefehl," 30.4.43, in NASM, FE750.
61. WD minutes of 4.8.43 meeting in NASM, FE750; Speer minutes, 8.7.43, in Boelcke, *Deutschlands Rüstung,* 280; WD draft letter to Saur, c. 29.7.43, in WD notebook #1, DM. Emphasis in the original.

7. *The Move Underground*

1. WD, *V-2,* 154–57; Middlebrook, *Peenemünde Raid,* 114–16; Schubert Aktennotiz, 25.7.40, in BA/MA, RH8/v.1207; Schubert chronicle, 19.9.42, in BA/MA, RH8/v.1209.
2. Middlebrook, *Peenemünde Raid,* 114–49.
3. *Ibid.,* 149–51, 217, 220; Irving, *Mare's Nest,* 104–15; Schubert chronicle, 18.8.43,; WD, *V-2,* 157–68; WD, "German V-2," 404; Steimel statement, 26.5.47, in NA, M-1079/4/897–98.
4. Hinsley *et al., British Intelligence,* 3/1:351–80; Irving, *Mare's Nest,* 33–83.
5. Schubert chronicle, 30.4, 21.5. and 2.6.43, in BA/MA, RH8/v.1210; WD Aktennotiz, 29.5.43, in BA/MA, RH8/v.1954; Middlebrook, *Peenemünde Raid,* 25–26, 133–34, 209–17.

6. WD, *V-2,* 165–67; Speer office chronicle, 18–19.8.43, in BA, R3/1738; Irving, *Mare's Nest,* 116–18.
7. Himmler Kalendernotizen, 15–20.8.43, in BA, NS19/1444; Speer minutes, 19–22.8.43, in Boelcke, *Deutschlands Rüstung,* 291; Freund, *Arbeitslager Zement,* 51–52; Hölsken, *V-Waffen,* 50–51.
8. Speer, *Infiltration,* 205–8; Himmler to Speer, 21.8.43, in BA, R3/1583.
9. Kammler files, BDC; Breitman, *Architect,* 199–203; Speer, *Inside,* 478; WD, *V-2,* 198–99.
10. Middlebrook, *Peenemünde Raid,* 195–201; Schubert chronicle, 21.8.43, in BA/MA, RH8/v.1210; WD to Sigismund von Braun, 28.6.47, in NA, RG319, Records of the Army Staff, G-2, Records Regarding Individuals, 1941–56, Box 77, 201 file of WD.
11. Schubert chronicle, 22.8.43. in BA/MA, RH8/v.1210; Reissinger to Fromm, 23.8.43, in NASM, FE341.
12. WvB(?) minutes, 25.8.43 meeting, in NASM, FE732.
13. Speer office chronicle, 26.8.43, in BA, R3/1738.
14. Bornemann, *Geheimprojekt,* 21–27, 36–43; Bornemann and Broszat, "Das KL," 156–60.
15. Schubert chronicle, 28.9.43, in BA/MA, RH8/v.1210; WD to Fromm, 31.5.44, in BA/MA, RH8/v.3730.
16. Schubert chronicle, 28.9., 8.12, 16.12.43, in BA/MA, RH8/v.1210; interview with Mr. and Mrs. Manfred Schubert, Dr. and Mrs. Hans Geipel, and Mr. and Mrs. Gerhard Zanssen, 1990; WD, *V-2,* 185, 209–10, 214; Rossmann on Kammler visit, 11.2.44, in NASM, FE694/f; HAP 11 Organisationsplan, 15.3.44, NASM, FE424.
17. WD to Fromm, 31.5.44, in BA/MA, RH8/v.3730; WD, *V-2,* 176–79; British interrogation of WD, 20.8.45, in NA, RG319, IRR files, WD dossier; Hölsken, *V-Waffen,* 62–63, 117–19; Kennedy, *Vengeance Weapon 2,* 32.
18. WD, *V-2,* 214; Reisig OHI, 1989, 99; Hölsken, *V-Waffen,* 55; Kammler to Brandt, 23.10.43, in NA, T-175/122/2647410; WD, *V-2,* 213–14, Himmler Kalendarnotizen, 28.9.43, in BA, NS19/1444.
19. Freund, *Arbeitslager Zement,* 61–74; Schubert chronicle, 12–20.9., 8.12., 27.12. and 28.12.43, 15.1., 22.1., 2.2. and 3–5.2.44, in BA/MA, RH8/v.1210; WvB, "*Vorschläge. . .* ," 2.10.43, in NASM, FE692/f; Kammler telegram to Brandt/Pers. Stab RFSS, 20.10.43, in BA, NS19/3546.
20. WvB, "*Vorschläge. . .* ," 2.10.43, in NASM, FE692/f; Schubert chronicle, 11.10.43, in BA/MA, RH8/v.1210; Wiesman OHI, 6–7; Huzel, *Peenemünde,* 61–62; Hermann, "Supersonic Wind Tunnel," 445–46; Kurzweg, "Aerodynamic Development," in Benecke and Quick, *History,* 52.
21. Middlebrook, *Peenemünde Raid,* 27–28, 152; Heimburg OHI, 58–59; Dahm OHI, 26–27; Wiesman OHI, 8–9; WD, *V-2,* 168.

22. Schubert chronicle, 17.6, 18.6., 11.7, 8.9., 13.10., 19.10., 20.10., 23.10, 16.11, and 20.11.43, in BA/MA, RH8/v.1210; WvB to Degenkolb, 12.11.43, in NASM, FE732; Rees to Storch, 6.7.44, and Storch to Kammler, 21.8.44, in NASM, FE692/f.
23. Degenkolb and WD minutes, 11.9.43 meeting, and Kunze minutes, 1.11.43 meeting, in NASM, FE732; WvB, "Bericht. . . ," 20.9.43, in NASM, FE692f; Freund and Perz, *Das Kz,* 84–88; Bornemann, *Geheimprojekt,* 46–62; Ordway and Sharpe, *Rocket Team,* 66–67. A 1.10.43 document in BA, NS4 Anhang/32, shows that Sawatzki originally planned for a Mittelwerk output of 1,800 a month, but that was quickly rejected as unrealistic.
24. Freund, *Arbeitslager Zement,* 58–59; Freund and Perz, *Das Kz,* 84–86; Heimburg OHI, 59–64; Ordway and Sharpe, *Rocket Team,* 84–87; WvB to Tessmann, 14.2.44, and WvB telegram to Justrow, 29.8.44, in NASM, FE692/f.
25. WvB pilot log, 1943–44, SRCH, WvB Papers; Kunze minutes, 1.11.43 meeting (with marginalia by WvB), and WvB to Degenkolb, 10.11.43 and 12.11.43, and WvB handwritten documents in NASM, FE732 (also in SRCH, WvB Papers); Freund and Perz, *Das Kz,* 86.
26. Bornemann, *Geheimprojekt,* 44–61; Rudolph testimony, 28.5.47, in NA, M-1079/1/401–2.
27. Bornemann, *Geheimprojekt,* 44–45; Rickhey statement, 25.6.47, and WvB defense questionnaire responses, 14.10.47, in NA, M-1079/4/144, 170; Bannasch testimony, 10.12.47, and Kunze testimony, 15.12.47, in NA, M-1079/10/808–10, 1060–61.
28. Kammler to Brandt, 16.10.43, in BA, NS19/3546; A-4 contract, 19.10.43, and Kettler and Förschner to Kühle, 20.11.43, in BA, NS4 Anhang/31; construction contract documents, 1943–44, in BA, NS4 Anhang/32; Kunze minutes, 1.11.43 meeting, in NASM, FE732; Bornemann, *Geheimprojekt,* 46–55.
29. Bornemann and Broszat, "Das KL," 165–66; testimony of SS-Col. Gerhardt Maurer, 13.11.47, in NA, M-1079/9/447–56.
30. Michel, *Dora,* 68; Bornemann, *Geheimprojekt,* 64–66; Bornemann and Broszat, "Das KL," 166–71.
31. Michel, *Dora,* 70; Bornemann, *Geheimprojekt,* 65.
32. Bornemann and Broszat, "Das KL," 168–71.
33. Speer office chronicle, 10.12.43, in BA, R3/1738; Speer, *Inside,* 474–75; Speer to Kammler, 17.12.43, in BA, R3/1585.
34. Speer office chronicle, 13.1. and 14.1.44, in BA, R3/1739; Speer, *Inside,* 474–75, 481; Speer, *Infiltration,* 211–12; Schmidt, *Albert Speer,* 181–95.
35. Béon, *La planète Dora;* interrogation of prisoner Verheyn, 14.4.45, in NA, M-1079/4/788, 790; Freund, *Arbeitslager Zement,* 121–28.

36. Franklin, *An American,* 78–79, and (for OSI interrogation of Rudolph, 4.2.83) 311.
37. Speer to Himmler, 22.12.43, in BA, R3/1583; Hölsken, *V-Waffen,* 61.
38. WvB, "Behind the Scenes," 30–31, SRCH, WvB Papers; WvB, "Reminiscences," 143.
39. WvB, "Behind the Scenes," 31, and WvB pilot log, 1943–44, in SRCH, WvB Papers; WvB affidavit, 18.6.47, in FOIA release, INSCOM; Irving, *Mare's Nest,* 205–6; Ordway and Sharpe, *Rocket Team,* 45; Himmler Kalendernotizen, 7–22.2.44, in BA, NS19/1444.
40. Kammler SS-officer file, BDC.
41. Speer, *Inside,* 421–28; Schmidt, *Albert Speer,* 85–99.
42. Jodl diary, 8.3.44, in NA, T-77/1429/144–45 (handwritten), T-77/1430/923 (typed); Irving, *Mare's Nest,* 207–8; HAP 11 Organisationsplan, 15.3.44, NASM, FE424.
43. Jodl diary, 8.3.44, in NA, T-77/1430/923; Irving, *Mare's Nest,* 206–8; Saur minutes, 5.3.44, in Boelcke, *Deutschlands Rüstung,* 341; Speer, *Infiltration,* 226–37.
44. Jodl diary, 8.3.44, in NA, T-77/1430/923; Irving, *Mare's Nest,* 207–8; Horeis, *Rolf Engel,* 24–25; Himmler Kalendernotizen, 22.2.44–19.3.44, in BA, NS19/1444.
45. WvB, "Behind the Scenes," 31, and pilot log, 1943–44, in SRCH, WvB Papers; WvB affidavit, 18.6.47, security questionnaire, 6.8.52, and CIC interview, 30.3.53, in FOIA release, INSCOM; Magnus von Braun CIC interview, 26.3.53, in FOIA release, INSCOM; Ordway and Sharpe, *Rocket Team,* 46–47; Bergaust, *Wernher von Braun,* 79; BzbV Heer minutes, 21.3.44 meeting, in NASM, FE692/f. The March 15 date in WD, *V-2,* 200, is certainly incorrect.
46. WD, *V-2,* 200–203.
47. Ibid., 203–7; Himmler Kalendernotizen, 23.3.44, in BA, NS19/1444; Hoffman, *History,* 294–95; Speer, *Inside,* 476; Speer office chronicle, 17–24.3.44, in BA, R3/1739.
48. WvB, "Behind the Scenes," 31, SRCH, WvB Papers; WvB, "Reminiscences," 143; WvB affidavit, 18.6.47, in FOIA release, INSCOM; Speer, *Inside,* 476; Speer minutes, 13.5.44, in Boelcke, *Deutschlands Rüstung,* 362; Riedel II death announcement, 5.8.44, in BA/MA, RH8/v.1941; Ordway and Sharpe, *Rocket Team,* 48.
49. For WvB's attitudes, see Heimburg OHI, 82–84; Huzel, *Peenemünde,* 119–20. For the mythology, see Ordway and Sharpe, *Rocket Team,* and Bainbridge, *Spaceflight Revolution.*
50. WvB(?), "Lagebericht A4," 5.12.43, in BA/MA, RH8/v.1954.
51. WD, *V-2,* 215–18.

52. *Ibid.;* WvB minutes, 4.4.44, in BA/MA, RH8/v.1278; Moser/BzbV Heer report, 17.4.44, in NASM, FE331.
53. WvB, "Behind the Scenes," 32, and WvB pilot log, 1943–44, SRCH, WvB Papers; WD, *V-2,* 220–22, 228–30; Steinhoff to T10, 23.6.44, in BA/MA, RH8/v.1278; Börgemann report, 6.10.44, in NASM, FE726; WvB to Arthur C. Clarke, 30.8.51, in SRCH, WvB Papers, BIS correspondence file.
54. Bühl/EW 2132, 31.10.44, in NASM, FE766.
55. *Ibid.;* minutes of 1.6.44 and 23.6.44 meetings in BA/MA, RH8/v.1278; Stegmaier report, 8.3.45, in NASM, FE763/c; Reisig Aktenvermerk, 13.3.45, in NASM, FE766; WD, *V-2,* 218–22.
56. WvB minutes, 4.4.44, in BA/MA, RH8/v.1278; Raithel OHI, 15–16.
57. WD, *V-2,* 222–24; WD to Wa Prüf, 6.12.44, in NASM, FE766; Bergaust, *Wernher von Braun,* 62.
58. Ordway and Sharpe, *Rocket Team,* 88–89; Bornemann, *Geheimprojekt,* 64; WvB(?), "Lagebericht A4," 5.12.43, in BA/MA, RH8/v.1954.
59. WvB(?), "Lagebericht A4," 5.12.43, in BA/MA, RH8/v.1954; Hölsken, *V-Waffen,* 61; minutes, 2.1.44 meeting of Führungsstab, and 2–3.44 A-4 failure reports in BA, NS4 Anhang/25; Rossmann on Kammler visit to Peenemünde, 11.2.44, in NASM, FE694/c.
60. Minutes of 9.9.43 Komm. f. Fernschiessen meeting, in NA, T-971/11/11; Speer office chronicle, 8.11.43, in BA, R3/1783; Rudolph OHI, 92, 96–97; WD and Degenkolb minutes, 11.9.43 meeting, in NASM, FE732; Stahlknecht/LZ telegram to WvB, 17.3.44, in NASM, FE694/a.
61. WD and WvB contributions in Zwicky, *Report,* 68, 73; BzbV Heer minutes, 21.3.44 meeting, in NASM, FE692/f; Reissinger minutes, 1.6.44 meeting, in BA/MA, RH8/v.1278; Steinhoff Denkschrift, 30.7.44, in NASM, FE766; WD to Rossmann, 20.8.44, in BA/MA, RH8/v.1278; Irving, *Mare's Nest,* 262.
62. WvB(?), "Lagebericht A4," 5.12.43, in BA/MA, RH8/v.1954; Leidreiter report, 15.2.44, in BA, NS4 Anhang/25; Sawatzki Aktenvermerk on 12.4.44 meeting, in NASM, FE694/a; Lindenberg to Fertigungsaufsicht, 15.5.44, in BA, NS4 Anhang/22; Ordway and Sharpe, *Rocket Team,* 73–74.
63. Bornemann and Broszat, "Das KL," 168, 176–80; Bornemann, *Geheimprojekt,* 72–73, 77–92.
64. Rickhey and Kettler, "Sonder-Directions-Anweisung," 22.6.44, and Förschner order, 30.12.43, in BA, NS4 Anhang/3; Bornemann and Broszat, "Das KL," 182–85; Pachaly and Pelny, *Konzentrationslager,* 144.
65. Rickhey, "Mittelwerke G.m.b.H.," 25.6.47, in NA, M-1079/4/146; Kahr testimony, 14.10.47, and Bannasch testimony, 10.12.47, in NA, M-

1079/8/586–87, 814–17; Simon to Rudolph and Haukohl, 18.4.48, in NA, M-1079/14/778; OSI interrogation of Rudolph, 13.10.82, in Franklin, *An American,* 266–69.

66. Stuhlinger and Ordway, *Wernher von Braun,* 103–5; Stuhlinger, "Gathering Momentum," 118.
67. Rickhey minutes, 6.5.44, in NASM, FE331; quotation from Boerner testimony, 10.12.47, in NA, M-1079/10/786–87; Rickhey, "Mittelwerke G.m.b.H.," 25.6.47, in NA, M-1079/4/145–46; Bower, *Paperclip Conspiracy,* 107.
68. WvB to Sawatzki, 15.8.44, in NASM, FE694/a.
69. Kettler and Förschner order, 8.1.44, in BA, NS4 Anhang/3 (emphasis in the original); Béon, *La planète Dora,* 78–80; Bornemann and Broszat, "Das KL," 194.
70. Bannasch testimony, 10.12.47, in NA, M-1079/10/820; interrogation of H. R. Friedrich, 28.5.47, in NA, M-1079/1/424, and same of G. Hobert, 14.4.45, in NA, M-1079/4/794-95; Archives of the United Nations War Crimes Commission, file no. 1192/Fr/G/523, obtained from OSI; OSI interrogation of Rudolph, 13.10.82, in Franklin, *An American,* 238–40, 276; Hunt, *Secret Agenda,* 74–75, 254; Bornemann and Broszat, "Das KL," 188–90.
71. Kloeckner testimony, 11–12.12.47, in NA, M-1079/10/943; Ordway and Sharpe, *Rocket Team,* 405–8; Irving, *Mare's Nest,* 220–22.
72. Hölsken, *V-Waffen,* 131–37; Irving, *Mare's Nest,* 233–39, 285–88; Speer minutes, 19–22.6.44 and 21–23.9.44, in Boelcke, *Deutschlands Rüstung,* 386, 413.
73. Storch to Kammler, 21.8.44, in NASM, FE692/f.
74. Von Bullion/L Flak 3 minutes, 14.1.43 meeting with Wa Prüf 11, in NASM, FE755/6; Von Renz/L Flak minutes, 16.1.43 meeting, in NASM, FE738/1; Flak dev. meeting, 1.3.43, in NA, T-321/160/185; "Untersuchung eines Vorprojektes für C1," 5.3.43, NASM, Arch. 63/32; Kurzweg, "Die aerodynamische Entwicklung der Flakrakete 'Wasserfall'," 15.3.45, NASM, Arch. 66/171.
75. Kurzweg, "Die aerodynamische Entwicklung. . . ," 15.3.45, NASM, Arch. 66/171; WvB and Grünewald minutes, 8.9.43 meeting, in BA/MA, RH8/v.1300; Reichel, "Die ferngesteuerte Flabrakete," 570.
76. Reichel, "Die ferngesteuerte Flabrakete," 571–73; Raithel OHI, 17–19.
77. Flak development meeting, 1.3.43, in NA, T-321/160/186; WD Aktennotiz, 29.5.43, in BA/MA, RH8/v.1954; minutes of 8–11.8.43 meetings in NASM, FE755/6; Oeckel/Henschel to WvB, 28.8.43, and WvB minutes, 19.11.43 meeting, in NASM, FE738/1.
78. Von Renz/L Flak to Milch, 13.1.43, in BA/MA, RH8/v.1235; Flak devel-

opment meetings, 1.3., 22.7., and 19.8.43, in NA, T-321/160/188, 245, 258; Hölsken, *V-Waffen,* 48–49.

79. Roth to Riedel III, 28.1.44, in NASM, FE754; WvB minutes, 19.11.43 meeting, in NASM, FE738/1.
80. Roth to Riedel III, 28.1.44, in NASM, FE754; Reichel, "Die ferngesteuerte Flabrakete," 571–72; Halder(?), "Vortragsnotiz," 10.1.44, in NA, T-321/154/235; WvB to Halder, 22.4.44, in NASM, FE738/1; Heller minutes, 21.7.44 meeting, in DM, GD664.6.2.
81. Roth to Riedel III, 28.1.44, in NASM, FE354; WvB to Kunze, 21.8.44, in BA/MA, RH8/v.1300.
82. WvB minutes, 26.9.42 meeting, and WD to Luftwaffe, 22.10.42, in BA/MA, RH8/v.1258; Zeyss/Wa Prüf 11 minutes, 8.2.43 meeting, in NASM, FE738/1; Thiry report, 24.7.43, in NASM, FE701; Roth minutes, 30.9.43 meeting, in NASM, FE755/6; Roth to Riedel III, 28.1.44, in NASM, FE754; E. W. Kutscher, ". . . Infrared Homing Devices," in Benecke and Quick, *History,* 201–17.
83. König, "Vorläufiger Einsatz der Fla.-R Wasserfall," 26.10.43, and prox. fuse documents in NASM, FE753/a; homing device documents in NASM, FE679, 680, 681, 682, 683; contracts list as of 15.8.44 in DM, (FE1224/1); F. v. Rautenfeld, ". . . Proximity Fuses," in Benecke and Quick, *History,* 218–37.
84. Roth to Riedel III, 28.1.44, in NASM, FE754; König (Grünewald), "Bericht Nr. 10," 17.8.43, and König, "Vorläufiger Einsatz. . . ," 26.10.43, in NASM, FE753/a; H. Wagner, ". . . A. Henschel Missiles," J. Dantscher, "Guided Missiles Radio Remote Control," and E. W. Kutscher, "Guidance of Surface-to-Air Missiles. . . ," in Benecke and Quick, *History,* 8–23, 109–124, 187–200.
85. Correspondence and meeting minutes on Wasserfall manufacturing, 14.5.43–27.5.44, in NASM, FE738/1; WvB to Kunze, 21.8.44, in BA/MA, RH8/v.1300; Schabel, "Wunderwaffen?" 75–76.
86. Flak development meeting, 18.1.43, in NA, T-321/160/175–76; Thiry report, 24.7.43, NASM, FE701; minutes of 30.9.43 meeting in NASM, FE755/6; Roth minutes, 18.12.43 meeting, in NASM, FE700/a; launch report, 9.3.44, Roth to König, 14.3.44, and reply, 24.3.44, in NASM, FE753/c; Klee and Merk, *Birth,* 68.
87. Hild minutes, 7.3.44 meeting, in BA/MA, RH8/v.1296; Roth minutes, 16.5.44 meeting, NASM, FE699/g; Dolezal report, 10.6.44, and Minnig(?) report, 23.8.44, in NASM, FE753/c; WD, "European Rocketry," 258; Roth minutes, 17.7.44 meeting, DM, GD660.0.10; Steinhoff to T10, 23.6.44, in BA/MA, RH8/v.1278; Irving, *Mare's Nest,* 263–69.
88. König to Roth, 1.5.44, and reply, 19.5.44, in NASM, FE753/c; Werrel, *Archie,* 41; Ludwig, "Flakraketen," 93–94.

8. Rockets, Inc.

1. Hoffmann, *History,* 377, 397–508, 528.
2. Speer office chronicle, 20.9.44, in BA, R3/1740. Schneider did not, however, get his job back.
3. Joachim Fest quoted in Hölsken, *V-Waffen,* 72; Himmler to Kammler, 6.8.44, in NA, T-175/122/2648275.
4. WD, *V-2,* 209; Hölsken, *V-Waffen,* 72.
5. Bornemann, *Geheimprojekt,* 68, 73–74.
6. WD, *V-2,* 208–10; Speer to Leeb, Kammler, Degenkolb, Rickhey, and MW Beirat, 12.5.44, and WD to Fromm, 31.5.44, in BA/MA, RH8/v.3730 (=IWM, M.I.14/861(V)).
7. WD, *V-2,* 208–9; WD to Fromm, 31.5.44, in BA/MA, RH8/v.3730.
8. WD to Fromm, 31.5.44, in BA/MA, RH8/v.3730; WD, *V-2,* 208-9; "Die Aufgaben der Elektromechanischen Werke G.m.b.H. (EMW)," 28.6.44, and Rees to Storch, 6.7.44, in NASM, FE692/f.
9. Rees to Storch, 6.7.44, and Storch to Kammler, 21.8.44, in NASM, FE692/f; International Tracing Service, *Verzeichnis,* 1:237.
10. Excerpt from Ordnance orders, 28.7.44, in DM, (FE1224/1); Rossmann to Storch, 12.11.44, in DM; WD, *V-2,* 209; interview with Mr. and Mrs. Manfred Schubert, Dr. and Mrs. Hans Geipel, and Mr. and Mrs. Gerhard Zanssen, 1990; Leo Zanssen, autobiographical chronology, c. 1966, courtesy of Gerhard Zanssen.
11. EW GmbH Organizational plan, 1.8.44, NASM, FE424, and same, 1.10.44, NASM, FE423; Huzel, *Peenemünde,* 123–26; Dannenberg OHI, 60–61; Rees OHI, 39–41.
12. Kammler to Thom, 16.8.44, and Kammler to WD, 18.8.44, in BA/MA, RH8/v.3730; WD, *V-2,* 210–11, 238.
13. WD, *V-2,* 236–39; Kammler (Thom?) order, 12.9.44, and Mayer "Vermerk" on Thom–Hübner phone conversation, 28.9.44, in IWM, M.I.14/861(V); Irving, *Mare's Nest,* 282–83; Villain, "France," 2; Hölsken, *V-Waffen,* 138–42, 259n. According to Hölsken, unit records show that two missiles were launched against London on September 7. The fact that the first impacts were recorded on September 8 near Paris at 11 A.M. and near London at 6:43 P.M. shows that the preceding day's attempts must have gone astray.
14. Thom "Vermerk," 15.9.44, on Jüttner meeting, in BA/MA, RH8/v.3730; Kammler to WD and Thom, 23.9.44, and Kammler "V-Befehl Nr. 1" to WD, 30.9.44, in IWM, M.I.14/861(V); WD to Kunze, Rickhey, Sawatzki, Storch, WvB, Rossmann *et al.,* 6.10.44, in NASM, FE692/f; Reisig OHI, 100; Hölsken, *V-Waffen,* 142.
15. Speer to Jüttner, 10.8.44 and 11.8.44, in BA, R3/1768; Speer, *Infiltration,* 213–14, 236–38; Mierzejewski, *Collapse,* 16–17, 93–97; WvB defense

questionaire, 14.10.47, in NA, M-1079/4/170 and 12/339–40; testimony of Bannasch, 10.12.47, and Kloeckner, 11–12.12.47, in NA, M-1079/10/811–13, 922–23; Jüttner order, 31.12.44, and drafts in NASM, FE333; Hölsken, *V-Waffen,* 72.

16. Himmler notes on Hitler meetings, 26–27.9.44, in BA, NS19/1474; Speer telegram to Himmler, 11.11.44, in BA, R3/1583; Riedel III to Rickhey, 11.12.44, in NASM, FE694/a.
17. WD, *V-2,* 210; WvB marginal notation on WD Aktenvermerk, 20.11.44, in DM, (FE1224/1).
18. WvB marginal notation on Contag Wasserfall report, 5.11.44, in NASM, FE738/1; Huzel, *Peenemünde,* 118–19; Heimburg OHI, 52, 82–83; Stuhlinger and Ordway, *Wernher von Braun,* 119–20; quote from Ward, *Wernher von Braun,* 28. For the atmosphere of the period, see WvB to Breé, 30.9.44, and reply, 10.10.44, in NASM, FE333.
19. Huzel, *Peenemünde,* 102–5, 114–15; report on 18.7. raid, 6.8.44, in NA, T-78/273/6220587; Craven and Cate, *Army Air Forces,* 3:537.
20. Schilling to Storch and Rees, 14.9.44, in NASM, FE735; Schilling to Storch, WvB, and Donaubauer, 19.8.44, Schilling notice, 21.9.44, and Storch order, 4.11.44, in NASM file "Peenemünde #2"; WvB to Kunze, 20.11.44, Grau (for Rickhey) to WvB, 15.1.45, and WvB to Rickhey, 23.1.45, in NASM, FE694/a; "Wichtige Punkte. . . ," 11.2.45, in NASM, FE731; Huzel, *Peenemünde,* 120–22.
21. Roth to WvB, 13.6.44, in NASM, FE738/1; WvB to Rossmann, 11.8.44, in DM, (FE1604); WD memorandum, 12.7.43, in NASM, FE333; WvB, "Vorschläge. . . ," 2.10.43, in NASM, FE692/f; Steinhoff to Zanssen, 7.3.44, in DM, (FE1224/1).
22. Rossmann to Kammler, 1.9.44, and Rossmann to Storch and Wa Prüf (BuM) 10, 3.9.44, in NASM, FE766; Roth, "Entwicklung einer geflügelter Fernrakete grosser Reichweite," 6.1.45, NASM, FE642; Irving, *Mare's Nest,* 280–89; Hölsken, *V-Waffen,* 135–45, 200.
23. Roth to Storch, 1.9.44, and Hellebrand minutes of 4.10.44 meeting in DM, (FE1604); Roth to Kurzweg/WVA, 6.9.44, in DM, (FE1081); Roth minutes of 28.9.44 meeting in NASM, FE699/r; Hellebrand minutes of 10.10.44 meeting and WvB minutes of 24.10.44 meeting in DM, GD636.0.1; Dahm OHI, 7.
24. Dahm OHI, 10–11, 16, 29.
25. Schilling report on 27.12.44 launch, 29.1.45, and Müller and Steinhoff to Rossmann, 3.1.45, in DM (FE1604); WD, *V-2,* 250–51.
26. A-4b meeting minutes and documents, 10.44–1.45, in DM, GD636.0.1 and (FE1604); Roth, "Entwicklung. . . ," 6.1.45, NASM, FE642.
27. Reisig, "Peenemünder 'Aggregaten'," 74; Heimburg OHI, 51; guidance documents, 7–12.44, in NASM, FE766, FE118, FE732, and DM,

GD632.11.9, and BA/MA, RH8/v.1971 and 1941; Hölsken, *V-Waffen,* 73; Reisig OHI, 1989, 99.

28. Rossmann to Kammler, 1.9., Rossmann (WvB) to Storch and Wa Prüf (BuM) 10, 3.9.44, and WD telegram to Kammler, 3.1.45, in NASM, FE766; Wa Prüf (BuM) 10/II Aktenvermerk, 2.12.44, in NASM, FE358; WD Aktenvermerk, 20.11.44, in DM, (FE1224/1); railroad launch reports, 12.44–1.45, in NASM, FE726; Huzel, *Peenemünde,* 127–28.
29. Hild minutes of 19.9.44 and 5.10.44 meetings in BA/MA, RH8/v.1296; Contag monthly reports, 5.11.44 and 1.12.44, in NASM, FE738/1; Contag minutes of 24.11.44 meeting in NASM, FE700/q.
30. Sources cited in preceding note; Haase and Peters, "Behälter-Entleerung," 29.11.44, NASM, FE738/10; Contag Wasserfall report, 27.1.45, in BA/MA, RH8/v.1300; contest documents courtesy of W. Dahm.
31. WvB to Kunze, 21.8.44, Steinmetz/RLM/Flak E5 to Halder, 24.11.44, and Contag report, 27.1.45, in BA/MA, RH8/v.1300; Hild minutes of 19.9.44 meeting in BA/MA, RH8/v.1296; WvB(?), "Übersicht und Stellungnahme. . . ," c. 12.44, NASM, FE738/2.
32. WvB to Kunze, 21.8.44, in BA/MA, RH8/v.1300; WD, *V-2,* 253–57; WvB(?), "Übersicht und Stellungnahme. . . ," c. 12.44, NASM, FE738/2; WvB to EW210, 29.12.44, and Hitler order of 4.11.44, in NASM, FE738/1.
33. WvB minutes of 20.11.44 meeting in DM, GD636.0.1; Storch to Kammler, 21.8.44, in NASM, FE692/f; Taifun documents, 31.8.44–31.1.45, in DM, GD660.0.33.
34. WvB to Rees, Maus *et al.,* 2.12.44, and Schilling minutes of 17.11.44 meeting in NASM, FE738/1; Hild minutes of 19.9.44 meeting in BA/MA, RH8/v.1296; Wegener and Orthmann minutes of 10.1.45 meeting at WVA, in DM, GD660.0.33; "Wichtige Punkte. . . ," 11.2.45, in NASM, FE731.
35. Test Stand XII documents in DM, GD639.3.4; Hüter minutes of 25.1.45 meeting in NASM, FE766.
36. Huzel, *Peenemünde,* 132–33; Steden notice, 24.10.44, in NASM file "Peenemünde #2"; WD Aktenvermerk, 20.11.44, on Kammler meetings in DM, (FE1224/1); EW227 to Steinhoff, 20.1.45, in DM, GD632.11.9; DeVorkin, *Science,* 30–37.
37. Freund, *Arbeitslager Zement,* 88–110; Speer minutes, 6–8.7.44, in Boelcke, *Deutschlands Rüstung,* 390; Schmid(?) to Steinhoff, 15.11.44, in DM; WvB to Kunze, 20.11.44, in NASM, FE694/a; Hüter to Schilling, 6.12.44, in NASM file "Peenemünde #2."
38. WD telegrams to WvB, 26.1.45, in DM, GD620.0.10.
39. Speer order, 13.1.45, in NASM, FE734; WD, *V-2,* 258–60; Hölsken, *V-Waffen,* 76–77, 161–62; Kammler orders, 6–7.2.45, in BA, R26III/52.
40. Rossmann orders, 30–31.1.45, in BA/MA, RH8/v.1941; Huzel, *Peenemünde,*

133–35; WvB minutes of 31.1.45 meeting in NASM file "Peenemünde #2"; "Wichtige Punkte aus Sch. Prof. v. B. an Aufsichtsrat v. 11.2.45. *Betr. Verlagerung,*" and Huzel Aktennotiz, 26.12.44, in NASM, FE731.

41. "Wichtige Punkte . . . v. 11.2.45," in NASM, FE731; WvB quoted in Ordway and Sharpe, *Rocket Team,* 254.
42. Ordway and Sharpe, *Rocket Team,* 254–55, 274; Rees OHI, 40–41; Dannenberg OHI, 68.
43. Huzel, *Peenemünde,* 137–39; Ordway and Sharpe, *Rocket Team,* 256–57; WvB, "Affidavit," 18.6.47, in FOIA release, INSCOM.
44. Huzel, *Peenemünde,* 138–42; Kaiser to Huzel, 24.2.45, in NASM, FE731; Ordway and Sharpe, *Rocket Team,* 261.
45. Huzel, *Peenemünde,* 139–48; WD, *V-2,* 265–70; McGovern, *Crossbow,* 94–95; Kammler to Storch *et al.,* 17.2.45, in BA/MA, RH8/v.1941; Rossmann, "*Notprogramm* für A-4 Entwicklung," 27.2.45, and Schneider minutes of 20.3.45 meeting at BzbV Heer, in NASM, FE766; WvB minutes of 2.3.45 meeting in NASM, FE738/1; WvB "Antrag," 6.3.45, in BA/MA, RH8/v.852; Kurz to WvB, 16.3., and Steinhoff to WvB, 27.3.45, in NASM, FE333; Wiesman OHI, 17–19.
46. WvB defense questionaire responses, 14.10.47, in NA, M-1079/4/169 and 12/348.
47. Bornemann and Broszat, "Das KL," 192–93; executions list, 1.11.44–27.3.45 (prosecution exhibit, 1947), in NA, M-1079/11/664–65.
48. Mierzejewski, *Collapse,* 125–61, 184–85; Bornemann and Broszat, "Das KL," 191–94; KL Mittelbau, prisoner list, 1.11.44, in NA, M-1079/11/580; Hein, "Lagerstaerke in KL.Dora" 1.12.44–3.4.45, c. 1947, in NA, M-1079/1/574.
49. Documents on Dora executions, 1947(?), in NA, M-1079/1/737 and 11/664–65; Bornemann and Broszat, "Das KL," 172–73, 194–95.
50. Béon, *La planète Dora,* 172 (translation courtesy of Yves Béon); interrogations of Ball, Friedrich, Rudolph, and Voss, 28.5–3.6.47, Fort Bliss, in NA, M-1079/1/385–428 (Ball quotes, 396–98). The Germans give various dates, but the tunnel hangings most probably occurred only in March 1945, when the large numbers correspond to the execution lists in the Nordhausen trial records.
51. Hölsken, *V-Waffen,* 79–80; Bornemann, *Geheimprojekt,* 135.
52. Hölsken, *V-Waffen,* 162–63; Herbert, *Fremdarbeiter,* 340.
53. Ordway and Sharpe, *Rocket Team,* 264–65; Reisig OHI, 1989, 109–10.
54. Bornemann and Broszat, "Das KL," 195–96; Bornemann, *Geheimprojekt,* 138–40.
55. Bornemann and Broszat, "Das KL," 197–98; Freund, *Arbeitslager Zement,* 328–29; Hölsken, *V-Waffen,* 162–63, 200–02.

56. Ordway and Sharpe, *Rocket Team*, 81, 265–66; telegrams to and from Kammler, 3–23.4.45, in NA, T-175/183/8718632–64; WD, *V-2*, 266–67; McGovern, *Crossbow*, 203.
57. Huzel, *Peenemünde*, 151–83; WD, *V-2*, 271; Ordway and Sharpe, *Rocket Team*, 261–67; Heimburg OHI, 77–78; Dahm OHI, 31–32.
58. Rees OHI, 43–44; Dannenberg OHI, 65–66; Huzel, *Peenemünde*, 187–88; Ordway and Sharpe, *Rocket Team*, 1–8, 268–69.

Epilogue: Peenemünde's Legacy

1. McGovern, *Crossbow*, 101–2, 151–86; Albrecht *et al.*, *Die Spezialisten*, 33–34.
2. Gimbel, "Project Paperclip"; McGovern, *Crossbow*, 125; Ordway and Sharpe, *Rocket Team*, 273–74.
3. Gimbel, "Project Paperclip"; Lasby, *Project Paperclip*, 66–92; McGovern, *Crossbow*, 185–97.
4. Ordway and Sharpe, *Rocket Team*, 318–20; McGovern, *Crossbow*, 205–6; Albrecht *et al.*, *Die Spezialisten*, 92–93.
5. Villain, "France," 5–8; McGovern, *Crossbow*, 200–204; Tessmann OHI, 28–32; FIAT Main Branch (British), "Memorandum for Dornberger P. File," 2.10.45, in NA, RG319, IRR files, WD dossier; Simpson, *Blowback*, 27–39.
6. McGovern, *Crossbow*, 207–15; DeVorkin, *Science*, 109–49; DeVorkin, "War Heads"; Ordway and Sharpe, *Rocket Team*, 310–17, 344–62.
7. Gimbel, "German Scientists"; Gimbel, *Science*, ch. 2; Bower, *Paperclip Conspiracy*; Hunt, *Secret Agenda*.
8. Hunt, *Secret Agenda*, 64–77.
9. Neufeld, "Guided Missile," discusses these issues more theoretically in terms of Thomas P. Hughes's "technological systems" approach.
10. Albrecht *et al.*, *Die Spezialisten*, 94–99; Bornemann, *Geheimprojekt*, 150–56; Ordway and Sharpe, *Rocket Team*, 318–43; Gröttrup, *Rocket Wife*.
11. WvB and WD contributions in Zwicky, *Report*, 67, 73. For other estimates, see Hölsken, *V-Waffen*, 79–80; Irving, *Mare's Nest*, 304, 315; Ordway and Sharpe, *Rocket Team*, 242–53, 405–8.
12. Hölsken, *V-Waffen*, 203–12; Irving, *Mare's Nest*, 304–6; Speer, *Inside*, 467–69.
13. Hölsken, *V-Waffen*, 163, 187–212; Murray, "Reflections," 90. Longmate, *Hitler's Rockets*, provides the other side: powerful eyewitness accounts of V-2 attacks in London.
14. Hölsken, *V-Waffen*, 208. Emphasis in original.

Bibliography and Archival Sources

ARCHIVAL SOURCES

This book is primarily based on what might be termed "the Peenemünde archive"—the documents that were salvaged by the U.S. Army and shipped to the Aberdeen Proving Ground, Maryland, for evaluation and exploitation. Some of the material was then forwarded to Fort Eustis, Virginia, for cataloguing and selective translation. Much of the archive, which included material from Wa Prüf 11 and its predecessors, was microfilmed, and a finding aid was created, numbering 2,683 pages. The first three volumes plus an Index were finished in 1946, followed by two supplementary volumes in 1948. The original paper documents appear to have followed the German rocket group to Fort Bliss, Texas, and Huntsville, Alabama, and then were returned to Germany (with minor exceptions) in the late 1950s. At that point the archive was artificially divided between the Deutsches Museum in Munich, which received the more "technical" files, and the Militärgeschichtliches Forschungsamt in Freiburg, which received the more "military" files. The latter collection then became part of the Bundesarchiv/Militärarchiv, when it was founded.

The American file system, and thus the microfilm as well, is divided into two bodies of material. The FE (Fort Eustis) files are administrative and technical documents that were renumbered, whereas the Peenemünde Archive Reports (the scientific report series) retained their original numbers. The Archive Reports, both in microfilm and in paper copy form, are to be found in a number of locations, because they were of greater value to postwar researchers than the FE material, which is more interesting to historians. Only one copy of the FE microfilm is known to me, and it is in the NASM Archives, but it covers only about 60 percent of the original files now found in Germany. In the notes I have cited this material according to the location and form in which it was used. (The same is true for National Archives microfilm of Captured German Records now found in Freiburg and Koblenz.) The BA/MA finding aid to the collection (RH8/II) has a concordance of FE file numbers in its possession to the new file numbers, but the situation in Munich is more confused, as is described below. The following are short descriptions of the archival collections consulted, including non-Peenemünde material.

Berlin Document Center (BDC)

This archive is currently run by the U.S. State Department but is being turned over to the Bundesarchiv when microfilming is complete. The filmed records will then be available at the National Archives. The BDC contains much of the central card file of the Nazi party, many SS officer files, and files of other Nazi organizations. Access is restricted to qualified persons, and records are obtained by supplying names and birthdates of relevant individuals.

Bundesarchiv (BA), Koblenz

Armaments Ministry files (record group R3) returned by the Imperial War Museum have a small amount of material relevant to the rocket program. It includes Speer's office chronicle, although the veracity of the retyped version has been questioned (see Schmidt, *Albert Speer*). A small number of Mittelwerk and Mittelbau-Dora camp documents are found under NS4/Anhang. I also consulted Reichsführer-SS Persönlicher Stab files in NS19, but all may be found in National Archives microfilm, although not always in a very readable form.

Bundesarchiv/Militärarchiv (BA/MA), Freiburg

Most of the really interesting administrative and political files from the Peenemünde archive are found here in record group RH8 (Army Ordnance) and are cited often in Hölsken, *V-Waffen*. I consulted only those files not found on FE microfilm at NASM, including the chronicle of the Peenemünde Production Plant, which never went to the United States at all. A small number of documents were later requested from RL3 (Reich Air Ministry).

Deutsches Museum (DM), Munich

The larger, more technical part of the Peenemünde archive was significantly damaged in the 1960s when a curator removed the most interesting files, often disassembling and reorganizing them. Those documents are thus no longer accessible through the five-volume Aberdeen list, which is the only finding aid the Museum possesses for the collection. At present these files are being reorganized, so it is not possible to give proper file numbers. When possible, I have given an FE number in brackets, in the hope that it will be of some use later on. The unaltered part of the collection is still organized by GD (German Document) numbers, which is a clumsy system created in Aberdeen to combine both FE and Archive Report material. A cross-reference list to Aberdeen volumes 1–3, published in a separate Index, gives GD number equivalents to FE and Archive Report numbers.

The Deutsches Museum also has two of the three notebooks of Walter Dornberger. The first covers 1938 to August 1943 but is scarcely more than notes on a few meetings until early 1943. After that point it begins to include draft documents and diarylike entries, but Dornberger's handwriting is so illegible that I have used the notebooks very sparingly. The second volume is missing, and the third, which most resembles a diary, covers from September 1944 to the end of the war.

Humboldt University Archive, Berlin

The Archive of the former University of Berlin has the 1934 doctoral examination records of Wernher von Braun, plus documents on the professorships of Erich Schumann and Karl Becker.

Imperial War Museum (IWM), London

Most German documents once possessed by the IWM have now been returned to Koblenz and Freiburg, but the Museum apparently retains microfilm copies. Dornberger's invaluable file on Rudolf Nebel, MI 14/801(V), is also to be returned to Freiburg. One item that will remain is "Papa" Riedel's memoir, which incorporates many pages of pictures from a 1933 Heylandt report to Ordnance. It is catalogued under German Misc. 148.

National Air and Space Museum (NASM), Washington

The NASM Archives Division has forty-three reels of PGM (Peenemünde Guided Missiles) microfilm of Archive reports and sixty-four reels of FE microfilm. In the latter case I created a finding aid of FE file numbers versus reel numbers. I have not listed reel numbers in the notes, because the film is available only at NASM in any case, where the finding aid may be consulted. A small number of miscellaneous original Peenemünde documents are also to be found in the file "Peenemünde #2," soon to be reorganized.

In addition, the NASM Department of Space History has transcripts of oral history interviews (OHI's) done by myself and by others in the Museum. The following transcripts are relevant to Peenemünde and are available to researchers, subject to restrictions imposed by the interviewees:

Interviewee	*Date*	*Interviewer*
Werner Dahm	1990	Neufeld
Konrad Dannenberg	1989	Neufeld
Walter Haeussermann	1990	Neufeld
Helmut Hoelzer	1989	Neufeld
Karl Heimburg	1989	Neufeld
Fritz Mueller	1989	Neufeld
Hermann Oberth	1987	Harwit/Winter
Herbert Raabe	1993	Neufeld/Winter
Wilhelm Raithel	1993	Neufeld
Eberhard Rees	1989	Neufeld
Gerhard Reisig	1985	DeVorkin/Collins
Gerhard Reisig	1989	Neufeld

Arthur Rudolph	1989	Neufeld
Bernhard Tessmann	1990	Neufeld
Georg von Tiesenhausen	1990	Neufeld
Walter Wiesman	1990	Neufeld

National Archives (NA), Washington

A small amount of material was drawn from Project Paperclip files in RG 319 and RG 330, as cited. A few of the files were actually housed in Suitland, but all modern military records will be moved to College Park, Maryland, by 1995. The great majority of the useful sources in this archive are to found, however, in microfilm publications of Captured German Records and war crimes trials, which may be read in Washington or purchased from the National Archives and Records Administration. I researched the following microfilm sets:

M-1079	U.S.A. v. Kurt Andrae *et al.* (Dora war crimes trial)
T-73	Reich Armaments Ministry (RmfBuM or RmfRuK)
T-77	Armed Forces High Command (OKW)
T-78	Army High Command (OKH)
T-84	Miscellaneous German Records
T-175	Reichsführer-SS and Chief of German Police
T-177	Reich Air Ministry (RLM)
T-178	Reich Research Council
T-321	Luftwaffe High Command (OKL)
T-581	Captured German Documents filmed by the Hoover Institution
T-971	Von Rohden Collection of Luftwaffe Documents

Redstone Scientific Information Center (RSIC), Huntsville, Alabama

The joint library of Redstone Arsenal and NASA Marshall Space Flight Center has a fairly complete paper set of Archive reports, including some missing from the PGM microfilm at NASM Archives.

Space and Rocket Center, Huntsville (SRCH)

The larger portion of the Wernher von Braun papers that reside in Huntsville is the SRCH's most important collection. Unfortunately, von Braun was able to bring very little with him to the United States, so that material relevant to the German period is rather limited. The other part of the von Braun papers is at the Library of Congress Manuscripts Division. It consists mostly of his public relations correspondence and contains very little that is relevant to the German period, other than copies of his scrapbooks.

U.S. Army Intelligence and Security Command (INSCOM), Fort Meade, Maryland

Army Project Paperclip files on specific German engineers and scientists may be obtained through Freedom of Information Act (FOIA) requests to INSCOM, subject to screening for privacy and security reasons.

BOOKS AND ARTICLES

Albrecht, Ulrich; Andreas Heinemann-Grüder; and Arend Wellmann. *Die Spezialisten: Deutsche Naturwissenschaftler und Techniker in der Sowjetunion nach 1945*. Berlin: Dietz Verlag, 1992.

Bainbridge, William Sims. *The Spaceflight Revolution: A Sociological Study.* New York: John Wiley & Sons, 1976.

Barth, Hans. *Hermann Oberth: Leben, Werk und Auswirkung auf die spätere Raumfahrtentwicklung.* Feucht: Uni-Verlag, 1985.

Bartov, Omer. *Hitler's Army: Soldiers, Nazis and War in the Third Reich.* New York and Oxford: Oxford University Press, 1991.

Baumgarten-Crusius, Artur. *Die Rakete als Weltfriedenstaube.* Leipzig: Verband der Raketen-Forscher und Förderer, 1931.

Benecke, Theodor, and A. W. Quick, eds. *History of German Guided Missile Development: AGARD First Guided Missile Seminar, Munich, Germany, April 1956.* Braunschweig: E. Appelhaus, 1957.

Béon, Yves. *La planète Dora.* Paris: Éditions du Seuil, 1985.

Bergaust, Erik. *Wernher von Braun.* Washington, D.C.: National Space Institute, 1976.

Boelcke, Willi, A., ed. *Deutschlands Rüstung im Zweiten Weltkrieg: Hitlers Konferenzen mit Albert Speer 1942–1945.* Frankfurt a.M.: Athenaion, 1969.

Boog, Horst. *Die deutsche Luftwaffenführung 1935–1945.* Stuttgart: Deutsche Verlags-Anstalt, 1982.

Bornemann, Manfred. *Geheimprojekt Mittelbau: Die Geschichte der deutschen V-Waffen Werke.* Munich: J. F. Lehmanns, 1971.

———, and Martin Broszat. "Das KL Dora-Mittelbau." In *Studien zur Geschichte der Konzentrationslager.* Schriftenreihe der Vierteljahreshefte für Zeitgeschichte, no. 21. Stuttgart: Deutsche Verlags-Anstalt, 1970.

Bower, Tom. *The Paperclip Conspiracy: The Battle for the Spoils and Secrets of Nazi Germany.* London: Michael Joseph, 1987.

Breitman, Richard. *The Architect of Genocide: Himmler and the Final Solution.* New York: Alfred A. Knopf, 1991.

Burger, Oswald. "Zeppelin und die Rüstungsindustrie am Bodensee." *1999: Zeitschrift für Sozialgeschichte des 20. und 21. Jahrhunderts,* 1987, no. 1, 8–49, and no. 2, 52–87.

Carroll, Bernice A. *Design for Total War: Arms and Economics in the Third Reich.* The Hague and Paris: Mouton, 1968.

Constant, Edward W., II. *The Origins of the Turbojet Revolution.* Baltimore and London: Johns Hopkins University Press, 1980.

Cranz, Carl. *Lehrbuch der Ballistik.* Vol. 1, *Äussere Ballistik.* 5th ed. Berlin: Julius Springer, 1925. Vol. 2: *Innere Ballistik.* Berlin: Julius Springer, 1926.

Craven, Wesley Frank, and James Lea Cate, eds. *The Army Air Forces in World War II.* Vol. 3, *Europe: Argument to V-E Day January 1944 to May 1945.* Chicago: University of Chicago Press, 1951.

Deutsch, Harold C. *Hitler and his Generals: The Hidden Crisis, January–June 1938.* Minneapolis: University of Minnesota Press, 1974.

DeVorkin, David H. *Science With a Vengeance: How the Military Created the U.S. Space Sciences After World War II.* New York, Berlin, and Heidelberg: Springer-Verlag, 1992.

———. "War Heads into Peace Heads: Holger N. Toftoy and the Public Image of the V-2 in the United States." *Journal of the British Interplanetary Society* 45 (1992): 439–44.

Dornberger, Walter. "European Rocketry After World War I." *Journal of the British Interplanetary Society* 13 (September 1954): 245–62.

———. *Peenemünde: Die Geschichte der V-Waffen.* (Reprint of *V 2.*) Frankfurt a.M. and Berlin: Ullstein, 1989.

———. "The German V-2." *Technology and Culture* 4 (Fall 1963): 393–408.

———. "The Lessons of Peenemünde." *Astronautics* 3 (March 1958): 18–20, 58, 60.

———. *V-2.* Trans. James Cleugh and Geoffrey Halliday. New York: Viking, 1954.

Ebert, Hans, and Hermann-J. Rupieper. "Technische Wissenschaft und nationalsozialistische Rüstungspolitik: Die Wehrtechnische Fakultät der TH Berlin 1933–1945." In *Wissenschaft und Gesellschaft: Beiträge zur Geschichte der Technischen Universität Berlin 1879–1979*. Ed. Reinhard Rürup. Berlin, Heidelberg, and New York: Springer, 1979, 469–81.

Ehricke, Krafft A. "The Peenemünde Rocket Center." *Rocketscience* 4 (1950): 17–22, 31–34, 57–63.

Engel, Gerhard. *Heeresadjutant bei Hitler 1938–1943: Aufzeichnungen des Majors Engel*. Ed. Hildegard von Kotze. Stuttgart: Deutsche Verlags-Anstalt, 1974.

Engelmann, Joachim. *Geheime Waffenschmiede Peenemünde*. Friedberg: Podzun-Pallas-Verlag, 1979.

Essers, Ilse. *Max Valier: Ein Vorkämpfer der Weltraumfahrt 1895–1930*. Technikgeschichte in Einzeldarstellungen, Nr. 5. Düsseldorf: VDI, 1968.

Fieber, Karl W. "Zur Geschichte der deutschen Raketensteuerung." Ms., Klaggenfurt, 1965. Deposited in the Siemens Archive, Munich.

Franklin, Thomas (pseud. for Hugh McInnish). *An American in Exile: The Story of Arthur Rudolph*. Huntsville, Ala.: Christopher Kaylor, 1987.

Freund, Florian. *Arbeitslager Zement: Das Konzentrationslager Ebensee und die Raketenrüstung*. Industrie, Zwangsarbeit und Konzentrationslager in Österreich, Bd. 2. Vienna: Verlag für Gesellschaftskritik, 1989.

———, and Bertrand Perz. *Das Kz in der Serbenhalle: Zur Kriegsindustrie in Wiener Neustadt*. Industrie, Zwangsarbeit und Konzentrationslager in Österreich, Bd. 1. Vienna: Verlag für Gesellschaftskritik, 1987.

Garlinski, Jósef. *Hitler's Last Weapons: The Underground War Against the V1 and the V2*. London: Julian Friedman, 1978.

Gartmann, Heinz. *The Men Behind the Space Rockets*. Trans. Eustace Wareing and Michael Glenny. New York: David McKay, 1956.

Germany, Militärgeschichtliches Forschungsamt. *Das Deutsche Reich und der Zweite Weltkrieg*. Bd. 1: *Ursachen und Voraussetzungen der deutschen Kriegspolitik*. Stuttgart: Deutsche Verlags-Anstalt, 1979.

———. *Das Deutsche Reich und der zweite Weltkrieg*. Bd. 5/1: *Organisation und Mobilisierung des deutschen Machtbereichs*. Stuttgart: Deutsche Verlags-Anstalt, 1988.

Gievers, Johannes, G. "Erinnerungen an Kreiselgeräte." *Jahrbuch der Deutschen Gesellschaft für Luft- und Raumfahrt E.V. (DGLR)*, 1971, 263–91.

Gimbel, John. "German Scientists, United States Denazification Policy, and the '*Paperclip* Conspiracy'." *International History Review* 12 (August 1990): 441–85.

———. "Project Paperclip: German Scientists, American Policy and the Cold War." *Diplomatic History* 14 (1990): 343–65.

———. *Science, Technology and Reparations: Exploitation and Plunder in Postwar Germany*. Stanford, Calif.: Stanford University Press, 1990.

———. "U.S. Policy and German Scientists: The Early Cold War." *Political Science Quarterly* 101 (1986): 433–51.

Goddard, Robert H. "A Method of Reaching Extreme Altitudes." In *The Papers of Robert H. Goddard.* Ed. Esther C. Goddard and G. Edward Pendray. New York: McGraw-Hill, 1970, 1:337–406.

———. "Liquid-Propellant Rocket Development." In *The Papers of Robert H. Goddard.* Ed. Esther C. Goddard and G. Edward Pendray. New York: McGraw-Hill, 1970, 2:968–964.

Gröttrup, Irmgard. *Rocket Wife.* London: André Deutsch, 1959.

Haeussermann, Walter. "Developments in the Field of Automatic Guidance and Control of Rockets." AIAA 81-4120. *Journal of Guidance and Control* 4 (May–June 1981): 225–39.

Halder, Franz. *Kriegstagebuch.* Ed. Hans-Adolf Jacobsen. 3 vols. Stuttgart: W. Kohlhammer, 1962–64.

Hallion, Richard. *Supersonic Flight: The Story of the Bell X-1 and Douglas D-558.* New York: Macmillan, 1972.

Hautefeuille, Roland. *Constructions speciales: Histoire de la construction par l'Organisation Todt, dans le Pas de Calais et le Cotentin, des neufs grands sites protégés pour le tir des V1, V2, V3 et las production d'oxygene liquide (1943–1944).* Paris: privately printed, 1985.

Heinkel, Ernst. *Stürmisches Leben.* Ed. Jürgen Thorwald. Stuttgart: Mundus, 1953.

Heitmann, Jan. "The Peenemünde Rocket Centre." *After the Battle,* no. 74, 1991, 1–25.

Henshall, Philip. *Hitler's Rocket Sites.* London: Robert Hale, 1985.

Herbert, Ulrich. *Fremdarbeiter: Die Politik und Praxis des "Ausländer-Einsatzes" in der Kriegswirtschaft des Dritten Reiches.* Bonn and Berlin: J. H. W. Dietz Nachf., 1985.

———. "Labour and Extermination: Economic Interest and the Primacy of *Weltanschauung* in National Socialism." *Past and Present,* no. 138, February 1993, 144–95.

Hermann, Rudolf. "The Supersonic Wind Tunnel Installations at Peenemünde and Kochel and Their Contributions to the Aerodynamics of Rocket-powered Vehicles." In *Space: Mankind's Fourth Environment,* London: Pergamon Press, 1982, 435–46.

Hillgruber, Andreas, ed. *Kriegstagebuch des Oberkommando der Wehrmacht (Wehrmachtführungsstab).* Vol. 2. Frankfurt: Bernard & Graefe, 1963.

Hinsley, F. H., *et al. British Intelligence in the Second World War: Its Influence on Strategy and Operations.* 3 vols. in 4. London: Her Majesty's Stationery Office, 1979–88.

Hoffmann, Peter. *The History of the German Resistance 1933–1945.* Trans. Richard Barry. Cambridge, Mass.: MIT Press, 1977.

Hölsken, Heinz Dieter. *Die V-Waffen: Entstehung—Propaganda—Kriegseinsatz.* Stuttgart: Deutsche Verlags-Anstalt, 1984.

———. "Die V-Waffen: Entwicklung und Einsatzgrundsätze." *Militärgeschichtliche Mitteilungen* 38 (1985): 95–122.

Homze, Edward L. *Arming the Luftwaffe: The Reich Air Ministry and the German Aircraft Industry, 1919–39*. Lincoln and London: University of Nebraska Press, 1976.

———. *Foreign Labor in Nazi Germany.* Princeton, N.J.: Princeton University Press, 1967.

Horeis, Heinz, ed. *Rolf Engel: Raketenbauer der ersten Stunde.* Munich: Lehrstuhl für Raumfahrttechnik, Techniche Universität München, 1992.

Hunt, Linda. *Secret Agenda: The United States Government, Nazi Scientists and Project Paperclip, 1945 to 1990.* New York: St. Martin's Press, 1991.

———. "U.S. Coverup of Nazi Scientists." *Bulletin of the Atomic Scientists,* April 1985, 16–24.

Hüttenberger, Peter. "Nationalsozialistische Polykratie." *Geschichte und Gesellschaft* 2 (1976): 417–42.

Huzel, Dieter K. *Peenemünde to Canaveral.* Engelwood Cliffs, N.J.: Prentice-Hall, 1962.

International Tracing Service. *Verzeichnis der Haftstätten unter dem Reichsführer-SS (1933–1945).* Arolsen: ITS, n.d. (1979?).

Irving, David. *Die Geheimwaffen des Dritten Reiches.* Gütersloh: Sigbert Mohn, 1965.

———. *The Mare's Nest.* Boston and Toronto: Little, Brown, 1965.

Karner, Stefan. "Die Steuerung der V2: Zum Anteil der Firma Siemens an der Entwicklung der ersten selbstgesteuerten Grossrakete." *Technikgeschichte* 46 (1979): 45–66.

Keilig, Wolf. *Das Deutsche Heer 1939–1945: Gliederung—Einsatz—Stellenbesetzung.* Bad Nauheim: Hans-Henning Podzun, 1956– .

Kennedy, Gregory P. *Vengeance Weapon 2: The V-2 Guided Missile.* Washington, DC: Smithsonian Institution Press for the National Air and Space Museum, 1983.

Kershaw, Ian. *The Nazi Dictatorship: Problems and Perspectives of Interpretation.* 2d ed. London: Edward Arnold, 1989.

Klee, Ernst, and Otto Merk. *The Birth of the Missile: The Secrets of Peenemünde.* Trans. T. Schoeters. New York: E. P. Dutton, 1965.

Klein, Heinrich. *Vom Geschoss zum Feuerpfeil: Der grosse Umbruch der Waffentechnik in Deutschland 1900–1970.* Neckargemünd: Kurt Vowinckel, 1977.

Koehl, Robert Lewis. *The Black Corps: The Structure and Power Struggles of the Nazi SS.* Madison: University of Wisconsin Press, 1983.

Krausnick, Helmut, and Martin Broszat. *Anatomy of the SS State.* Trans. Dorothy Lang and Marian Jackson. London: Granada, 1970.

Kunze, Harald. "Zur Zusammenarbeit von Hugo Junkers und Johannes Winkler." *NTM* 24 (1987): 63–82.

Lasby, Clarence G. *Project Paperclip: German Scientists and the Cold War.* New York: Atheneum, 1971.

Leeb, Emil. *Aus der Rüstung des Dritten Reiches (Das Heereswaffenamt 1938–1945).* Wehrtechnische Monatshefte, Beiheft 4. Berlin and Frankfurt: E. S. Mittler & Sohn, 1958.

Lehman, Milton. *Robert H. Goddard: Pioneer of Space Research.* (Reprint of *This High Man,* 1963.) New York: Da Capo, 1988.

Ley, Willy. "Count von Braun." *Journal of the British Interplanetary Society* 6 (June 1947): 154–56.

———. *Rockets, Missiles and Space Travel.* Rev. and enl. for 1960s. New York: Viking, 1961.

Longmate, Norman. *Hitler's Rockets: The Story of the V-2's.* London: Hutchison, 1985.

Ludwig, Karl-Heinz. "Die deutschen Flakraketen im Zweiten Weltkrieg." *Militärgeschichtliche Mitteilungen,* no. 1, 1969, 87–100.

———. "Die 'Hockdruckpumpe', ein Beispiel technischer Fehleinschätzung im 2. Weltkrieg." *Technikgeschichte* 38 (1971): 142–57.

———. *Technik und Ingenieure im Dritten Reich.* Königstein and Dusseldorf: Anthenäum/Droste, 1979.

Mackenzie, Donald. *Inventing Accuracy: A Historical Sociology of Nuclear Missile Guidance.* Cambridge, Mass., and London: MIT Press, 1990.

McDougall, Walter A. . . . *the Heavens and the Earth: A Political History of the Space Age.* New York: Basic Books, 1985.

McGovern, James. *Crossbow and Overcast.* New York: William Morrow, 1964.

Mehrtens, Herbert. "Die Naturwissenschaften im Nationalsozialismus." In *Wissenschaft und Gesellschaft: Beiträge zur Geschichte der Technischen Universität Berlin 1879–1979.* Ed. Reinhard Rürup, Berlin, Heidelberg, and New York: Springer, 1979, 427–43.

Michel, Jean, with Louis Nucera. *Dora.* Trans. Jennifer Kidd. New York: Holt, Rinehart & Winston, 1979.

Middlebrook, Martin. *The Peenemünde Raid: The Night of 17–18 August 1943.* London: Penguin, 1988.

Mierzejewski, Alfred C. *The Collapse of the German War Economy, 1944–1945: Allied Air Power and the German National Railway.* Chapel Hill and London: University of North Carolina Press, 1988.

Moore, Thomas M. "German Missile Accelerometers." *Electrical Engineering* (November 1949): 996–99.

Mueller, Fritz K. "A History of Inertial Guidance." *Journal of the British Interplanetary Society* 38 (1985): 180–92.

Müller, Rolf-Dieter. "World Power Status Through the Use of Poison Gas? German Preparations for Chemical Warfare, 1919–1945." In *The German Mili-*

tary in the Age of Total War. Ed. Wilhelm Deist. Leamington Spa: Berg, 1985, 171–209.

Murray, Williamson. *German Military Effectiveness.* Baltimore: Nautical & Aviation Publishing, 1992.

———. "Reflections on the Combined Bomber Offensive." *Militärgeschichtliche Mitteilungen* 51 (1992): 73–94.

Nebel, Rudolf. *Die Narren von Tegel: Ein Pioneer der Raumfahrt erzählt.* Düsseldorf: Droste, 1972.

Neufeld, Michael J. "Hitler, the V-2, and the Battle for Priority, 1939–1943." *The Journal of Military History* 57 (July 1993): 511–38.

———. "The Guided Missile and the Third Reich: Peenemünde and the Forging of a Technological Revolution." In *Science, Technology and National Socialism.* Ed. Monika Renneberg and Mark Walker. Cambridge: Cambridge University Press, 1993, 51–71, 352–56.

———. "Weimar Culture and Futuristic Technology: The Rocketry and Spaceflight Fad in Germany, 1923–1933." *Technology and Culture* 31 (October 1990): 725–52.

Nuss, Karl. "Einige Aspekte der Zusammenarbeit vom Heereswaffenamt und Rüstungskonzernen vor dem zweiten Weltkrieg." *Zeitschrift für Militärgeschichte* 4 (1965): 433–43.

Oberth, Hermann. *Die Rakete zu den Planetenräumen* (1923). Reprint. Nuremberg: Uni-Verlag, 1960.

———. *Wege zur Raumschiffahrt* (1929). Reprint. Bucharest: Kriterion, 1974.

Ordway, Frederick I., III, and Mitchell R. Sharpe. *The Rocket Team.* New York: Thomas Y. Crowell, 1979.

Overy, Richard J. "From 'Uralbomber' to 'Amerikabomber': The Luftwaffe and Strategic Bombing." *Journal of Strategic Studies* 1 (1978): 154–78.

———. *Goering: The "Iron Man."* London and New York: Routledge & Kegan Paul, 1984.

———. "Hitler's War and the German Economy: A Reinterpretation." *Economic History Review,* 2d Series, 35 (1982): 272–91.

———. "Mobilization for Total War in Germany 1939–1941." *English Historical Review* 103 (1988): 613–39.

Päch, Susanne. "Rolf Engel: Fifty Years of Activity in Rocketry and Space Flight." *Spaceflight* 22 (June 1980): 231–36.

Pachaly, Erhard, and Kurt Pelny. *Konzentrationslager Mittelbau-Dora: Zum antifaschistischen Widerstandskampf im KZ Dora 1943 bis 1945.* Berlin (East): Dietz, 1990.

Philipps, W. "Karl Becker," obituary. *Zeitschrift des Vereines deutscher Ingenieure* 84 (May 4, 1940): 293–94.

Reichel, Rudolf H. "Die ferngesteuerte Flabrakete C2 'Wasserfall.'" *Interavia* 6 (1951): 569–74.

Reisig, Gerhard H. R. "Das kongeniale Vermächtnis Hermann Oberth's und Wernher von Braun's für die Raumfahrt-Entwicklung." *Astronautik,* 1987, 103–6, and 1988, 10–11.

———. "Von den Peenemünder 'Aggregaten' zur amerikanischen 'Mondrakete': Die Entwicklung der Apollo-Rakete 'Saturn V' durch das Wernher-von-Braun-Team an Hand der Peenemünder Konzepte." *Astronautik,* 1986, 5–9, 44–47, 73–77, 111.

Riedel, W. H. J. "A Chapter in Rocket History." *Journal of the British Interplanetary Society* 13 (July 1954): 208–12.

Ruland, Bernd. *Wernher von Braun: Mein Leben für die Raumfahrt.* Offenburg: Burda, 1969.

Rürup, Reinhard, ed. *Topographie des Terrors: Gestapo, SS und Reichssicherheitshauptamt auf dem "Prinz-Albrecht-Gelände"—Eine Dokumentation.* Berlin: Willmuth Arenhövel, 1987.

Sänger-Bredt, Irene. "The Silver Bird Story: A Memoir." In *History of Rocketry and Astronautics.* Ed. R. Cargill Hall. AAS History Series, vol. 7, Part I. San Diego: Univelt, Inc., for the American Astronautical Society, 1986, 195–228.

———, and Rolf Engel. "The Development of Regeneratively Cooled Liquid Rocket Engines in Austria and Germany, 1926–1942." In *First Steps Toward Space.* Ed. Frederick C. Durant III and George S. James. AAS History Series, Vol. 6; IAA History Series, Vol. 1. Reprint. San Diego: Univelt, Inc., for the American Astronautical Society, 1985.

Schabel, Ralf. "Wunderwaffen? Strahlflugzeuge und Raketen in der Rüstungspolitik des Dritten Reiches." PhD dissertation, University of Augsburg, 1989.

Schmidt, Matthias. *Albert Speer: The End of a Myth.* Trans. Joachim Neugroschel. London: Harrap, 1985.

Schneider, Erich. "Technik und Waffenentwicklung im Kriege." In *Bilanz des Zweiten Weltkrieges.* Oldenbourg and Hamburg: Gerhard Stalling, 1953, 225–47.

Schumann, Erich. "Wehrmacht und Forschung." In *Wehrmacht und Partei.* Ed. Richard Donnevert. 2d ed. Leipzig: Johann Ambrosius Barth, 1939, 133–51.

Seidler, Franz W. *Fritz Todt: Baumeister des Dritten Reiches.* Munich: Herbig, 1986.

Simpson, Christopher. *Blowback: America's Recruitment of Nazis and the Its Effects on the Cold War.* New York: Weidenfeld & Nicholson, 1988.

Smelt, R. "A Critical Review of German Research on High-Speed Airflow." *Journal of the Royal Aeronautical Society* 50 (1946): 899–934.

Späte, Wolfgang. *Top Secret Bird: The Luftwaffe's Me-163 Comet.* Trans. Richard E. Moore. Missoula, Mont.: Pictorial Histories, 1989.

Speer, Albert. *Infiltration*. Trans. Joachim Neugroschel. New York: Macmillan, 1981.

———. *Inside the Third Reich*. Trans. Richard Winston and Clara Winston. New York: Avon, 1970.

Steinhoff, Ernst A. "Development of the German A-4 Guidance and Control System 1939–1945: A Memoir." In *History of Rocketry and Astronautics: Proceedings of the Third Through Sixth History Symposia of the International Academy of Astronautics*. Ed. R. Cargill Hall. San Diego: Univelt, Inc., for the American Astronautical Society, 1986.

Stubno, William J., Jr. "The Von Braun Rocket Team Viewed as a Product of German Romanticism." *Journal of the British Interplanetary Society* 35 (1982): 445–49.

Stuhlinger, Ernst. "Gathering Momentum: Von Braun's Work in the 1940s and 1950s." In *Blueprint for Space: Science Fiction to Science Fact*. Ed. Frederick I. Ordway III and Randy Liebermann. Washington, D.C., and London: Smithsonian Institution Press, 1992, 113–33.

——— and Frederick I. Ordway. *Wernher von Braun: Aufbruch in den Weltraum*. Esslingen and Munich: Bechtle, 1992.

Tarter, Donald. "Peenemünde and Los Alamos: Two Studies." *History of Technology* 14 (1992): 150–70.

Tomayko, James E. "Helmut Hoelzer's Fully Electronic Analog Computer." *Annals of the History of Computing* 7 (July 1985): 227–40.

Trischler, Helmuth. *Luft- und Raumfahrtforschung in Deutschland 1900–1970: Politische Geschichte einer Wissenschaft*. Frankfurt and New York: Campus, 1992.

United States, Army Ordnance. *The Story of Peenemünde, or What Might Have Been*. (Also known as *Peenemünde East, through the Eyes of 500 Detained at Garmisch*.) Mimeographed. 1945.

Villain, Jacques. "France and the Peenemünde Legacy." IAA-92-0186. Paper presented at the 43d Congress of the International Astronautical Federation, Washington, D.C., August–September 1992.

Von Braun, Magnus Freiherr. *Weg durch vier Zeitepochen*. Limburg an der Lahn: Starke, 1965.

Von Braun, Wernher. "Konstruktive, theoretische und experimentelle Beiträge zu dem Problem der Flüssigkeitsrakete." PhD dissertation, University of Berlin, 1934. Reprint: *Raketentechnik und Raumfahrtforschung*, Sonderheft 1, n.d. (c. 1960).

———. "Reminiscences of German Rocketry." *Journal of the British Interplanetary Society* 15 (May–June 1956): 125–45.

Walker, Mark. *German National Socialism and the Quest for Nuclear Power 1939–1949*. Cambridge: Cambridge University Press, 1989.

Ward, Bob. *Wernher von Braun Anekdotisch*. Esslingen: Bechtle, 1972.

Weinmann, Martin, ed. *Das nationalsozialistische Lagersystem (CCP)*. Frankfurt a.M.: Zweitausendeins, 1990.

Werrell, Kenneth P. *Archie, Flak, AAA and SAM: A Short Operational History of Ground-Based Air Defense*. Maxwell AFB, Ala.: Air University Press, 1988.

———. *The Evolution of the Cruise Missile*. Maxwell AFB, Ala.: Air University Press, 1985.

Winter, Frank H. *Prelude to the Space Age: The Rocket Societies: 1924–1940*. Washington, D.C.: Smithsonian Institution Press, 1983.

———. *Rockets into Space*. Cambridge, Mass., and London: Harvard University Press, 1990.

———, and Michael J. Neufeld. "Heylandt's Rocket Cars and the V-2: A Little Known Chapter in Rocket History." IAA-92-0185. Paper presented at 43d Congress of the International Astronautical Federation, Washington, D.C., September 1992.

Young, Richard Anthony. *The Flying Bomb*. New York: Sky Books Press, 1978.

Zwicky, Fritz. *Report on Certain Phases of War Research in Germany*. Vol. 1. Pasadena: Aerojet Engineering Corporation, 1945.

Index

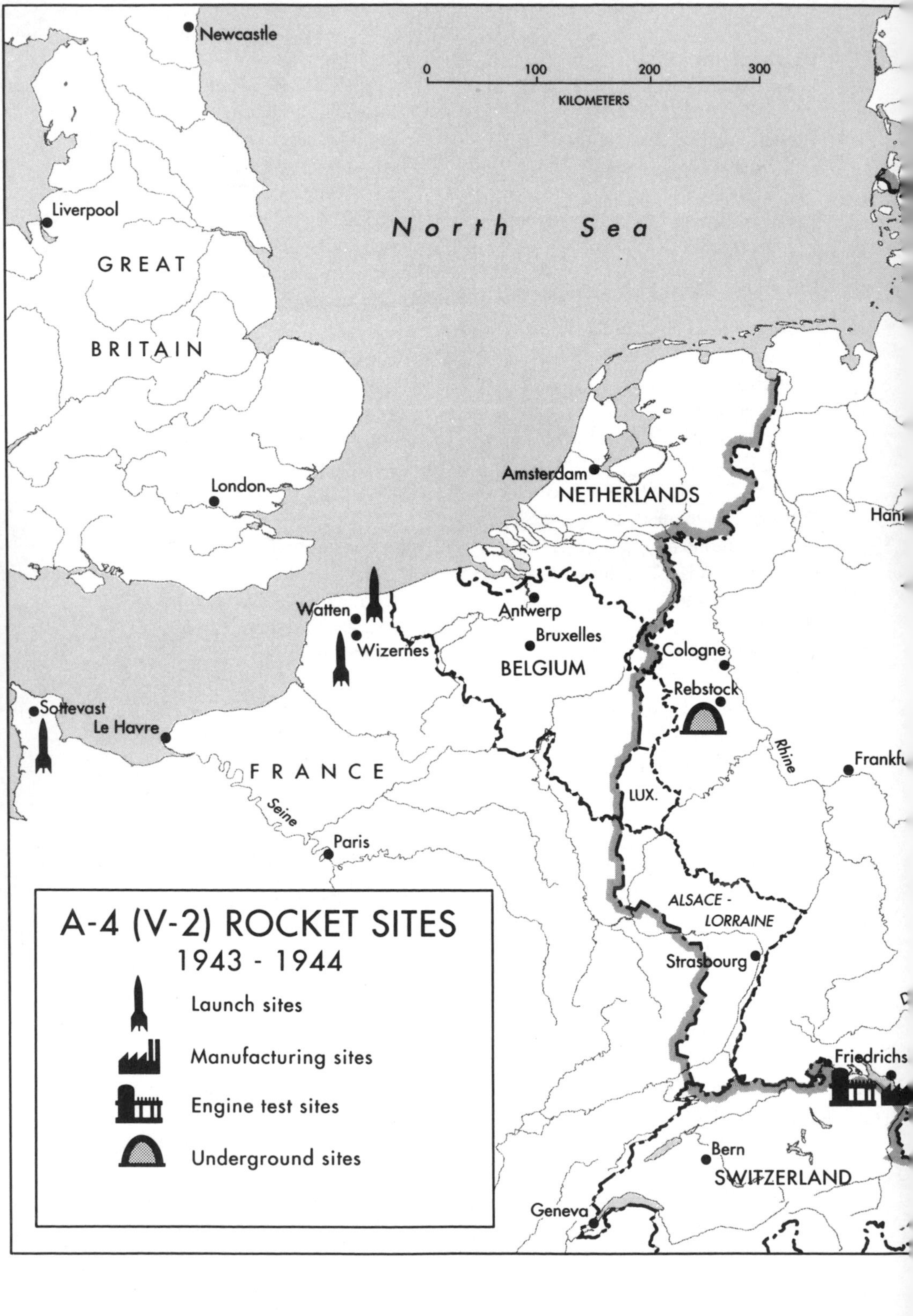
A-4 (V-2) ROCKET SITES
1943 - 1944
Launch sites
Manufacturing sites
Engine test sites
Underground sites
0
100
200
300
KILOMETERS
North Sea
Newcastle
Liverpool
GREAT
BRITAIN
London
Amsterdam
NETHERLANDS
Antwerp
Bruxelles
BELGIUM
Watten
Wizernes
Cologne
Rebstock
Rhine
Sottevast
Le Havre
FRANCE
Seine
Paris
LUX.
ALSACE -
LORRAINE
Strasbourg
Bern
SWITZERLAND
Geneva

SWEDEN
MARK
LITHUANIA
Copenhagen
Baltic Sea
Königsberg
Danzig
Rostock
Peenemünde
Heidekraut
burg
Stettin
Elbe
Vistula
Berlin
Posen
Warsaw
GREATER GERMAN REICH
ittelwerk
Nordhausen
Leipzig
GENERAL GOVERNMENT OF POLAND
Oder
Bliszna
(Heidelager)
Lehesten
Prague
Krakow
PROTECTORATE OF BOHEMIA AND MORAVIA
SLOVAKIA
Bratislava
Munich
Vienna
Salzburg
Schlier
Wiener
Neustadt
Ebensee
(Zement)
Oberammergau
Budapest
HUNGARY
Danube
Graz
Drava
ITALY
Pecs